The Resilient Earth

Science, Global Warming and the Future of Humanity

Doug L. Hoffman

Allen Simmons

http://www.theresilientearth.com

ISBN-10 1-4392-1154-X
ISBN-13 9781439211540

Published in the USA by
BookSurge Publishing
http://www.booksurge.com

First Edition

Table of Contents

Preface

The Resilient Earth had its genesis in a number of events spanning several years. The authors have been friends and colleagues for more than three decades and, while they have often discussed writing a book together, the timing never seemed quite right. Then, at the start of 2007, the debate surrounding human-caused global warming reached a crescendo. Those who questioned the extent and causes of global warming, other than human CO_2 emissions, were labeled "climate criminals," "industry stooges," and "traitors" by ecological activists. Those on the other side of the issue used terms like "hoax" and "scam."

The shrill level of the "debate" was driven home when Hoffman attended a business meeting. A co-worker asked a seemingly innocent question: "Doug, you're a scientist, what do you think about this global warming thing?" Hoffman framed a fairly neutral reply—"I don't think it's as bad as portrayed in the media, certainly we shouldn't ruin our economy in a panic." Hearing this, the senior executive present made a sarcastic, scatological remark regarding the offered opinion and stormed out of the room. Taken aback by this emotional reaction, Hoffman resolved to look more deeply into the subject of global warming.

In a matter of days, Hoffman was on the phone to Simmons suggesting that the time to write that often talked about book had arrived, and the topic should be the science of global warming—the real science, not the pseudo-science being reported in the popular media. Simmons immediately agreed and a long distance collaboration, linking coastal Texas and a log cabin in Arkansas, began. The more deeply we delved into the "facts" portrayed in the media the more concerned we became—not from fear of impending ecological disaster, but from the total lack of scientific objectivity, rationality and detachment exhibited by those on both sides of the global warming issue.

As the months past, the viciousness of the rhetoric used by activists and deniers continued unabated, reaching almost religious proportions. News anchors, never noted for their deep scientific insights, deliriously reported wild speculation about global ecological catastrophes as though they were established scientific fact. Also troublesome was the use of the term "scientific consensus" as a debate stopping argument by both overzealous fanatics and people who should know better.

Having both worked for years on numerous engineering and scientific projects, we resolved to uncover the actual scientific underpinnings of climate science and communicate our findings to a non-scientific audience. During the process of researching and writing *The Resilient Earth,* we were continually amazed with how little of the real science made it into the public debate. Further surprise came from the lack of knowledge among the general public and scientists, some of whom were involved in climatology. We discovered that global warming is a topic much discussed but little understood.

In conversations with colleagues about our progress writing *The Resilient Earth,* we found them astounded when told certain facts—facts such as the Earth had no ice caps for much of its history—or, who was the first person to comment on global warming due to greenhouse gases. At a dinner with colleagues, several participants expressed astonishment when the actual facts and figures regarding CO_2 and greenhouse warming were revealed. "I didn't know any of this!" exclaimed one senior and very respected scientist. How can the public clamor about global warming be so omnipresent while not only laymen, but even scientists seem to be unaware of the facts?

As the facts unfolded, the form and tone of *The Resilient Earth* changed many times. We were constantly amazed by the complex and interrelated nature of Earth's environment. We learned how fundamentally incomplete humanity's actual level of scientific understanding is regarding how our planet's climate system works. We have tried to maintain an even-handed approach while presenting the information contained in this book—to present an undistorted view of the science behind Earth's changing climate. We hope that our passion for both science and protecting the natural world are evident in our words.

Units and Measurements

This is a book about science, and that makes the presentation of facts and figures a necessity. Most of the measurements in this book are given in metric units, since all measurements and quantities found in the scientific literature are expressed using metric units. Since a sizable portion of our target audience resides in the United States we have also frequently stated measurements in the more familiar American units; pounds, feet, miles and temperatures in degrees Fahrenheit (°F). In most cases we give metric translations in parentheses, except when the number of alternate measurements would detract from the readability of the text.

Dealing with scientific subjects ranging from the life-cycle of stars to the chemistry of carbon on an atomic scale, very large and very small numbers must frequently be dealt with. In most situations we have managed to avoid scientific or engineering notation by using common prefixes. These prefixes are used to indicate powers of ten:

peta	1 000 000 000 000 000	P
tera	1 000 000 000 000	T
giga	1 000 000 000	G
mega	1 000 000	M
kilo	1 000	k
milli	0.001	m
micro	0.000 001	μ
nano	0.000 000 001	n

On occasion we have had to fall back on scientific notation, but only rarely.

About The References

Throughout this book you will find numbered references. Many of these references are to scientific articles from refereed journals that are the source for statements and assertions made in the text. Were this book a scientific treatise all the references would come from such sources. However, *The Resilient Earth* is intended for a wider, general readership audience so we have also included references to sources that non-scientists may find more accessible—magazine and newspaper articles and URLs for online web sites.

Acknowledgments

The authors would like to thank Dr. Terry Talley, Professor Amy Apon, Dr. Rik Faith, Brandon Willis and Alan Rainey for reading over the many early versions of this book and for their helpful comments and criticisms. Thanks to Bob Arrington for his many emails and data sources. We would also like to thank NASA legend, Dr. Rudolf Hanel, for his comments and advice on Chapter 7, Changing Atmosphere Gases, regarding the absorption spectra of CO_2. Thanks also to Dr. Nir Shaviv for kindly answering our inquiries regarding cosmo-climatology. Though we have sought the advice of many, we are solely responsible for any errors or inaccuracies in the text.

Special thanks to Eleanor Simmons for her diligence in editing and reediting the raw text. It took endurance handling two authors 650 miles apart. The cover art is a composite of two photo images; a NASA Hubble Space Telescope image of the giant nebula NGC3603, one of the most massive young star clusters in the Milky Way Galaxy, and an image of Earth and the moon, taken by the Galileo spacecraft while on its way to Jupiter. The composite was made by Hoffman using the GIMP.

Allen Simmons, Rockport, Texas

Doug L. Hoffman, Conway, Arkansas

December 5, 2007

Chapter 1 Introduction

"Scientists observe nature, then develop theories that describe their observations. Science is driven by nature itself, and nature gives us no choice. It is what it is."

— *Meg Urry*

A million years after the birth of our sun, the violent explosion of a nearby supernova nearly ended life on Earth before it began.[1] Over the next four and a half billion years, forces of nature shaped our planet and the life it harbored. Barely surviving the traumatic birth of the Moon, buffeted by supernovae, and bombarded by asteroids, the resilient Earth endured. And despite planet-freezing ice ages, devastating mass extinctions, and ever changing climate, life not only survived, it thrived.

Today, we are told all life on Earth is threatened by a new peril. Some say people are the cause of this impending crisis: human activity creates an increase in the amount of certain gases in the atmosphere, mainly carbon dioxide (CO_2), which cause the Earth's temperature to rise.[2] The catchphrase for this is *human-caused global warming* and we are warned, if something is not done immediately to stop it, our planet is doomed.

Earth *is* doomed, and science tells us how our world will end. In a billion years, the Sun's output will have increased by ten percent above today's levels, causing runaway greenhouse heating and the end of life on Earth.[3] In 5 billion years, the Sun will start to run out of hydrogen and begin to swell. In 6 billion years, the Sun will become a *red giant,* engulfing Mercury, Venus and Earth. Finally, in 7 billion years, the Sun will eject its outer layers and slide into retirement as a *white dwarf.* The Sun will spend its final days quietly cooling—a dimming ember in space. Earth, and its vibrant ecology, will have long vanished.[4] This happens to stars and their planets all the time, it is the fate decreed by the laws of nature.

The prophets of ecological doom are not warning of truly apocalyptic events, such as those outlined above. They say humans are destroying Earth's ecosystem on a much more immediate and personal scale. Their tale of impending destruction goes something like this:

> The average temperature of the Earth has been rising in recent decades and will keep rising in the future. Most of the observed increase in globally averaged temperatures since the mid-20th century is very likely due to the observed increase in anthro-

1

pogenic greenhouse gas concentrations. Anticipated effects include rising sea levels, repercussions to agriculture, slowing of ocean circulation, reductions in the ozone layer, increased intensity and frequency of hurricanes and extreme weather events, lowering of ocean pH, and the spread of diseases such as malaria and dengue fever.

The statement above summarizes the main points being made by those backing human-caused global warming. Some of the terms used sound technical and scientific, particularly the phrase "anthropogenic greenhouse gas concentrations." What the statement means is: people are adding so much carbon dioxide to the air that Earth's climate is affected. Is this the truth? How can a non-scientist winnow fact from fiction, truth from exaggeration?

Science Obscured

The debate over global warming and its possible human causes has become the defining scientific controversy of our time. Opinions vary regarding the severity of the problem among both scientists and lay people, though there are some who claim the threat is so immense and so immediate that all doubters should be silenced. The arguments presented to the public are mostly simplistic and cursory, usually accompanied by images of calving glaciers, melting icebergs and smokestacks belching clouds of pollution. The public debate has become vicious and nasty, filled with personal attacks and insults. As disturbing as this shift from reasoned scientific discourse to acrimony is, it is not the most troubling aspect of the global warming debate.

The most troubling aspect of the global warming controversy is what it reveals about the level of scientific understanding among the general populace. There is a growing disconnect between the scientific community and the general population. This is a consequence of the ever-widening knowledge gap between scientists and non-scientists. Even the separation between engineers and the public has grown to the point where the workings of everyday devices has become incomprehensible. For comparison, consider the state of technology fifty years ago.

In the United States, during the late 1950s, the shift from the war time economy of the 1940s was complete—a new, consumer driven economy was in full bloom. Every American family worked to own a house, a new car, modern appliances and a television set. But the inner workings of all these shiny, modern marvels were still understood by the average consumer.

Most people performed simple maintenance on their own automobiles. Changing the motor's oil and filter, replacing the spark plugs, and rotating the tires were part of car ownership. Many owners tackled more complicated maintenance and repair work; rebuilding the brakes, changing a water pump, or cleaning a carburetor. Today, most people never open the hoods of their autos. If they do, they are greeted by a featureless engine cover that effectively prevents any owner maintenance more complicated than checking the fluid levels.

A new wonder of the modern age was the television set. In the 1950s, these were large pieces of furniture filled with wires and softly glowing vacuum tubes. Since tubes had a rather short life expectancy they were installed in sockets for easy replacement. When a TV set malfunctioned, a thrifty and enterprising TV owner could remove any suspicious looking tubes and take them to the local supermarket where there was a testing unit available. Any tubes that didn't pass the tester were replaced with new ones. Today, no one works on their own TV set when it breaks. More than likely, if it is out of warranty, the unit is discarded and a new one is bought.

Even a 1950s era telephone could be disassembled and its parts examined. There was a recognizable speaker and microphone in the handset, and simple circuitry in the body. Today, a modern cellphone contains a color display screen, a camera and an electronic memory for storing addresses, ring tones, and mp3s. It is no longer connected to the phone system by wires, and the consumer can talk, send images, text message and even browse the Internet. Opening up a cellphone would gain the owner nothing, other than a voided warranty.

The point is that all of these "high tech" devices—cars, televisions and telephones—were accessible and understandable by their owners. Most people could describe how an internal combustion engine, a telephone, or television set worked. Perhaps not all of the details, but the general principles involved. Can the same be said today? In an age of high tech miniaturization, tubes are replaced by integrated circuits and flat-screen displays, carburetors by electronic fuel injectors and engine management computers, and everyone owns a multi-function cellphone. We all use these devices, but do we understand how they work? Modern life is filled with increasingly sophisticated devices that are increasingly incomprehensible. Even scientists don't work on their own cars and hardly any engineers try to fix their own television sets. As Arthur C. Clarke[†] said, "any sufficiently advanced technology is indistinguishable from magic."

If we are befuddled by everyday technology, imagine the confusion surrounding modern science. Five hundred years ago, Earth was still

---

† Arthur Charles Clarke (born 1917) British science-fiction author and inventor.

3

believed to be the unmoving center of the Universe. Two hundred years ago it was accepted that Earth circled the Sun, but the Sun was surely the center of all things. In short order, it was discovered that the Sun was just a star, and not a particularly remarkable one. Just an average star among the hundreds of billions in the Milky Way galaxy, which itself was one of billions of galaxies in the Universe. Also during this time, the Earth's estimated age changed from a few thousand years to a few million, and then to over four and a half billion years. And we now know that mankind has existed on Earth for such a short time that we can scarcely claim residency.

In chemistry, amazing new materials are now invented, not discovered. Advances are being made in organic synthesis, computationally aided molecular design, nanotechnology and space chemistry. In biology, scientists thought they had life figured out when they discovered DNA in the 1950s, and mapped the human genome at the end of the twentieth century. Now, they have discovered that RNA may play a role in biology as important as DNA. Hybrid plants and transgenic animals abound. Physics has moved from the certainty of Newton and Descartes to the warped space-time of Einstein and the quantum uncertainty of Heisenberg and Bohr. Instead of atoms, physicists talk about quarks, gluons, quantum gravity and string theory. Astronomers ponder black holes and quasars, the life and death of stars, and the ultimate fate of the Universe.

Most people struggle with science during their basic schooling and gladly abandon it to others upon graduating. For a while, this approach seemed to work: what did accountants, businessmen and lawyers need to know about science? But science has infiltrated every aspect of human existence. More and more, business means science and technology. As a result, judges and lawyers are faced with increasingly complex cases rooted in technology. Major criminal cases are decided by DNA evidence, and fingerprints seem so old-fashioned. Governments struggle to keep pace with scientific development, wrestling with the rights of frozen embryos, human cloning, genetically engineered crops, network neutrality and email spam. Science and technology cannot be avoided or ignored—our world is built on them.

This technological bewilderment is an indication of fundamental problems in modern education. Even in technologically advanced countries, the knowledge gap is growing, leaving average citizens adrift in a world that is becoming harder and harder to understand. Closing this knowledge gap was one of the main motivations for writing *The Resilient Earth*. Without knowledge, citizens cannot make intelligent decisions about technological problems; without knowledgeable citizens democracy cannot function. A case in point is the global warming debate.

Climate science is one of the most complicated fields of study in the history of science. Comparing Earth's climate to the workings of a star like the Sun is like comparing the workings of a formula one racer with a forest fire. A forest fire is large, dangerous and impressive but the processes involved are fairly simple and well understood. A race car engine is also based on fire, but it contains many individual parts, all interacting to turn expanding gas into linear motion and then rotational motion. Some of the rotational movement is transferred to the cam shafts, which translate that motion back to linear motion of the valves. The rest is passed through the transmission to the wheels to be translated into movement of the vehicle. A complicated and improbable machine where the relationships among the various parts are not straightforward or obvious.

Similarly, Earth's climate is made up of thousands of mechanisms and processes, all interacting in ways neither obvious nor fully understood. A modern internal combustion engine is highly refined—smooth, efficient and powerful. This is because engineers have been improving such engines for 120 years. Nature has been refining Earth's ecosystem for more than 4,000,000,000 years. It is not surprising that Earth's climate, which is intimately tied to and regulated by life, should be complicated in ways that escape current human understanding.

All disciplines from the natural sciences are involved in climate study, to the point where gaining a detailed overall understanding of climate is impossible. Most scientists lack a clear understanding of our imperfect knowledge of Earth's climate. Yet the public is asked to make decisions about climate policy based on televised shouting matches between pundits and politicians.

Global Warming Confusion

With classic Russian bluntness, Fedor Dostoevsky once said, "A man who lies to himself, and believes his own lies, becomes unable to recognize truth, either in himself or in anyone else." We're not saying that people are intentionally lying about the global warming issue, but there is a great deal of misinformation and numerous skewed conclusions in circulation around the planet. Call it human-caused global confusion.

In April, 1975, *Newsweek* reported, "there are ominous signs the earth's weather is changing dramatically and these changes portend a drastic decline in food production with serious political implications for every nation on earth." The article further reported delayed growing seasons, drought, devastation and warnings about increased severe weather activity, and "the most devastating outbreak of tornadoes ever recorded." The catastrophic scenario described was blamed on climate **cooling**. Scientists suggested melting the Arctic ice caps by covering them with soot—thus blunting the oncoming ice age.

Three decades later, our world faces similar ominous predictions that it will suffer floods, pestilence and starvation. Only now, the planet will experience catastrophic climate **warming**. Just as the cooling crisis spawned a number of far-fetched technological fixes, *IEEE Spectrum* reported in May 2007, that "Space Shields" have been proposed to cool the planet: "Steerable micrometers-thick refractive screens could divert a portion of the sun's energy away from Earth, thus cooling the atmosphere. The screens would orbit between the sun and the Earth." But salvation comes at a high price. *Spectrum* states: "Even using futuristic launching technology, the 20 million metric tons of mesh would cost US $4 trillion to deploy."[5]

Scientists are not the only ones making outlandish suggestions. A pop-music star stated we could help stop the warming if the whole world used just "one sheet of toilet paper per restroom visit."[6] *Time Magazine*, guilty of sensationalist reporting of a cooling climate in the 1970s, published in April, 2007—*The Global Warming Survival Guide/51 Things You Can Do to Make a Difference*. One web site proposed "going vegetarian" as a way to avert disaster.[7]

The main proponent of the human-caused global warming scenario is the Intergovernmental Panel on Climate Change (IPCC), an agency of the United Nations. Established in 1988, the panel has many experts on climatology and ecology. This panel, consisting of thousands of scientists and bureaucrats, was formed to evaluate "the risk of climate change brought on by humans."

Not surprisingly, the IPCC concluded the entire world is at risk and humans are to blame. This has been stated repeatedly in the IPCC reports over the last decade. The latest report, the fourth in the series, states that the world's scientists are "90% sure" that humans are the cause for the "unprecedented" rise in Earth's temperature.[8] Scientists dealing with climate change always use percentages because they know they cannot be certain of their predictions. Unfortunately, uncertain numbers from scientists quickly become firm predictions in the media.

The IPCC reports rallied many eco-conscious people and brought global warming to the attention of politicians worldwide. In 1997, the international community came together to draft the Kyoto Protocol,[9] an international treaty, which describes the actions needed to defeat the global warming menace. The treaty signing was announced with great fanfare, but its terms were far from universally accepted.

The United States and Australia refused to be bound by its restrictions. This was for a number of reasons—a reduced GDP and lower standard of living for both countries among them. The United States experienced unprecedented economic growth in the 1990s. Interestingly, the Kyoto emission goals for developed nations are set back to 1990 levels. But the Kyoto treaty gives *developing countries*, such as China and India, a free pass. This ignores claims by numerous sources that China has passed the US in annual emissions of CO_2.[10] At least one UN source states that India has also surpassed North America in CO_2 emissions and that emissions from the region are growing 5-6 times faster than in the developed countries.[11] The "brown clouds" of Asia have become a permanent fixture, ozone levels in Mexico City exceed air-quality standards 284 days per year, and airborne pollutants from Moscow can be found more than 600 miles away.[12]

Some countries that did sign the Kyoto Protocol are now reneging. Canada, for example, after a change in government, decided to abandon the emissions reduction targets of the Kyoto Treaty.[13] And Canada is not alone. Austria, Belgium, Japan, Portugal, Greece, Italy, Ireland, Finland, Norway, Denmark, and New Zealand have failed to meet their Kyoto goals for CO_2 emissions reduction. Acting in concert, the countries of the European Union found loopholes, which gave them a huge advantage relative to other countries.

The reunification of Germany led to the elimination of much dirty, polluting industry in what was formerly East Germany—though this was done for economic, not environmental reasons. Similarly, in the United Kingdom, the discovery of natural gas in the North Sea facilitated the phase-out of the coal industry. Coal, a major source of greenhouse gas emissions, had been a major source of fuel in both Britain and Germany. Reduced coal use meant the EU could reapportion emissions allotments no longer needed by Britain and Germany to other big polluters—awarding large net increases in some cases—thereby obtaining flexibility that no individual country had.[14] Even so, overall greenhouse emissions in Europe have *increased* since 1990.[15]

In January, 2007, the European Union nations announced a new agreement to limit their emissions to 20% below 1990 levels by 2020, and support for dropping global emissions to 50% below 1990 levels by 2050. The EU communique stated: "The European Union's objective is to limit global average temperature increase to less than 2°C compared to pre-industrial levels." But, according to a study from Canada's University of Victoria, stopping Earth's rising temperature will require going well beyond the reduction of industrial emissions discussed in international negotiations. "There is a disconnect between the European Union arguing for a 2°C threshold and calling for 50% cuts at 2050 - you can't have it both ways," says Andrew Weaver, leader of the Canadian study, adding: "If you're going to talk about 2°C you have got to be talking 90% emissions cuts." According to the study, only the total elimination of industrial emissions will succeed in limiting climate change to a 2°C rise in temperatures.[16]

The Kyoto Treaty is due to expire in 2012, having resulted in much hand wringing and shouting, but no noticeable impact on greenhouse gas emissions or global warming. In light of the Kyoto Treaty loopholes, the manipulation of emissions allotments by treaty signatories, and failure in reducing greenhouse gas emissions, the question arises: was Kyoto an effective way to respond to global warming, or were there hidden agendas? According to independent analysis by scientists at the Massachusetts Institute of Technology:

> "The 1990 level of emissions that is used in the Protocol, as the base from which the reductions would be made, and the reductions targets themselves, are quite arbitrary and not based on a specific target for the future world climate. In addition, the particular allocations of greenhouse gas emissions restrictions among countries do not have a principled logic. This arbitrariness has led to allocations that impose sharply different costs on the participating countries that have no consistent relation to their income or wealth."[17]

Are people, particularly those in the United States and other developed nations, responsible for worldwide climate change, or is this all a geo-political smokescreen? A rational person might ask a few questions before rushing headlong into crash programs to de-industrialize the world. Questions such as:

> Is global warming a real crisis?
> How do we know mankind is responsible?
> Has Earth had higher temperatures before?
> If it has, what caused the temperatures to cool?
> What can we actually do about a warming Earth?

These questions are reasonable from a scientific point of view, but the global warming debate is hardly about science anymore. Global warming has become a *cause célèbre,* championed by numerous politicians, activists and celebrities. To veer from the alarmists' view of human-caused global warming—that man's CO_2 emissions are causing global warming with catastrophic effect—is to invoke the wrath of those who blindly believe in the *cause*. As Louis Brandeis[†] wrote, "The greatest dangers to liberty lurk in insidious encroachment by men of zeal, well-meaning but without understanding."

The Three Pillars of Science

Global warming, therefore, is the central theme of our book, but not the only theme. Global warming is what is called the *driving problem* in this scientific investigation. By studying the science behind climate change, all manner of useful and interesting information will be revealed. This book will not inundate the reader with pages of equations, though there will be some of the Greek characters so beloved by scientists—they will not appear without plain language explanations. Along with the theories, facts and figures, each chapter also presents some of the history of scientific discovery. But the central character is nature itself.

The natural processes that create and regulate Earth's climate form a gigantic, extraordinarily complex heat engine powered by energy from the Sun. Earth's climate is perhaps the most complicated natural system science has ever tried to understand. The fundamental sciences—chemistry, physics and biology—are all intimately involved, along with a host of more specialized scientific disciplines. To gain an understanding of climate change requires knowledge of geology, archeology, anthropology, oceanography, meteorology, astrophysics, paleontology, glaciology and computer science. Practically every natural science has a role to play.

This is a tremendous amount of science to try and understand, even for trained scientists. But, in order to understand the claims and counterclaims bandied about regarding human-caused global warming, these topics must be explored and their connection to Earth's climate system understood. To provide a logical framework for understanding this most complex subject, we turn to the fundamental tenets of science.

When discussing the essential aspects of science, scholars often mention the *three pillars of science*. Many different formulations of the three pillars can be found and they have changed over the decades. One well accepted statement defines the three pillars as theory, experimentation, and computation.[18] For climate science these pillars translate to the following:

1. Theoretical understanding of Earth's climate.

2. Methods of collecting climate data both past and present.

---

† Louis Dembitz Brandeis (1856-1941) American Supreme Court Justice.

9

3. Use of computer climate models.

In the next several chapters we will use these definitions to examine the theory of human-caused global warming. We will address each pillar in turn, analyzing the strength and weaknesses of the IPCC's case.

The Path Ahead

Our investigation begins with a brief history of global warming and an examination of the claims in the IPCC reports. In the rest of Chapter 2, Global Warming–The Crisis Defined, the IPCC's own confidence levels in their predictions of a warming world, and the impact that warming could have on nature, are examined. We will show that their case rests on the output of global climate models (GCM), complex computer programs that attempt to model Earth's environment and predict the future.

From there we will look to Earth's past for proof that the current warming trend is "unprecedented" in Chapters 3, We are in an Ice Age?, and 4, Unprecedented Climate Change? We will investigate the causes of ice ages, the most dramatic of climate changes, in Chapter 5. The causes of mass extinctions will be examined in Chapter 6, since widespread extinctions are a predicted consequence of global warming.

After placing climate change into historical perspective we will then examine the various scientific theories that offer explanations for climate change. There are a number of different factors involved and no simple, single explanation will suffice. In order to keep the volume of scientific facts and figures from becoming overwhelming, we have structured the chapters for each topic to start with historical background behind the science. The scientists behind the important theories are introduced, along with the circumstances that led to their discoveries and their theories' paths to acceptance.

Since the IPCC models are driven primarily by carbon dioxide, we begin the review of the science behind climate change with carbon dioxide and the greenhouse effect. In Chapter 7, Changing Atmospheric Gases, we investigate the amount of CO_2 in the atmosphere, the terrestrial sources of carbon, how the greenhouse effect works, and its impact on climate. Looking beyond the greenhouse effect, the other major scientific causes of climate change are examined in detail, starting with the changing face of our planet and the ever flowing ocean waters in Chapter 8, Moving Continents & Ocean Currents.

Earth's changing path around our local star is a major factor, as is the Sun itself. These influences are discussed in Chapter 9, Variations In Earth's

Orbit, and Chapter 10, Varying Solar Radiation. Venturing farther out into the Universe, we present the recent theory of how supernovae and the solar system's path through the Milky Way influence Earth's climate in Chapter 11, Cosmic Rays.

After a discussion of how science and the scientific method of investigation developed in Chapter 12, How Science Works, we will pull all the evidence together and evaluate each of the three pillars of climate science. Chapter 12 ends with a summary of the first pillar, theory. The remaining two pillars, experimentation and computation, are examined in Chapter 13, Experimental Data and Error, and Chapter 14, The Limits of Climate Science, respectively.

We also examine something unscientific—the media, who are often wrong, but never in doubt. Media coverage and the public pronouncements of politicians, special interest groups, and celebrities will be scrutinized in Chapter 15, Prophets of Doom. After reviewing the role of non-scientists in the global warming controversy, the predicted consequences of global warming will be studied in Chapter 16, The Worst That Could Happen.

We finish with an examination of the IPCC's proposed solutions to the global warming threat in Chapter 17, Mitigation Strategies. We will identify the strategies that will work, help a bit, and not help at all. Then we will offer several rational proposals of our own to help defuse the crisis in Chapter 18, A Plan for the Future. We close on a hopeful note in the final chapter, The Fate of Planet Earth.

Our investigation will report if human-caused global warming is myth, or founded in fact. Don't misunderstand, we want people to stop polluting, stop wasteful consumption, recycle garbage, plant more trees, save energy— that's plain common sense. We wish to show that you can question the details of global warming as reported in the media and not be anti-nature or a tool of the oil industry. More importantly, we wish to provide non-scientists with enough background knowledge to enable them to question global warming, because all scientific theories must be questioned—that is how science works.

Anatomically modern humans have been on Earth about 130,000 years.[19] A mere blink of an eye on nature's time scale. Only in recent years have we moved from ignorance and superstition to a rudimentary, scientific understanding of nature. As you read about our resilient Earth, think of humanity's place in the grand scheme of life. Always keep in mind the wise words of Meg Urry, head of the Physics Department at Yale University— nature is what it is.

Chapter 2 Global Warming–The Crisis Defined

"We live in the midst of alarms; anxiety beclouds the future; we expect some new disaster with each newspaper we read."
— *Abraham Lincoln*

It seems that global warming suddenly appeared on the scene, rapidly spread by media warnings of impending peril. What is Global Warming? Where did it come from and how long have we known about it? This chapter examines where the idea of global warming came from, particularly global warming tied to human generated CO_2 emissions.

The History of Global Warming

Discovery of the *greenhouse effect,* the ability of certain gases to trap heat, is attributed to Joseph Fourier,[†] in 1829. Sixty-five years later, Swedish scientist Svante Arrhenius became the first investigator to report on the effects of heat-absorbing gases in the atmosphere. According to NASA, Arrhenius was "the first person to investigate the effect that doubling atmospheric carbon dioxide would have on global climate."[20]

Arrhenius had not started out in the fields of climatology and geophysics. He was awarded the Nobel Prize for Chemistry for his work on the electrolytic theory of dissociation. As a hobby, Arrhenius began studying the increased emission of atmospheric carbon by humans and its effects. In a paper presented in 1895, he stated that "the slight percentage of carbonic acid in the atmosphere may, by the advances of industry, be changed to a noticeable degree in the course of a few centuries."[21]

Arrhenius suggested that this increase could be beneficial, making Earth's climate "more equable," while stimulating plant growth and food production. Most scientists thought the idea that humans could actually affect average global temperatures far-fetched.

In 1938, G. S. Callendar, a British amateur meteorologist, made a bold claim that may sound familiar. He argued that man was responsible for heating up the planet with CO_2 emissions. Despite the previous work of Arrhenius, it still wasn't a common notion at the time. He published an article in the *Quarterly Journal of the Royal Meteorological*

Illustration 1: Svante August Arrhenius (1859-1927).

[†]Joseph Fourier (1768-1830), famous French mathematician and physicist.

13

Society on the subject stating, "In the following paper I hope to show that such influence is not only possible, but is actually occurring at the present time." He went on the lecture circuit describing carbon-dioxide-induced global warming, similar to Al Gore's present day crusade.

The 1940s saw significant developments in infrared spectroscopy, particularly for measuring long-wave IR radiation. It was empirically demonstrated that increasing the amount of atmospheric carbon dioxide resulted in greater absorption of infrared radiation. Also, it was discovered that water vapor, ozone and carbon dioxide absorbed radiation at different wavelengths. This allowed scientists to measure the increase in atmospheric CO_2 by studying spectroscopic data.

In 1956, Gilbert Plass, a Canadian physicist working in the United States, concluded that adding more carbon dioxide to the atmosphere would capture infrared radiation that is otherwise lost to space.[22,23] He further concluded that human activities were raising the average global temperature. Plass also pioneered the use of electronic computers in climate studies, a foreshadowing of modern climate science's dependency on computer models.

The American government first became involved in 1965, when Roger Revelle, a member of the President's Science Advisory Committee Panel on Environmental Pollution, helped publish the first high-level government document to mention global warming.[24] The report identified many of the environmental troubles the nation faced, including the potential for global warming by carbon dioxide. Revelle would continue to popularize the notion of human-caused global warming, publishing a widely-read article in *Scientific American* in 1982. The article linked the rise in global sea level and the "relative role played by the melting of glaciers and ice sheets versus the thermal expansion of the warming surface waters."[25]

Revelle, an oceanographer by training, convinced himself that he was the "grandfather" of global warming theory, conveniently ignoring the work of Arrhenius, Chandler, and Plass. On his death in 1982, his hometown paper, the San Diego County edition of the *Los Angeles Times*, began its front page coverage as follows: "Roger Revelle, the internationally renowned oceanographer who warned of global warming 30 years before greenhouse effect became a household term, died Monday of complications related to a heart attack." Even if his role in discovering global warming wasn't as central as he and the press thought, Revelle did play a major part in popularizing the subject.

Increased political awareness of global warming issues led to the establishment of the Intergovernmental Panel on Climate Change (IPCC) in 1988. The IPCC was established by two UN organizations, the World Meteorological Organization (WMO) and the United Nations Environment Programme (UNEP), to evaluate the risk of climate change brought on by humans. The IPCC was the first international effort of this scale to address environmental issues and has become the primary source of global warming information quoted by government officials and the media.

The remaining portion of this chapter will examine the case for human-caused global warming as advanced by the IPCC. We will show how a scientific observation by one man evolved into a political controversy, involving thousands of scientists, politicians and bureaucrats. First, we will examine changes in Earth's temperature over the past century or so.

Is Earth Really Warming?

There is little doubt that, over the past 100 years or so, Earth's climate has been warming. There may still be a few scientists out there who would disagree, but the fact that the planet is warming comes as close to universal acceptance (i.e. consensus[†]) as anything in science. But, as we shall see, consensus is not a valid scientific argument. Despite claims of consensus, opinions vary widely when the rate of warming and its causes are discussed.

The changing global temperature shown in Illustration 2 reveals a warming trend of around 1.8°F (1°C) per century. What is also apparent is the fact that the temperature didn't rise in a smooth, straight line. There were a number of ups and downs, particularly when traced from year to year. The five year mean temperature provides a smoother curve, but even it has had a bumpy rise. While Earth's climate is constantly changing, the change is anything but constant.

After a slight dip at the beginning of the last century, temperatures rose for 30 years, until around the start of World War II. Then the temperature sank for four decades, resuming an upward trend only after the mid-1970s. Because of this capricious temperature fluctuation, the time frame used to calculate the temperature trend is tremendously important.

When measured from 1975 to the 2005, a temperature rise of 0.8°C is observed in 30 years, a rate of +2.7°C per century. But, if the past 20 years were measured in 1964, a 20 year decrease of 0.4°C would be recorded, a cooling trend of -1.3°C per century. The temperature difference between 1900 and 2000 shows a rise for that century of 0.4°C, significantly lower

---

† Consensus (noun): agreement in the judgment or opinion reached by a group as a whole. Source Word Net V2.0, on http://www.dict.org.

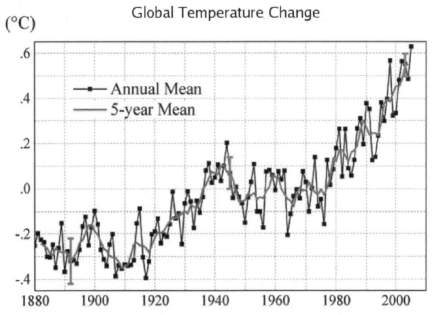

Illustration 2 Change in global mean temperature since 1880. Source NASA.

than the accepted 1°C per century. Clearly, how the measuring points are chosen can distort the perceived temperature trend greatly.

Judging a trend on yearly fluctuations is also an unsound practice. There was a 0.2°C drop from 1998 to 1999, in the midst of a 30 year warming period, while from 1956 to 1957 there was a 0.2°C rise during a 30 year period of temperature decline. Claiming that a year is the "hottest year in a decade" is totally meaningless in terms of the overall trend, and claiming that it is proof of global warming is an outright lie.

The truth is that Earth's climate is always changing. It varies from year to year, from decade to decade, and from century to century. What is seldom mentioned in the global warming debate is that there are also longer trends, spanning thousands and even millions of years. The forces affecting Earth's climate are colossal, long-acting, and only poorly understood by science.

How should we judge climate change? Assuming that the previous century's trend will continue for the next 100 years is naive, but probably a good place to start. If you extend the warming trend of the past century out into the future, a *linear prediction,* Earth's climate should grow warmer by another 1.8°F by the year 2100, but there is no guarantee this will happen. This continued, modest increase falls below the lowest increase predicted by the

IPCC, which uses much more complicated means of estimating future temperature change.

The IPCC Reports

Since its creation, the IPCC has issued four major, and numerous special, reports on the threat of human-caused global warming. Each installment has reported more dreadful consequences with greater certainty than the last. The latest report, the fourth in the series, states that the world's scientists are "90% sure" that humans are to blame for the unprecedented rise in Earth's temperature.[26]

1990: First assessment report (FAR)
1995: Second assessment report (SAR)
2001: Third assessment report (TAR)
2007: Fourth assessment report (AR4)
~2013: Fifth assessment report (AR5)

Table 1: Time-line for IPCC reports.

The previous report, the Third Assessment Report (TAR), had been released in 2001. Its findings were dire enough. In the words of Dr. R. K. Pachauri, Chairman of the IPCC at the High Level Segment, in an address given at the 11th Conference of the Parties to the United Nations Framework Convention on Climate Change and 1st Conference of the Parties serving as Meeting of the Parties to the Kyoto Protocol, in Montreal, Canada, 7 December 2005:

"The Earth's climate system has demonstrably changed on both global and regional scales since the pre-industrial era, and there is new and stronger evidence that most of the warming observed over the last 50 years is attributable to human activities...

Projections using the SRES[†] emissions scenarios based on a range of climate models point to an increase in globally averaged surface temperature of 1.4 to 5.8°C over the period 1990 to 2100. This is about two to ten times larger than the central value of observed warming over the 20th century, and the projected rate of warming is very likely to be without precedent during at least the last 10,000 years, based on paleo-climate data."[27]

So, here is a prediction of unprecedented climate change over the next century, a change not seen in "at least the last 10,000 years." Temperatures

---

† The IPCC Special Report on Emissions Scenarios.

could rise by as much as 2.5°F to 10°F (1.4°C to 5.8°C) around the world. This change could be bad for the climate, living things, and human civilization.

But the TAR isn't all doom and gloom. It mentions several courses of action that could reduce the level of threat. There is also some carefully worded bureaucratic bet-hedging about the effect of these "change mitigation efforts" on the ultimate outcome. What the report says is, if we cut down on CO_2 emissions, the temperature won't go up as much. So, the sooner we stop the emissions the better.

Dr. Pachauri noted the AR4 would contain additional and new information to update the older report. He said, "It will address some specific cross-cutting themes, which cover, in addition to other issues, scientific and technical aspects of Article 2 of the Convention."

The Fourth Assessment Report

The accompanying text box contains the press release from the IPCC announcing the release of the Fourth Assessment Report (AR4). The AR4 consists of four distinct sections: The Physical Science Basis; Impacts, Adaptation and Vulnerability; Mitigation of Climate Change and The Synthesis Report (SYR). Each of the first three sections has its own set of experts, called a Working Group. The three Working Group report sections were released by early May, 2007.

The most concise statement of the IPCC's findings is found in the Summary for Policy Makers (SPM). Unfortunately, the language found in all the IPCC reports is anything but clear and understandable. A strange brew of scientific jargon and bureaucratic doublespeak, the SPM is intended for use by non-scientists and is the most accessible of the lot.

What does the IPCC say about Earth's temperature? It's going up, of course; that conclusion was not really in doubt. What is questionable are the IPCC's presentation of historic data, methodology and estimated severity of temperature increase.

Illustration 3, taken directly from the IPCC report, contains two graphs. The upper graph shows average temperatures for the past 140 years. The lower graph extends the temperature readings to cover the past 1000 years. The reason for splitting the data up this way is that only temperatures for the past 140 years are based on direct measurements (e.g. thermometer readings). All the older temperatures shown are inferred from other measurements, called *proxies*. An interesting feature of the lower graph in Illustration 3 is the dramatic upswing, shown at the far right of the graph.

Paris, 2 February 2007 – Late last night, Working Group I of the Intergovernmental Panel on Climate Change (IPCC) adopted the Summary for Policymakers of the first volume of "Climate Change 2007", also known as the Fourth Assessment Report (AR4).

"Climate Change 2007: The Physical Science Basis", assesses the current scientific knowledge of the natural and human drivers of climate change, observed changes in climate, the ability of science to attribute changes to different causes, and projections for future climate change.

The report was produced by some 600 authors from 40 countries. Over 620 expert reviewers and a large number of government reviewers also participated. Representatives from 113 governments reviewed and revised the Summary line-by-line during the course of this week before adopting it and accepting the underlying report.

The Summary can be downloaded in English from www.ipcc.ch and http://ipcc-wg1.ucar.edu. A webcast of the final press conference has also been posted. The Summary will be available in Arabic, Chinese, French, Russian and Spanish at a later date.

This graph is known as the "hockey stick," because of its shape. The long, almost straight portion is the shaft and the up-swing at the end is the blade. For reasons we will discuss later, this graph has become infamous among those who question the IPCC's methods and conclusions.

The sharp up-tick right around the year 2000 is not part of the heavily drawn line that meanders across the rest of the graph. This is because the heavy line represents a moving average, a way to mathematically smooth out choppy data like Earth's yearly temperature. There are not enough data points in the steepest portion of the graph to smooth out the peaks. The lack of smoothing helps make the final up-tick all the more dramatic.

Note the wide, dark gray swath that encloses the temperature curve up until 1900. This band represents the uncertainty in the proxy temperatures used to create the graph. This means that even though the averaged temperature curve is shown consistently below the straight line, representing the average temperature for 1960-1990, some proxy data indicated that the temperatures were above that line. The IPCC takes uncertainty very seriously—to the point where they have published a number of guidelines on exactly what the

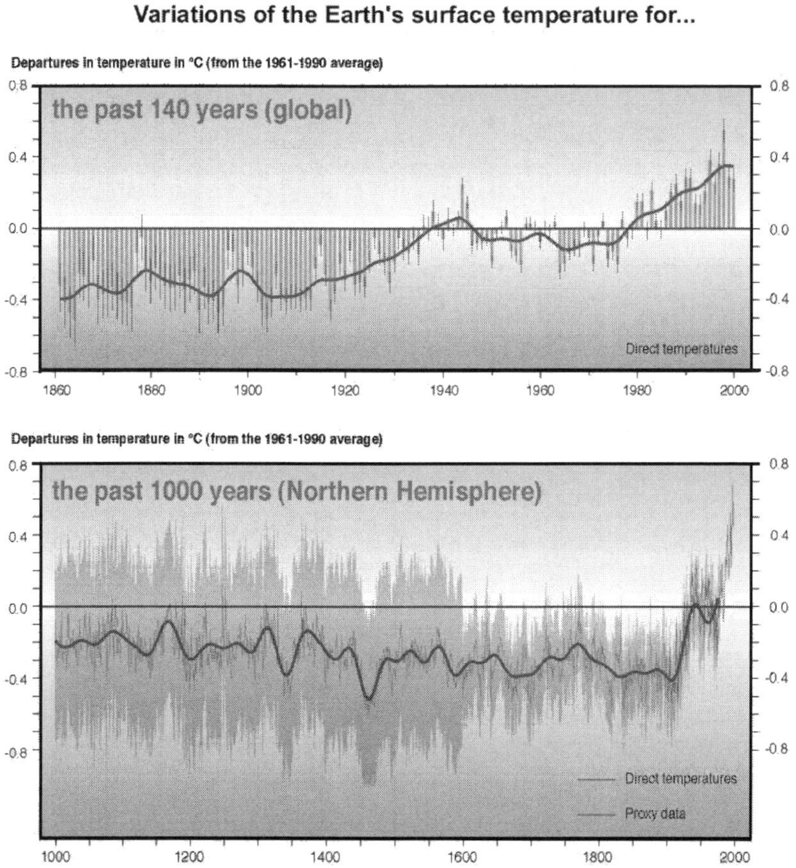
Variations of the Earth's surface temperature for...
Departures in temperature in °C (from the 1961-1990 average)
the past 140 years (global)
Direct temperatures
0.8
0.4
0.0
-0.4
-0.8
1860
1880
1900
1920
1940
1960
1980
2000
Departures in temperature in °C (from the 1961-1990 average)
the past 1000 years (Northern Hemisphere)
Direct temperatures
Proxy data
1000
1200
1400
1600
1800
2000

Illustration 3 Temperature change for the past 140 and 1,000 years. Source IPCC Fourth Assessment Report.

terms used in their reports mean. These include: *Uncertainties in Guidance Papers on the Cross Cutting Issues of the Third Assessment Report of the IPCC,*[28] published in 2000; *IPCC Workshop on Describing Scientific Uncertainties in Climate Change to Support Analysis of Risk and of Options: Workshop report,*[29] published in 2003; and the *IPCC Uncertainty Guidance Note,*[30] published in 2005. In the introductory letter to the guidance note, use of the terms "confidence" and "likelihood" as alternative ways of expressing uncertainty are given "particular attention."[31]

Multiple scales are provided to allow IPCC authors to confidently identify their level of uncertainty. Table 2 defines the meanings of phrases used when expressing levels of confidence, while Table 3 defines phrases used when talking about likelihood. It is not uncommon for scientists to use statistics and probabilities when they are unsure of their conclusions. This might seem confusing, but at least it is well defined confusion.

Terminology	Degree of confidence in being correct
Very High confidence	At least 9 out of 10 chance of being correct
High confidence	About 8 out of 10 chance
Medium confidence	About 5 out of 10 chance
Low confidence	About 2 out of 10 chance
Very low confidence	Less than 1 out of 10 chance

Table 2: Quantitatively calibrated levels of confidence

When the terms "high confidence" or "about as likely as not" appear in an IPCC report they have specific meanings. "High confidence" means that there is an eight in ten, or 80%, chance of being correct. "About as likely as not" means there is between a one in three and two in three (33%-66%) chance of being correct. All the conclusions presented in the IPCC reports, especially those in the Summary for Policymakers, must be viewed through this haze of statistical uncertainty. As Mark Twain said, "There are three kinds of lies: lies, damn lies, and statistics."

Terminology	Degree of confidence in being correct
Virtually certain	>99% probability of occurrence
Very likely	>90% probability
Likely	>66% probability
About as likely as not	33% to 66% probability
Unlikely	<33% probability
Very unlikely	<10% probability
Exceptionally unlikely	<1% probability

Table 3: IPCC Likelihood Scale.

An example of how these definitions are used to interpret figures and statements in the IPCC reports is shown in Illustration 4. Warmer days and nights are rated a 99% sure bet. More warm spells and heavy rain come in at 90% while drought, more hurricanes, and rising sea levels are a two in three chance at 66%.

But even these tentative predictions are conditional, based on the total amount of actual temperature rise. According to the IPCC, "The categories defined in this table should be considered as having 'fuzzy' boundaries."[32] Fuzzy indeed. With these definitions in mind, we will examine the effects of global warming as predicted by the IPCC's latest report.

Predicted Effects of Global Warming

IPCC Working Group II (WGII) is responsible for accessing the impact of global warming. Their section of the AR4 document contains "current scientific understanding of impacts of climate change on natural, managed and

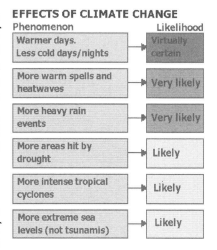

EFFECTS OF CLIMATE CHANGE

Phenomenon	Likelihood
Warmer days. Less cold days/nights	Virtually certain
More warm spells and heatwaves	Very likely
More heavy rain events	Very likely
More areas hit by drought	Likely
More intense tropical cyclones	Likely
More extreme sea levels (not tsunamis)	Likely

Illustration 4: Likelihood of Climate Effects.

human systems, the capacity of these systems to adapt and their vulnerability."[33] This report claims that global warming will cause significant change in many physical and biological systems. As an example of the actual text in the report, we quote from the summary report.

> With regard to changes in snow, ice and frozen ground (including permafrost), there is high confidence that natural systems are affected. Examples are:
>
> • enlargement and increased numbers of glacial lakes [1.3]†;
>
> • increasing ground instability in permafrost regions, and rock avalanches in mountain regions [1.3];
>
> • changes in some Arctic and Antarctic ecosystems, including those in sea-ice biomes, and also predators high in the food chain [1.3, 4.4, 15.4].
>
> Based on growing evidence, there is high confidence that the following effects on hydrological systems are occurring:
>
> • increased run-off and earlier spring peak discharge in many glacier- and snow-fed rivers [1.3];

---

†Sources to statements in the report text are given in square brackets. For example, [3.3] refers to Chapter 3, Section 3.

- warming of lakes and rivers in many regions, with effects on thermal structure and water quality [1.3]...

There is very high confidence, based on more evidence from a wider range of species, that recent warming is strongly affecting terrestrial biological systems, including such changes as:

- earlier timing of spring events, such as leaf-unfolding, bird migration and egg-laying [1.3];

- poleward and upward shifts in ranges in plant and animal species [1.3, 8.2, 14.2].

To save space, here is a summary of the other major outcomes and predictions from subsequent sections:

- There could be more fish farther north, more algae growing in lakes, and fish runs will start earlier in the year.

- Spring will come earlier. Warmer higher latitudes will mean animals and plants will be found in places that were previously too cold for them.

- Crops *might* be affected, also, there *might* be more forest fires and pests. People *may* experience a worse hay fever season.

- The hot summer weather *might* be dangerous to some. Also, snow sports and winter hunting *might* be diminished.

- Warmer days and nights are rated a 99% certainty.

- More warm spells and heavy rain are rated 90%.

- Drought, more frequent and severe hurricanes, and rising sea levels receive 66%.

The report also cautions that warming *might* cause floods from glacier melt-water. *Some* places will experience drier conditions, which could affect agriculture. Water released from melting glaciers *could* make sea-levels higher and flood low-lying areas. It is noted, however, that these effects have not "become established trends."

The report goes on to break out various effects for different categories based on a sliding scale according to the amount of temperature increase. The table from the original report is reproduced in Illustration 5. A close inspection reveals a mixed bag of results.

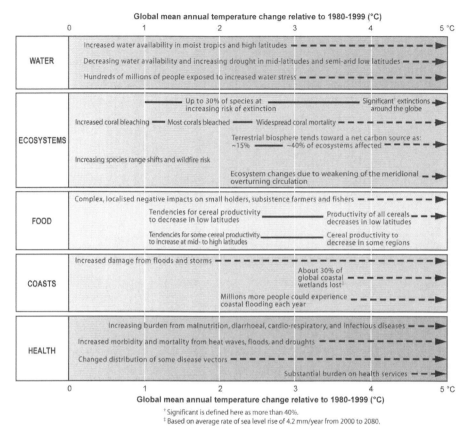

Illustration 5: Figure SPM-1 from the AR4 WGII Summary.

Take food production, for instance; some areas are hurt because of drought while others may see increased productivity. As discussed in later chapters, moderate global warming may actually benefit some areas due to greater rain-fall and higher agricultural production.

The impact on wildlife is also mixed. Some species will have expanded ranges while others may go extinct. Coral seems to have a particularly bleak future. If the deep ocean currents that transport heat around the globe weaken, the effects are unpredictable. These currents are called the meridional overturning circulation (MOC), also referred to as the thermohaline circulation or the "great ocean conveyor belt."

The Causes of Global Warming

What are the causes of the rise in temperature? There are a number of possible causes or contributing factors listed in AR4. Climate scientists refer

24

to the causes as forcing factors or *forcings*. The forcings identified by the IPCC's fourth report are shown in Illustration 6.

Notice that the factors are shown in order of their "level of scientific understanding," from left to right. By level of understanding the IPCC means how confident they feel that their conclusions are correct. If we apply the confidence level definitions from Table 2, the true meaning of this illustration becomes clear. The only forcing factor given a high level of confidence is the leftmost column in the greenhouse gases section. IPCC scientists claim to be 80% confident that they understand the effects of the gases listed in this column, the major one being CO_2. The other two columns in this section are devoted to ozone, O_3.

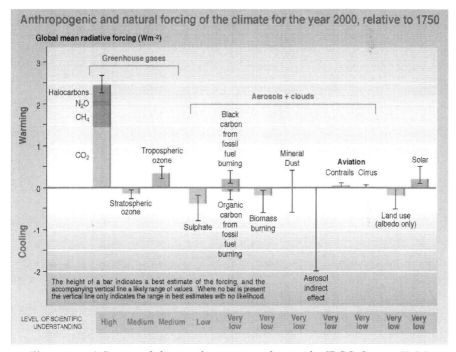

Illustration 6 Causes of climate change according to the IPCC. Source IPCC.

Ozone gets two columns because it can either cause warming if it appears in the troposphere (the lower atmosphere), but can cause cooling if it is present in the stratosphere (part of the upper atmosphere). Either way, the confidence in understanding is rated medium, only five in ten. It's a coin toss that the IPCC scientist's knowledge is correct.

The remaining nine forcing factors are rated low or very low. Understanding of eight of the twelve categories shown are rated as having less than a one in ten chance of being correctly interpreted. So, using the IPCC's own figures and confidence scale, they really only have high confidence in their understanding of less than 10% of the causes of global warming. They admit to not understanding 75% of the causes they list.

Interestingly, the cooling effects shown almost cancel out the warming effects if you add them all up. Notice how the possible cooling effect of aerosols alone could cancel most of the greenhouse gas effect. The bars for the poorly understood factors are drawn as simple lines, not broad distinctive bars. This purposefully obscures their possible impact. Since all of the cooling effects are given low levels of certainty, they have been mostly ignored.

This lack of understanding of the fundamental factors of climate change explains why the case for global warming presented to the public always focuses on greenhouse warming. To further simplify the presentation, carbon dioxide, the largest part of the greenhouse gas column, becomes *the cause*. And because humans emit large amounts of CO_2 into the atmosphere the case is complete—our planet is threatened by human-caused global warming.

Global Climate Models and CO_2

A 1.8°F temperature rise is lower than future temperature forecast by the IPCC. The IPCC has predicted a range of future temperature increase of 2.0°-11.5°F.[34] Rather than simply estimating future temperature increase from the past trend, the IPCC relies on *global climate models* (GCM).[†] These models are complex computer programs that have been under development for several decades. Using these models, climate scientists try to simulate the effects of various environmental factors on global climate. Of the many predictions made by these models, the result that gets the most attention is, of course, temperature.

The complexity of modeling global climate is addressed in best seller, "The Skeptical Environmentalist," by author Bjørn Lomborg.

> "Essentially, answering the question about temperature increase from CO_2 means predicting the global temperature over the coming centuries—no mean feat, given that Earth's climate is an incredibly complex system. It is basically controlled by the Earth's exchange

---

† Also called *global circulation models*, because a major component of these models is atmospheric and oceanic circulation. We will use the more inclusive term.

of energy with the sun and outer space. The calculations comprise five important basic elements: the atmosphere, the oceans, the land surface, the ice sheets and the Earth's biosphere."[35]

Because Earth's climate is amazingly complex, modelers try to find a minimum of input parameters to simplify their programs. Fewer inputs make the programs easier to write and debug. Simpler models also take less computer time to run. A successful model accurately captures the effects of all the important inputs to a system and ignores the unimportant ones.

For a model to be considered accurate it must undergo a process called *validation*. Validation compares a model's results with real world measurements from the past, a process called *backcasting*. During validation, scientists set the values of the input variables to a known state. The model is adjusted (called tweaking or tuning) until the outputs of the model match the known answers for the initial conditions.

But, as modelers will tell you, a model can always be adjusted to give the right output values for a single set of test conditions.[36] Passing validation, or even making one successful prediction, is no guarantee that the model is correct, or that it will provide accurate future predictions. As modeling experts have stated, "in complex natural systems, successful prediction of one event doesn't mean that it will work the next time the model is applied."[37]

Knowing they are on shaky scientific ground, the IPCC doesn't rely on a single model. Their figures are the results of hundreds of different *scenarios*, each based on different assumptions. Perhaps this is done with the hope that hundreds of unsupportable answers will, by sheer luck, stumble upon a correct one. Even that doesn't really matter. The IPCC reports are written by hundreds of authors, modified by hundreds of reviewers, and then submitted to a committee of political representatives for a final edit. The numbers in the IPCC reports are those that are politically acceptable. Politics have created *consensus* science.

Climate modelers have been tweaking their programs for decades, trying to get their models to produce valid answers. As stated, models of Earth's climate are extremely complex. The more complex the system being modeled, the more complex the model, and the longer it takes to get good results. Accordingly, many modeling teams have simplified their models by choosing a dominant input—carbon dioxide. Because computer models are at the heart of the IPCC's climate predictions, and because the GCMs are being driven primarily by atmospheric CO_2 levels, we must examine why this should be so. The subjects of CO_2 induced greenhouse warming and computer climate models will be covered in detail in later chapters.

A Summary of the Problem

In total, the consequences of global warming are rather moderate, at least at the lower end of the temperature increase scale. The IPCC data does not support the strident warnings of impending disaster portrayed in some media stories. Increase in storm activity, famine, drought and epidemics are mentioned but not given prominence. The more extreme effects are only likely to happen if Earth's temperature rises by more than a few degrees over the next century.

As stated earlier, science is not based on political consensus. It is based on real data and provable facts. Unfortunately, the truth about global warming has been obscured by a number of exaggerated, dumbed down, sensationalized claims made mostly by non-scientists for reasons that have nothing to do with the scientific search for knowledge.

Here are the facts that are not in dispute:

- Since around 1850, Earth has been experiencing a general warming trend with temperatures rising about 1.8°F (1°C) per century.

- Human beings are adding large amounts of carbon dioxide to the atmosphere each year by burning fossil fuels that have remained buried in the Earth for hundreds of millions of years.

- Carbon dioxide in the atmosphere has a warming effect on the planet as a whole.

Those are the *facts*, everything else is speculation. All the projected apocalyptic disasters are based on rising temperatures, with the worst damage requiring the most drastic temperature elevation. In turn, the temperature predictions are based, not on empirical evidence, but on CO_2 driven computer climate model programs.

As we will see, these computer programs are incomplete, error prone stand-ins for real experimental science. This fact is well known by the IPCC and the scientists working on the climate modeling programs. Unfortunately, the inaccuracy of climate models is seldom discussed in the media. An exception is this statement in *New Scientist*:

> "Most modelers accept that despite constant improvements over more than half a century, there are problems. They acknowledge, for instance, that one of the largest uncertainties in their models is

how clouds will respond to climate change. Their predictions, which they prefer to call scenarios, usually come with generous error bars."[38]

All a model can do is project our present, limited understanding of Earth's climate into the future. There is no way for climate scientists to test their theories about global warming directly. The only way to do that would be to have a second Earth to run experiments on. Perhaps in the distant future mankind will become so powerful that we can use planets as playthings, but for now, this is only fantasy. Scientists are left having to wait for the passage of time to prove them right or wrong. The best that can be done is to make rational projections based on how Earth's climate has acted in the past.

Filling the gap between what science can and cannot prove is global warming hysteria, manufactured by a scientifically illiterate press urged on by a cadre of special interest groups. Every cause that can possibly establish a link to global warming has done so because that ensures they will get media attention. If you are anti-industrialist, anti-globalist, anti-American, or even vegetarian, the smart move is to jump on the global warming bandwagon. Politicians and celebrities, instinctive seekers of publicity, are totally captivated.

How do the people who work on these IPCC reports see their task? In the words of Chairman R. K. Pachauri, "we are privileged to perform by bringing together the world's best experts and scientists on an ongoing basis to serve you and to serve the interests of the human race and all life on this planet."[39] A noble sentiment, but high-minded ideals are meaningless if the results of your work are distorted, or used improperly.

The Working Group III (WGIII) report on "mitigation" states that "with current climate change mitigation policies and related sustainable development practices, global GHG[†] emissions will continue to grow over the next few decades."[40] In Chapter 17, we will discuss suggestions in the WGIII report that addresses lowering Earth's temperature. One important point to know is that greenhouse gas levels will not stabilize for 100-150 years. This IPCC conclusion is the result of running 177 simulation scenarios. The most aggressive scenarios are based on *negative emissions* of greenhouse gases, achieved using technologies that don't currently exist.

The only way to significantly reduce greenhouse gas emissions right now is to ban the automobile and severely curtail industrial activity worldwide. As hard as this would be on people in developed countries, it would be devastating for underdeveloped nations. Other "cures" proposed by NGOs

---

† Greenhouse Gas.

and special interest groups, such as respecting indigenous peoples, reducing third-world infant mortality, and empowerment of women, while they may be good and noble ideas, have nothing to do with Earth's climate warming.

In the following chapters we will contrast the case made by the IPCC with other theories and opinions expressed by scientists. Despite claims of consensus, many scientists do not agree with the IPCC reports. We will examine Earth's past to see if the current "crisis" is, in fact, unprecedented. We will also determine how much credibility we should give the prophets of doom.

An examination of climate changes in the past, cooling periods as well as warming periods, might tell us something about the forces controlling Earth's temperature. If the current warming trend proves to have historical precedents, particularly during times when Man wasn't adding sizable amounts of greenhouse gas to the atmosphere, then we may gain some insight into the major causes of climate change happening now.

In the next chapter we will examine climate changes in the recent past. We will look back over history, both recorded and from before humans learned to write. Back to a time when the glacial ice sheets retreated northwards, ushering in the relatively warm period we are in today. To understand our present climate we must understand the *ice age* we live in.

Chapter 3 We are in an Ice Age?

"Well, why don't they call it The Big Chill? Or The Nippy Era? I'm just sayin', how do we know it's an Ice Age?"

— *Macrauchenia #1 in the movie Ice Age*

An ice age is a period when Earth's climate cools, leading to the formation of persistent ice sheets, called glaciers, at the poles and on continental land masses. The current ice age has been going on for the last 3 million years. In the movie *Ice Age,* the question was posed, "how do we know it's an ice age?" The answer given was "because of all the ice!" Scientists who study glaciers, *glaciologists*, would agree. They consider any time period when there are permanent, year-round ice sheets in both the northern and southern hemispheres, to be an ice age. That includes the really cold periods, as well as the warmer ones. During the past 3 million years, there have been times when there has been less ice than the present, and times when there has been much, much more.

The Iceman

Around 3300 BC, a lone traveler struggled to make his way across an alpine mountain pass. His joints ached from the cold, caused by what modern people call arthritis. Hunger was beginning to gnaw at him, the last time he had eaten was more than eight hours earlier. That had been a good meal of deer meat, grain and fruit. He had not realized when he set out on the journey that it was to be his last.

A well-dressed resident of the bronze age, the traveler wore a cloak made of woven grass with a vest, a belt, and a pair of leggings. Underneath was a loincloth and on his feet were shoes, all made of leather. On his head was a bearskin cap with a leather chin strap. The shoes were waterproof and wide, seemingly designed for walking across the

Illustration 7: Ötzi the Iceman.

31

snow. Their soles were made of bearskin for added traction with deer hide top panels and a netting made of tree bark. Within the shoes, soft grass wrapped his feet like warm socks.

His belt had a sewn pouch that contained his personal possessions: a scraper, a drill, flint flakes, and tinder for starting a fire. A copper axe with a yew handle, a flint knife, a quiver of arrows with dogwood shafts. Only two of the arrows were finished, twelve were not. A three foot (1 m) yew longbow, unstrung, rounded out his kit. The traveler was well-armed and able to defend himself. According to blood traces found on his clothes and weapons, the bronze age man had recently killed or injured at least four other people. Unfortunately, they were not the only ones injured.

A deep wound in the man's hand throbbed and the arrow he had taken in the back had bled him almost to unconsciousness. He had pulled out the arrow shaft but the head remained stuck in his shoulder. He reached the top of the pass, almost 10,000 ft (3200 m) above the valley, but was exhausted and weakened from bleeding. He could go no further. Stacking his possessions neatly on a snow bank, the traveler sat down, bowed his head and died. He had lived about 46 years—he was to remain buried in the ice for 5,300 years.

The traveler was found by two German tourists, Helmut and Erika Simon, on September 19, 1991. His ice-preserved body was found in the Ötztal valley in Italy, less than 100 yards from the Austrian border. At first, the body was thought to be a modern corpse, like several others that had recently been found in the region. In the Alps, people have been known to die in accidents and become encased by the ice. When it was discovered that the victim dated to pre-historic times he was christened Ötzi the Iceman by the news media.

What prompted the Iceman to leave the hospitable valley below the pass, with no food or water to speak of, and try to cross the mountain at a time of year when several feet of snow obscured the steep, rocky Alpine ridge remains a mystery. It is known that he died from the arrow wound. At first, it was thought that he might have been the victim of a hunting accident. The discovery of other evidence strongly suggests that he was harried to his death by enemies. He must have escaped his attackers or he would not have had his weapons. Above all, he had his precious bronze axe with him when he died alone in the mountain pass. Perhaps the killers attacked and the Iceman managed to fight back, wounding or driving them off before escaping into the mountains. We will never know for sure.

We do know from finding the Iceman's body, and his well-preserved possessions, that thousands of years before the rise of the Roman empire, there were people traveling over passes in the Alps. They were well-dressed for the weather and able to use trails that only recently have become passable to hikers again. Ötzi is called the Iceman, but he lived during a time when Europe's climate was as warm as today's.

We are in what scientists call an *interglacial* period, a time of warming when the ice sheets around the world retreat. The current interglacial is called the Holocene. The last *glacial* period, a time when ice sheets advance, ended about 12,000 years ago. Illustration 8 shows the variation in average temperature over the past million years. For most of that time it was much colder than the present day, indicated by the dashed line, but it has also been hotter than the present on several occasions.

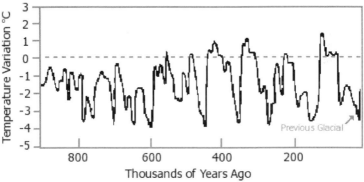

Illustration 8 Temperature variation over the last million years.

The previous interglacial, called the Eemian, ended around 120,000 years ago. It had lasted about 20,000 years, which is the average length for interglacials during the last million years. The temperatures during the Eemian were as high or higher than they are today.[41] CO_2 levels were also about what they are today, though they fell considerably when the last glacial period began. The most recent glacial period is called the Wisconsin in the US. In Europe, it is often referred to as the Weicksel glaciation.

How cold did it get during the last icy period? Actual low temperatures are hard to determine, but ice core data indicates that the average temperature in Antarctica was about 12°F (7°C) colder than today. That may not sound like a lot, but the impact it had on the ice caps was huge. In North America, ice covered all of Canada and the northernmost parts of the United States. The extent of northern hemisphere ice coverage is shown in Illustration 9.

At its peak, some 20,000 years ago, the last glacial advance locked up so much water in ice sheets that sea levels were 400 ft (120 m) below today's levels. Proof of this can be found in the channel of the Hudson river in New York.

Most river valleys end at the ocean's edge, where the lighter, fresh river water disperses on top of the denser, salty ocean water. The power of a river to cut a channel ends there as well. Oceanographers have discovered that the channel of the Hudson extends far beyond the New York-New Jersey coast.

Some parts of the Hudson Canyon rival in size the Grand Canyon of the Colorado. Wide and deep, the Hudson Canyon continues over 450 miles across the continental shelf.[42] This magnificent and mysterious canyon lies next to one of the largest metropolitan areas in the world, yet remains unseen under hundreds of feet of ocean. This hidden valley, one of the largest submarine canyons in the world, bears testimony to a time when the oceans were much shallower than today.

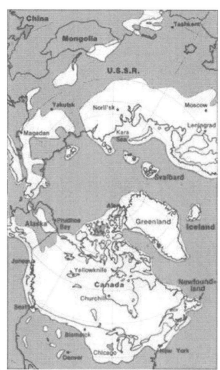

Illustration 9 Ice sheets during last glacial period.

The fact that alternating cold and warm periods have been occurring for a long time indicates that Earth can experience significant climate change without man's interference. The latest warming period, the Holocene, started before human civilization took root, and well before the industrial revolution. This tells us that humans were not the cause of the Holocene warming.

Since the Ice Started Melting

Even though we are in a warm period, that doesn't mean there haven't been recent swings in temperature. Just as temperatures vary from day to day, month to month, and year to year, there are longer term cycles that occur over centuries. Sometimes these changes are significant enough and happen fast enough for people to notice. When this occurs, the episodes are often given names: the Roman Warm Period, the Dark Ages Cold Period, the Medieval Warm Period, the Little Ice Age, and so on.

Illustration 10 shows variation in average temperature since the beginning of the Holocene. Notice the long period starting about 9,000 years ago, when the temperatures were at or above today's level. This period is called the *Holocene Climate Maximum* (HCM), or sometimes the Holocene Climate Optimum, and it lasted for 4,500 years. The Iceman's ill-fated journey through the Alps took place near the end of the HCM.

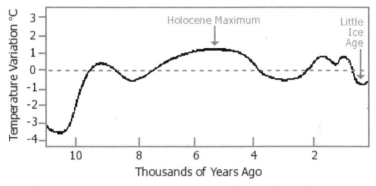

Illustration 10 Temperature variation during the Holocene.

During this time, temperatures at the North Pole rose 7°F (4°C). Some studies suggest that winter temperatures rose more than summer temperatures, and that the warming was limited to the northern hemisphere.[43] These studies claim that temperatures in the southern hemisphere remained close to modern levels or a bit cooler. Some authorities claim this lack of worldwide change proves that warming during the HCM was fundamentally different than global warming today.

Reliable temperature records have existed for only the last 150 years, worldwide records for 50 years, and satellite data for 35 years. We observe greater variation today, in part, because there are more sources of data, providing details about more sources of change. For example, possible sources of temperature change are variation in *insolation* (the energy from the Sun), increase in greenhouse gases due to rock weathering or volcanic activity, the extent of sea ice, and variation in cloud formation rates. There are also a number of temperature exchange mechanisms that affect Earth's climate; ocean currents, deep water up-welling, air currents, floating sea ice, and ice shelf base melting, to name a few. Today all these sources and mechanisms are probed and measured worldwide. We have no way to recreate a similar wealth of data for days long past.

Scientists have discovered that "the Holocene climate developed differently in different regions of the North Atlantic due to different dominant forcing factors."[44] It also appears that different parts of the world experienced the warming at different times, with the southern hemisphere having warming periods before and after the northern hemisphere.[45] According to data taken from Antarctic ice cores:

> "All the records confirm the widespread Antarctic early Holocene optimum between 11,500 and 9000 yr; in the Ross Sea sector, a secondary optimum is identified between 7000 and 5000 yr, whereas all eastern Antarctic sites show a late optimum between 6000 and 3000 yr."[46]

Accordingly, ice core data indicates that not all parts of the world were "hot or cold" at the same time, or by the same amount. Today, not all parts of the world experience the same temperature trends at the same time. The same was true during the HCM.

Recent studies have revealed that, despite the general worldwide warming trend, Antarctica is actually growing colder. From 1986 to 2000, central Antarctic valleys cooled 1.2°F (0.7°C) per decade with serious ecosystem damage from cold.[47] Most of Antarctica now experiences a longer sea-ice season, lasting 21 days longer than it did in 1979.[48] Satellite radar altimetry measurements from 1992 to 2003 indicate that, on average, the elevation of about 8.5 million square kilometers of the Antarctic interior has been increasing. The rising elevation has been linked to increases in snowfall, which translates into a mass gain of 45 billion tons per year, tying up enough moisture to lower sea level by 0.005 inches per year.[49]

Criticisms that claim no global warming took place during the HCM are not substantiated by current available data. Both hemispheres experienced warming during the early Holocene, but not simultaneously. Today, similar conditions exist. The northern hemisphere is warming, while the Antarctic regions are growing colder. Claiming the HCM warming was not global because both hemispheres did not warm simultaneously implies that the current spate of warming isn't global either.

Historic Climate Changes

So far, we have addressed climate change in the pre-historic past, long before there were written records or thermometers. All of the data we have discussed come from what scientists call *proxies*. Proxies are measurements like oxygen isotope ratios extracted from the tiny skeletons of dead sea creatures, pollen counts from lake bed sediment layers, or tree ring widths. These proxies stand in for actual measurements of other parameters, like temperature or CO_2 levels. Proxies are second-hand data. There are many different proxies and we will explain how they are used in Chapter 13, Experimental Data and Error.

The task of compiling climate data becomes easier during recorded human history. Even if our ancestors did not have thermometers, we can collect anecdotal evidence from written documents. Several warm periods and cold periods, that lasted for centuries, were noted by commentators in the past: The Roman Warm Period, 400 BC - 100 AD, the Dark Ages Cold Period, 100-800 AD, and the famous Medieval Warm Period, 800-1300 AD.

Though these historic periods were named by European scientists, their effects were global. From a study of climate in China during the same times comes evidence that China experienced the same climate variations. K. F. Yu, et al. state, "historic records show that it was relatively warm and wet in China during 800-300 BC (Eastern Zhou Dynasty), but was significantly colder and drier in east China during the period of 386-589 AD."[50]

The Chinese researchers continue: "it was so warm during the early Eastern Zhou Dynasty (770-256 BC) that rivers in today's Shangdong province (35-38°N) never froze for the whole winter season in 698, 590, and 545 BC," but that "the period of Southern-Northern Dynasties (42-550 AD) was so dry and cold ... that Beiwei Dynasty (386-534 AD) was forced to move its capital from Pingcheng (40.10°N) to Luoyang (34.67°N) in 493 AD after a series of severe droughts."

Research into the collapse of ancient civilizations in the Americas have revealed links to these shifts in climate. A paper in *Science* observed, "the collapse of Maya civilization in the Terminal Classic Period occurred during an extended regional dry period, punctuated by more intense multi-year droughts centered at approximately 810, 860, and 910 AD."[51] The paper concludes, "new data suggest that a century-scale decline in rainfall put a general strain on resources in the region, which was then exacerbated by abrupt drought events, contributing to the social stresses that led to the Maya demise."

Sometimes, the change in climate was viewed as beneficial. In Europe and the rest of the northern hemisphere, the Medieval Warm Period was considered a period of plenty, with warm summers and mild winters. Crops were good and living was easy, unlike the Dark Ages Cold Period that preceded it, and the Little Ice Age that followed.

The Little Ice Age

One of the most notorious periods of climate variation during the Holocene Epoch is the Little Ice Age. This worldwide cooling trend began at different times in different parts of the planet. There was a great deal of local temperature variation, but it was not just a cold snap—its depths lasted for nearly four centuries. This frigid time coincided with a reduction in the Sun's activity, as indicated by sunspots, known as the Maunder Minimum. This observation has lead many scientists to investigate the link between solar activity cycles and climate.[52,53,54,55] The swing in temperature from the Medieval Warm Period to the Little Ice Age is shown in Illustration 11.

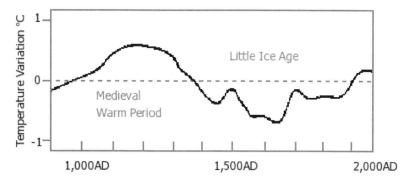

Illustration 11 Temperatures over the last 1,000 years.

Europe experienced a general cooling between the years 1150 and 1460 and very cold conditions between 1560 and 1850. This climate shift, which brought late harvests and wet summers, is blamed for poor health, crop failures, and bad years for wine making. Prior to this period, England had a significant wine industry but, after the 1400s, temperatures became too cold for widespread grape production. Poor grape harvests in continental Europe, particularly in Switzerland and Austria, resulted in a change in drinking preference from wine to beer.

The Little Ice Age caused many disasters: the disappearance of the orange groves from Jiang-Xi province in China, the failure of the Viking colonies in Greenland and North America, and the Black Death in Europe. The disappearance of cliff-dwelling Anasazi culture in the American Southwest, the bitter cold at Valley Forge during the American Revolution, and the Irish Potato Famine have all been attributed to the Little Ice Age.[56,57] There are even claims that the storm that destroyed much of the Spanish Armada, during its disastrous attempt to circumnavigate the British Isles in 1588, was due to shifting climatic conditions.[58]

The harsh climate played an unmistakable role in the course of human events. During the late 13th century, Kublai Khan turned his attention to the Islands of Japan. Mongolian invaders from the Asian mainland were pushed back twice in the span of only a few years. Not by Japanese soldiers, but rather by a devastating natural phenomenon—the Japanese typhoon season. The grateful Japanese named the storms "kamikaze," meaning "divine wind," for protecting their homeland.

In the early 1300s, the old sea route from Iceland, directly west to Greenland, became impassable due to southward expansion of sea ice. Floating ice in the North Atlantic adversely impacted fisheries in Iceland and Scandinavia while Eskimos were seen paddling kayaks off the coast of Scotland. By 1350, this had caused a serious decline in the Viking Greenland settlements. The last reliable account of Norsemen still living in Greenland was around 1410. When German merchants visited Greenland in 1510, they found Inuits living among ruins of Norse settlements.[59]

Mountain glaciers advanced in Scandinavia, New Zealand and the Alps during the 1600s and early 1700s, threatening villages and pasture lands. Today, the Chamonix valley of France is a popular resort area and tourist destination, nestled in the Alps in the shadow of Mont Blanc. But starting in the late 1500s, advancing alpine glaciers, accompanied by massive floods and sudden landslides, began a reign of terror. The rivers of ice descended to the lowest levels in recorded history. In the Chamonix valley, the glacial onslaught entirely destroyed three villages; Bonnenuit, Le Chatelard and La Bonneville. A fourth village, La Rosiere, was severely damaged. Le Chatelard, the oldest village, had been continuously inhabited since the 13th century.[60]

Around 1700, the Mer de Glace (sea of ice) glacier, the largest of the seven Chamonix glaciers, threatened to obstruct the Arve river valley. Local inhabitants of the village of Les Bois resorted to an "exorcism" of the evil glacier spirits. The Bishop was summoned, a procession made its way to the base of the glacier, singing hymns and praying all the way. The Bishop blessed the glacier, sprinkling it with holy water, in the hope of halting the great wall of ice. A statue of Saint Ignatius was erected barring the glacier's path. These religious exhortations only provided temporary relief; as late as 1825, when Les Bois village had to be evacuated, the ice was still advancing.[61]

There were a few positive aspects; in London, the Thames River froze during the winters, leading to big winter carnivals on the ice. In New York City, the harbor froze solid, allowing people to walk to Manhattan and Staten Island. Some have suggested that the cold temperatures increased the density of tree wood, helping to create the distinctive sound of Stradivarius violins.[62] The variable climate forced changes in agricultural methods that ushered in land reform in several parts of Europe.

In 1812, the French Emperor Napoleon Bonaparte, flush from victories across Europe, fixed his gaze on Russia. Napoleon invaded Russia with more than 600,000 men and, though he looted and burned Moscow, the invasion turned into one of the worst military disasters of all time. The coldest temperature recorded on the army's return from Russia was -36°F (-38°C), freezing men and horses to death as they marched. Fewer than 20,000 of Napoleon's men survived the Russian campaign. Many were lost in battle and typhus killed even more, but the enemy that destroyed Napoleon's Grand Armée, the largest army Europe had ever seen, was the cold.

In 1816, New England experienced its "year without summer," when cold, rainy weather during the growing season caused many crops to fail. This low point has been linked to several volcanic eruptions around the world, their volcanic ash further chilling the already cold climate. The art and literature of the time reflected the menacing climate. A survey of historical paintings has revealed that images of the sky painted during the Little Ice Age show significant cloud cover. Depiction of low clouds increased sharply after 1550 and did not decline until after 1850.[63] When Dickens wrote "A Christmas Carol" in 1843, it reflected the cold, snowy winters Britain experienced at that time. By the 1850s, the climate relented and average yearly temperatures slowly began to rise—a trend that continues to this day.

These well-documented historical episodes have been presented here to underline the fact that these periods of climate change actually happened. The evidence for these periods is clear and can be found in the written histories of every nation on Earth. Yet, some modern scientists have cast doubt on their existence.

The IPCC Changes Climate History

Early IPCC reports reflect these well-established historical climate periods,[64] but since the publication of two papers by Michael E. Mann, et al.[65,66] that has all changed. These papers presented a new climate history for the past thousand years based on a reinterpretation of selected proxy temperature records, primarily tree ring data. They are the source of the famed "hockey stick" temperature graph (page 20) that has caused so much controversy.

This revised climate history totally ignores a century of scholarship and scientific study. Both the Medieval Warm Period and Little Ice Age are eliminated—replaced with a slight, almost linear temperature decrease from 1000 A.D. to the early 1990s. All that's left is an abrupt rise in temperature over the last decade, as shown in Illustration 3, on page 20. Bob Carter,

Research Professor of Geology at James Cook University, in an appearance on Radio National, had this to say about the IPCC's revised history:

> "That's where the big argument is centered over the 'hockey stick' because that purports to summarize accurately the temperatures, particularly for the northern hemisphere, over that time period. And the answer is that there is good evidence that in the Medieval warm period and also prior to that in the Roman warm period before the birth of Christ, temperature was at least as warm as it was today, if not indeed a little bit warmer still. Then, in between those warm periods, of course, we've had cold periods such as the famous Little Ice Age experienced in Europe in the 17th, 18th centuries."[67]

There are any number of studies that contradict the IPCC findings by reporting temperatures during the historical warm periods at least as warm as today. For example, scientists have reported that the mean temperature of the Medieval Warm Period in northwest Spain was 1.5°C warmer than it was over the preceding 300 years. During the Roman Warm Period the average temperature was 2°C warmer than today, several decades were more than 2.5°C warmer than today, and one interval longer than 80 years was more than 3°C warmer than today.[68] Many other examples have been reported.[69,70,71,72,73]

As a result of these contradictions, the IPCC's new climate history has caused a furor in scientific circles.[74,75,76] Refusal to release source data and analysis programs prevented other researchers from reproducing Mann's results. This recalcitrance led to charges that Mann and company had "manufactured" the hockey stick by selectively "censoring" datasets that did not agree with their desired outcome.[77,78] In science it is acceptable, even encouraged, to introduce new theories that re-interpret the meaning of data and observations from the past. Changing the data itself, in effect rewriting history in order to better match a theory, is not science. It is deception.

Another Ice Age?

The Little Ice Age came to an end around 1860, and since then Earth has been experiencing a general warming trend. This does not mean that temperatures have been steadily increasing. As we have seen in the previous chapter, there was a significant dip in temperature during the middle of the 20th century. There have always been variations in Earth's climate, and history has shown that turning short-term observations into long-term predictions is not a sound idea. Knowing this, however, has not stopped people from doing just that.

In the 1970s, Earth was entering a fourth decade of decreasing temperatures. Scientists were concerned what this might imply for the future and warned that the climate might be returning to the chilly temperatures of the Little Ice Age, or worse. The popular press ran articles filled with dire warnings of an impending ice age. When the first Earth Day was celebrated on April 22, 1970, a major concern was the coming new ice age. *Science News* pictured rampaging glacial ice knocking over skyscrapers on its March 1, 1975, cover.

Illustration 12 The cover of Science News, March 1 1975.

While the media frenzy in the '70s didn't quite reach the levels of today's global warming "crisis," the story of an impending ice age, covering major northern cities with glacial ice and forcing mankind to retreat to the tropics, was widely reported at the time. The peril of the coming ice age was wide spread in the popular media. In an article, dated November 13, 1972, *Time* magazine reported on the looming crisis:

> "The arrival of another ice age has long been a chilling theme of science fiction. If the earth's recent history is any clue, says Marine Geologist Cesare Emiliani of the University of Miami, a new ice age could become a reality."[79]

Quite a different story from the one presented in the press today. It is hard to fathom, in this time of global warming anxiety, that scientists 30 years ago were worried about global cooling. Like today, the scientific evidence was not conclusive, even if the news media chose to interpret it that way. Dr. Emiliani expressed his uncertainty as the article continued:

> Writing in *Science,* Emiliani reports that the earth has undergone at least eight periods of extreme cold and seven of torrid heat in the past 400,000 years.

> In what direction will the earth's climate then turn? Emiliani refuses to speculate. But if man continues his "interference with climate through deforestation, urban development and pollution," says Emiliani in typical scientific jargon, "we may soon be confronted with either a runaway glaciation or a runaway deglaciation, both of which would generate unacceptable environmental stresses."[80]

The reporter disparages Emiliani's "typical scientific jargon" while at the same time sensationalizing his rather equivocal predictions. Similar circumstances exist today, with scientific possibilities restated in headlines as certainties. This tendency on the part of the press has not diminished, as we will see in Chapter 15.

This cooling crisis produced charts and graphs, much like the ones we are bombarded with today. Illustration 13 shows the alarming downward temperature trend in 1972. Then, as now, concerned climatologists were "pessimistic that political leaders will take any positive action to compensate for the climatic change, or even to allay its effects." They gloomily opined, "the longer the planners delay, the more difficult will they find it to cope with climatic change once the results become grim reality."[81]

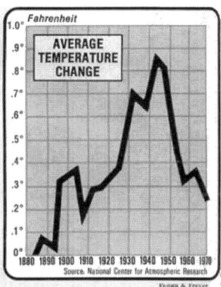

Illustration 13 Temperature trend in 1972, source Time Magazine.

Newsweek, in April of 1975, reported, "there are ominous signs that the earth's weather patterns have begun to change dramatically and that these changes may portend a drastic decline in food production—with serious political implications for just about every nation on earth."[82]

The Newsweek article went on to report delayed growing seasons, disease, drought and devastation. Echoing contemporary warnings about increased severe weather activity, the 1974 tornado season was called, "the most devastating outbreak of tornadoes ever recorded." The increase in tornado activity was blamed on the cooling climate.

Scientists went so far as to suggest melting the arctic ice caps by covering them with soot. Others suggested diverting the flow of northern rivers or building mirrors in space to capture extra sunlight, all in an effort to *warm* the planet. "As for the present cooling trend a number of leading climatologists have concluded that it is very bad news indeed," *Fortune* announced in February 1974.

This type of environmental alarmism has a long and colorful history. On at least three prior occasions the *New York Times* has reported impending climate disaster. Here are some headlines from the past:

- "Geologists Think the World May Be Frozen Up Again." February 24, 1895.

- "America in Longest Warming Spell Since 1776," March 27, 1933.

- "Scientists Ponder Why World's Climate is Changing; A Major Cooling Widely Considered to be Inevitable," May 21, 1975.

Take your choice, fire or ice.

Today, of course, the *Times* headlines proclaim "Science Panel Calls Global Warming 'Unequivocal.'" The article describes the IPCC report as "a grim and powerful assessment of the future of the planet."[83] *Time, Newsweek* and others show similar historical patterns of sensationalist climate reporting. No thinking person confuses the *news* with the truth.

The drum beat of conjecture continues today with wild speculation in journals that report on scientific matters. The British weekly, *New Scientist,* is onboard the impending disaster bandwagon, reporting elevated storm frequency in the North Atlantic[84] and warning of worldwide flooding, "the most deadly of all natural disasters."[85] On the cover of its July 28, 2007, issue, the state of Florida suffers the calamitous effects of rising sea levels, accompanied by a headline in red reading "Goodbye Miami."

Illustration 14: New Scientist, 28 July 2007.

What can we expect in the future?

While scientists can't be 100% positive, most think that we are still in an active ice age. This means that Earth will probably return to glacial conditions, with much colder temperatures, lower sea levels, and advancing ice sheets. The return to glacial conditions assumes global warming hasn't disrupted the natural rhythms of Earth's climatic system.

Some think that humanity's greenhouse gas emissions are sufficient to derail the current ice age and drive the planet into catastrophic hot house conditions. Is humanity really so powerful that our profligate burning of fossil fuels will irreversibly alter the course of climatic development—or is this just hubris?

Another point to consider is this: scientists defining an ice age as a period of time when there are permanent ice sheets in both the Northern and Southern Hemispheres, implies that there have been times when Earth was not in the grip of an ice age. During those times there was no permanent glacial ice—no Greenland ice sheet and no Antarctic ice cap. In the public debate, the disappearance of the world's glaciers is touted as one of the catastrophic effects of global warming. If Earth has been ice free for long periods in the past, how catastrophic can the disappearance of the ice sheets be?

As Winston Churchill said, "The farther backward you can look, the farther forward you are likely to see." To understand the true story of Earth's changing climate, and the magnitude of the forces that shape it, we must dig deeper. We need to understand more of Earth's history: Where our planet came from, how life developed, the interaction of life and the environment, the effect of the Sun, the planets and even the stars.

Chapter 4 Unprecedented Climate Change?

"What has been is what will be, and what has been done is what will be done; there is nothing new under the sun."

— Ecclesiastes 1:9

To read about global warming in newspapers or listen to pundits' sound bites on TV, you can't avoid the impression that the current rate of global warming has no historical precedent, that the increase in temperature is something new and unparalleled in the history of life on Earth. Indeed, part of the justification for taking urgent, drastic action is predicated on the unprecedented nature of this change in Earth's climate. But is this really true? Has Earth's average temperature never reached current or higher levels before? Because if it has, then a number of other questions arise. Questions such as: If the temperature has been higher than it is today, why is this a crisis? What caused the climate to change before? What happened to bring the temperature back down the last time?

To answer these questions we must look at the history of Earth and of life on Earth because Earth's climate is intimately tied to the life it hosts. We will see that Earth's climate has changed life, but that life has also changed Earth's climate. This chapter will trace the history of our planet from its first formation along with the Sun and the other planets of the solar system. We will examine the development of life on Earth from its first appearance, when Earth more closely resembled Hell than a natural paradise. We'll see how Earth's atmosphere was transformed by the smallest of life-forms. Special attention will be given to the time since complex life arose, the geologic time period called the Phanerozoic Eon.

From the Cambrian Explosion, 545 million years ago (mya), through five major extinction events—including the Permian-Triassic extinction that nearly destroyed all life on our planet—to the asteroid that ended the reign of the dinosaurs, and down to the present, life's struggle will be examined. Along the way, despite wide variation in climate and any number of natural cataclysms, an increasing trend toward greater species diversity will be uncovered. In fact, without climate shifts and mass extinctions in the past, mankind would not exist today.

In The Beginning

Scientists believe that our planet was formed about 4.5 billion years ago when our sun and the other planets of the solar system coalesced out of interstellar gases and debris left over from the explosions of other, older stars. At some point, the Sun collected enough hydrogen from space and

47

deep in its heart nuclear fires ignited—our star was born. There was a period called the "Great Bombardment," some 4 billion years ago, when the last rubble left over from the formation of the solar system was swept up by the larger planetary bodies.[86]

Illustration 15 The Earth during the Great Bombardment.

During that time, Earth's crust solidified and was remelted over and over. If we could peer back in time to view our planet, we would not recognize it and we certainly wouldn't want to live there. A collision with a planetesimal (or small planet) the size of Mars nearly destroyed Earth, ejecting large volumes of matter. A disk of orbiting material was formed and this matter eventually condensed to form the Moon in orbit around the planet. This off-center impact also gave Earth its spin leading to today's 24 hour day-night cycle.

First proposed in the 1970s, this *giant impact theory*[87,88] was not believed by most scientists for nearly a decade. However, in 1984, a conference devoted to the origin of the Moon prompted a critical comparison of existing theories. The giant impact theory emerged from this conference as the best explanation of how the Moon formed. Scientists adopted this model of planet formation in which large impacts were common events in the late stages of terrestrial planet formation. This explanation remains the dominant theory of the Moon's origin today.

How long the bombardment lasted is not known, but it must have affected every planet in the solar system. On Earth, the signs of this terrific pounding have long since been erased by weather and Earth's active geology, but the effects of the bombardment can still be seen today in the cratered, pockmarked face of the Moon.

Illustration 16 The Moon's surface as seen from Apollo 11. Source NASA.

Scientists who study ancient life (palaeontologists) and those who study the physical structure and processes of our planet (geologists), classify periods of time based on changes in the layers of rock that make up Earth's crust—the *rock record*. For them, time is broken into Eons, Eras, Periods and Epochs. Eons, of which there are four, are the longest intervals into which *Geologic Time* is divided. The earliest three, the Hadean, Archean, and Proterozoic Eons, are frequently lumped together and referred to as Pre-Cambrian Time or just the Precambrian.

Table 4 shows the time spans of the major geologic time periods and their names. It is important to realize that, when new information about boundaries in the rock record or new measurements dating geological formations become available, the time scale can change. Revisions to the geologic time scale, also called *deep time,* have occurred since its inception in the late 1700s. As scientific tools rapidly improved since the 1930s, the dates attached to the various time spans have been continuously refined.[89] However, the magnitude of adjustments made with each revision have become smaller over the decades.[90]

Eon	Era	Dates (mya)
Phanerozoic	Cenozoic	66-0
	Mesozoic	251-66
	Paleozoic	542-251
Proterozoic	Neoproterozoic	1000-542
	Mesoproterozoic	1600-1000
	Paleoproterozoic	2500-1600
Archean	Neoarchean	2800-2500
	Mesoarchean	3200-2800
	Paleoarchean	3600-3200
	Eoarchean	3900-3600
Hadean		4600-3900

Table 4 Geological Time Periods. Source ICS[91]

The Hadean and Archean Eons

The Hadean, or pre-geologic, Eon is the time period during which Earth was transformed from a gaseous cloud into a solid body. The name Hadean comes from Hades, the mythological underworld of the Greeks. Hadean time is not a true geologic period because no rock that old has survived intact to the present day. The beginning of the rock record available to scientists dates back to the start of the second oldest eon, the Archean.

The oldest rock, dating from about 3.9 billion years ago, shows that there were volcanoes, continents, oceans, and Life on Earth even in those days. Not that life back then was the same as we see around us today. Life started as simple, single-celled organisms, called *prokaryotes* by biologists. *Karyose* comes from a Greek word which means "kernel" or "nucleus," and *pro* means "before." So, prokaryote means "before a nucleus." Though

50

prokaryotes were single-cell organisms of simple construction, they were the most complicated life on Earth for more than a billion years.

These microscopic organisms were to have a significant impact on Earth's environment. Prokaryotes are responsible for building the atmosphere that surrounds the modern Earth and makes our planet a suitable home for higher life forms. The early atmosphere contained nitrogen (N_2), water vapor (H_2O), carbon dioxide (CO_2), carbon monoxide (CO), methane (CH_4), ammonia (NH_3), hydrogen (H_2), helium (He) and trace gases. Hydrogen and helium are very light and tend to escape the atmosphere into space. Oxygen was present, but was bound to other elements. This atmosphere is thought to have been quite similar to the present day atmosphere of Titan, one of the larger moons of Saturn. To humans, and most other life-forms that inhabit Earth today, this atmosphere would have been toxic.

The tiny one-celled organisms, that were the primitive Earth's only inhabitants, labored for more than a billion years to convert the poisonous atmosphere into a more life-friendly form. Over time, the tiny bacteria—in particular, those called *cyanobacteria*—developed ways of fixing nitrogen and freeing oxygen. Cyanobacteria, also called *blue-green alga,* developed a way to use light as an energy source (see Illustration 17). This natural process is called photosynthesis, and it is the same mechanism used by green plants today. Energy from sunlight is used to split CO_2 into carbon and oxygen. The carbon is absorbed, becoming part of the growing plant, and the oxygen is released into the atmosphere.

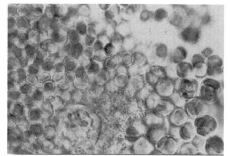

Illustration 17 Chroococcus sp., a type of cyanobacteria. Source NASA.

The climate during the Archean Eon was hot and wet with very warm oceans. Worldwide, volcanic eruptions ejected volcanic ash and dust into the atmosphere causing violent lightning storms and continual rain. In the oceans, reefs were built by *stromatolites,* colonies of photosynthesizing cyanobacteria that formed in shallow waters.[92] There were no corals as there are in today's oceans—they would not appear for billions of years.

The Proterozoic Eon

The Proterozoic Eon was from 2,500 mya to 540 mya. In rock deposited during the Proterozoic Eon, the first evidence of multi-celled organisms is seen. Stromatolites remained common in the shallow waters of Proterozoic Eon oceans. Many other life-forms were also present, such as *archaea,* another type of single-celled organism that have only recently been placed in their own biological *domain.* Archaea may be the only organisms that can live in extreme habitats, such as thermal vents or extremely acidic, alkaline or salty water. Many types of archaea produce methane (CH_4). Today, archaea can be found in the stomachs of cows and the guts of termites where they manufacture more greenhouse gases than all of man's planes, trains, and automobiles combined.[93]

The *Oxygen Catastrophe* was a dramatic environmental change believed to have happened about 2.4 billion years ago, near the beginning of the Paleoproterozoic Era. Measurable amounts of free oxygen appeared in Earth's atmosphere for the first time. Before this significant increase in atmospheric oxygen, caused by the cyanobacteria mentioned previously, almost all life did not require oxygen. In fact, to these types of *anaerobic* organisms, large amounts of free oxygen (O_2) are poisonous. This event is called a catastrophe because for the first time, but by no means the last, most life on Earth went extinct.

This was a positive event for our own species. If the cyanobacteria had not "polluted" the atmosphere with toxic oxygen, the world we live in would not exist. With the advent of an oxygen-rich atmosphere, Earth gained something else that helped Life survive and expand; the ozone layer.

The ozone layer consists of a particular molecular form of oxygen, O_3, that is created when ultraviolet (UV) light strikes free oxygen in the atmosphere. By absorbing UV light, the ozone layer protects life from cellular damage. Photosynthesis, the process that emits oxygen making the ozone layer possible, depends on CO_2. Labeling carbon dioxide a pollutant is an error—CO_2 is essential to life on Earth. So, stealing a phrase from the movie *Men In Black,* we owe our existence today to "barely evolved pond scum."

These developments are the first examples of some common themes that reoccur throughout the history of life on Earth. Life evolves and diversifies into new forms, often with significant impact on the environment. Then, some dramatic event occurs, killing off a significant number of living species. After such extinction events, life struggles back, taking new forms. These new forms expand into ecological niches previously occupied by the unfortunate species that died out; a process called *adaptive radiation* by biologists.

With the dramatic changes in Earth's atmosphere, brought about by the ancient prokaryotes, the evolution of life quickened. About 1.6 billion years ago, during the Mesoproterozoic era, new, larger and more complex forms of life began to appear. These newcomers, called *eukaryotes,* also started as single-cell organisms.

Eukaryotes differ from their more primitive prokaryotic cousins by having a nucleus, a sort of cell within a cell, to contain their genetic material. As mentioned earlier, *Karyose* comes from a Greek word, which means "kernel" or "nucleus." The prefix *eu* means "true" so, as prokaryotic means "before a nucleus," eukaryotic means "having a true nucleus". These differences are shown in Illustration 18.

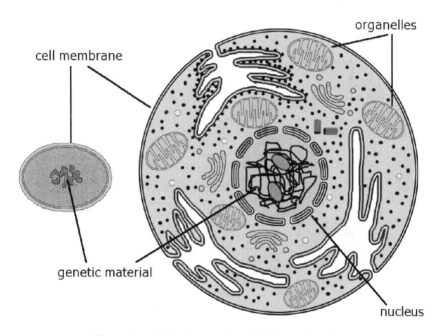

Illustration 18 Prokaryotic and Eukaryotic cells.

The new eukaryotic organisms were much larger than their more primitive cousins, so large that it is believed they surrounded and absorbed some species of the smaller prokaryotes.[94] These absorbed cells became part of the larger organism, often maintaining their own genetic code separate from the host cell's DNA. In this way, it is believed that early plants developed by capturing cyanobacteria, enabling them to reap the benefits of photosynthesis.

Animal cells are also a type of eukaryotic cell, though much evolved from single-cell organisms. Human mitochondria are examples of *organelles,* miniature organs providing specialized services to the cells that contain them while maintaining their own DNA.

The Phanerozoic Eon

The next major milestone for life on Earth occurred at the beginning of the Phanerozoic Eon with the *Cambrian Explosion*. This event, with new multicellular organisms popping up in great profusion, resulted in an explosion of life. It marked the end of the Proterozoic Eon and the beginning of the Phanerozoic, Greek for "visible life." This eon signals the rise of truly complex life, where individual organisms are large enough to be recognized without a microscope.

From this time forward, the pace of development really speeds up, at least when time is measured using geologic time scales. Though time spans at this point are still hundreds of millions of years long, it is prudent to examine subsequent events with a finer grained time scale. To do this we need to change from observing Eons to observing Eras and Periods of geologic time.

Table 5 shows the three Eras that make up the Phanerozoic Eon; the Paleozoic, the Mesozoic, and the Cenozoic. The root word *zoic* comes from the Greek word *zoion* and means "animals." The prefixes *paleo*, *meso*, and *ceno* mean "old, middle, and new," respectively. That makes the three eras of the Phanerozoic the times of "old animals, middle animals, and new animals."

Era	Period	Dates (mya)
Cenozoic	Neogene	23-0
	Paleogene	65-23
Mesozoic	Cretaceous	146-65
	Jurassic	200-146
	Triassic	251-200
Paleozoic	Permian	299-251
	Carboniferous	359-299
	Devonian	416-359
	Silurian	444-416
	Ordovician	488-444
	Cambrian	542-488

Table 5 Geological Eras and Periods of the Phanerozoic Eon. Source ICS[95].

Different geologic time periods are marked by significant changes in the types of creatures living on Earth. The rock deposited during the Phanerozoic Eon contains evidence of fossilized hard body parts from living things and it is this *fossil record* that is used to date rock layers from the three eras. By reading the fossil record, scientists have constructed an outline of the development of life during the time following the Cambrian Explosion.

The Paleozoic Era

The Paleozoic spanned from roughly 542 mya to 251 mya, and is subdivided into six geologic Periods. From oldest to youngest these are: the Cambrian, Ordovician, Silurian, Devonian, Carboniferous, and Permian. The Paleozoic covers the time from the first appearance of abundant, hard-shelled fossils to the time when large reptiles and modern plants began their domination of life on land. From the great explosion of life in the Cambrian, until the catastrophic, worldwide extinction at the end of the Permian, life struggled to evolve. Starting in the oceans, life at the beginning of the era was confined to bacteria, algae, sponges and a few other types of early multicellular animals. With the colonization of the land, many new life-forms emerged, some with relatives that survive to this day. Many others went extinct, disappearing from Earth forever.

The poster child for this early part of the Paleozoic is the Trilobite, a primitive type of arthropod that thrived during the Cambrian and Ordovician. Arthropods are members of the largest *Phylum* of animals, a classification that contains modern insects, spiders and crustaceans. Trilobites were an extremely successful family of animals, with an estimated 10 to 15 thousand different types living over a period of 300 million years.[96] It is thought that their numbers began to decrease with the arrival of the first sharks and other early fish in the Silurian and Devonian periods.

Illustration 19 An ancient Trilobite.

Trilobites went into sharp decline when, during the Late Devonian Extinction, all trilobite orders except one died out. The last of the trilobites disappeared in the Permian-Triassic Extinction, 251 million years ago. The closest living relatives of trilobites might be horseshoe crabs,[97] that are themselves throwbacks to a more primitive time.

There is some evidence that simple life-forms had already invaded the land by the start of the Paleozoic, but large plants and animals did not gain a firm hold on land until the Silurian, 444–416 mya. Although primitive vertebrates (animals with backbones) are known to have existed near the start of the Paleozoic, animal forms were dominated by invertebrates (animals without backbones) until well into the middle of the era. Life on land did not really start to thrive until the Devonian, 416–359 mya.

The Devonian is named after Devon, a county in southwestern England, where rocks from this Period were first identified and studied. During the Devonian, fish populations exploded and life continued to colonize the land. Early bony fish grew legs and learned to walk. Insects spread across the continents along with seed-bearing plants.

Illustration 20 An artist's view of the Devonian.

In 2005, paleontologists working in upstate New York discovered the fossilized remains of the oldest known trees. These plants would look strange, almost alien, next to a modern oak or spruce tree. They had 19-30 ft (6-9 m) tall trunks with pronounced fibrous strands running from bottom to top, like giant stalks of celery. At the top, there were no limbs with leaves or needles. Instead, there were a number of bifurcating branches ending in structures resembling bottle brushes. The branches were about 3 ft (1 m) long, which were shed as the tree grew taller. This forest is thought to date from 385 mya.[98]

The continents of the time would not have been recognizable to us today; in the south was the supercontinent of Gondwana, to its north, Euramerica, with Siberia forming a smaller continent of its own even farther north. All of these land masses were in the process of a slow-motion collision that would form a single super continent, Pangaea, by the end of the era. The movement of the continents will be presented in greater detail in Chapter 8, Moving Continents & Ocean Currents.

Illustration 21 A Carboniferous coal forest.

In the late Paleozoic Era, during the Carboniferous Period, great forests of primitive plants thrived on land, forming extensive peat-swamps. These huge masses of plant matter were buried with sediment, eventually forming the great coal deposits found in North America, Europe and around the world. A global drop in sea level at the end of the Devonian reversed early in the Carboniferous, creating large shallow seas and huge deposits of carbonate minerals. These deposits trapped large quantities of atmospheric carbon that would later form vast beds of limestone.[99]

During the later part of the Carboniferous, the amount of oxygen in Earth's atmosphere was about 35%, nearly double what it is today. At the same time, global CO_2 went below 300 parts per million—a level which is now associated with glacial periods.[100] The abundance of O_2 led to the existence of the largest insects ever seen on Earth. Hawk-sized dragonflies, with 29 inch (75 cm) wing spans, spiders the size of house plants, 5 foot (1.5 m) long centipedes and soup bowl-sized crawling bugs.[101] It was truly a time when insects ruled the planet. Perhaps it's a good thing the atmospheric oxygen level is only 21% today.

Carboniferous plants resembled the plants that live in tropical and mildly temperate areas today. From fossils, we know that many of them lacked growth rings, suggesting a uniform climate. But the climate was changing. By the middle of the Carboniferous, Earth was sliding into an Ice Age, the Permo-Carboniferous. The growth of large ice sheets at the southern pole locked up large amounts of water as ice. Because so much water was taken out of the environment, sea levels dropped, leading to a mass extinction of shallow marine invertebrates, the gradual decline of the swamps, and an increase in dry land.[102]

Many times, these conditions were reversed when the glaciers receded. Glacial melt water was released back into the oceans, and again flooded the swamps and low-lying plains.[103] Carboniferous rock formations often occur as a pattern of stripes, with alternating shale and coal seams indicating the cyclic flooding and drying of the land. Even under these stressful conditions, or perhaps because of them, life continued to develop. By the end of the era, the first large reptiles and the first modern plants, ancestors of today's conifers, had appeared.[104]

At the end of the Permian, life had just weathered the extreme cold temperatures and glaciation of the Permo-Carboniferous Ice Age. Most of the world's dry land was still concentrated in a single super continent, *Pangaea*, meaning "all lands" in Greek. But Pangaea was starting to break up, eventually forming continents we are familiar with today. Events seemed to be progressing when disaster struck. Approximately 251 mya a sudden event, the nature of which is still a topic of hot debate among scientists, caused 95% of all life on Earth to become extinct.[105]

We still don't know the cause of the Permian-Triassic Extinction, or "Great Dying," but we do know that it was of cataclysmic proportions.[106] Was it a meteor? A sudden release of methane from the oceans, or the outbreak of an intense period of volcanic eruptions? The possible causes of this worst of all mass extinctions will be discussed in Chapter 6. Continuing with the history of Earth, the impact this extinction had on the development of life will be examined next.

The Mesozoic Era

The Mesozoic Era spans the time from 251 mya to 65 mya and has three geologic Periods: the Triassic, the Jurassic, and the Cretaceous. It is the era that most people think of when someone mentions ancient life on Earth, the time of the dinosaurs. But things didn't start out with giant beasts roaming the plains in scenes reminiscent of Steven Spielberg's "Jurassic Park" movies.

Life, at the beginning of the Mesozoic Era, must have been very hard. The great Permian-Triassic Extinction had set life back to a state not seen since the early Cambrian. Stromatolites were once again the main builders of reefs in the oceans and life on land was greatly diminished. At least nine of every ten species in the oceans went extinct. This extinction, now thought to consist of two separate events occurring about 10 million years apart, was as disastrous as the next two largest extinctions combined.[107]

Illustration 22 Life in the Jurassic.

Given the unusual severity of the Permian-Triassic Extinction, it might seem reasonable that recovery would be slow—it was. The fossil record from this time shows a period of five to ten million years when life was greatly impoverished, with diversity levels lower than any time since the Cambrian. But slowly life rebounded with the appearance of new and more dynamic life-forms. During this time, the first primitive mammals made their debut along with a new form of reptilian life destined to rule Earth for the next 185 million years. By the end of the Triassic, the first dinosaurs had evolved and the stage was set for the coming Jurassic Period.

The Triassic was also a time of great change in the terrestrial vegetation. The forests were dominated by giant ferns, cycads, ginkgophytes, and other unusual plants. Modern trees, such as conifers, first appeared in their current recognizable forms in the early Triassic. The Triassic Period closed with an extinction event that primarily affected marine life, including most marine reptiles.

By the Jurassic, 200-146 mya, the Age of Reptiles was in full swing. Dinosaurs proliferated, dominating life on land and in the seas. Sauropods, immense plant-eating dinosaurs, were ubiquitous and were preyed on by large flesh-eaters, including *Allosaurs* (the wolves of the Jurassic) and *Megalosaurs* (great lizard). *Pterosaurs* (winged lizard), also known as *Pterodactyls* (wing finger), flew in the Jurassic skies. During the later Jurassic, the first true birds appeared, as did more modern mammals.

Illustration 23
Archaeopteryx, an early bird ancestor. Source Berlin Museum.

The entire Mesozoic Era is thought of as being warm and arid, but there are some conflicting signs. Coral reefs did not have an extended range beyond what they have today, an indication that the climate was more temperate. On the other hand, ferns, which do not thrive in cold climates, are found at higher latitudes than today, a sign of warmer temperatures. One thing is certain, however, there is no evidence of glacial conditions during most of the era. Earth had no polar ice caps.

Geological evidence indicates that the warmth of the Triassic and Jurassic Periods continued into the Cretaceous, 146–65 mya. Global sea levels rose during the Jurassic and Cretaceous, possibly because of increased sea-floor spreading as Pangaea began breaking up.[108] The elevated sea levels created an equatorial seaway, called the Tethys Sea, that circled the globe bringing warmth and moisture to low latitude regions. There is also evidence of warmer temperatures in higher latitudes during the mid-Cretaceous.[109] Earth was a warmer planet by 11°F to 18°F (6°C to 10°C). CO_2 levels were between two and eight times higher than preindustrial values.[110]

The Cretaceous Period brings the height of the dinosaurs' rule on Earth, the first flowering plants, and the oldest known ants. The land was covered with primitive coniferous forests and surrounded by shallow seas. Seasonal variation increased causing mass animal migrations. Several of the more exciting dinosaur species, including the ferocious Velociraptors and

Tyrannosaurus Rex portrayed in the movie "Jurassic Park," were actually residents of the Cretaceous. But somehow "Cretaceous Park" doesn't have the same Hollywood zing.

Finally, the Cretaceous Period, and the entire Mesozoic Era, came to a sudden end with the second-most extensive mass extinction in the history of Earth. Scientists had long speculated about what caused the dinosaurs to die out. Then, in 1980, a research team led by Nobel-prize-winning physicist Luis Alvarez, geologist Walter Alvarez (Luis' son) and chemists Frank Asaro and Helen Michels discovered that sediment layers found worldwide at the Cretaceous-Tertiary (KT) boundary contained high concentrations of the rare metal iridium. This concentration, hundreds of times greater than normally found on Earth, led them to conclude that an asteroid, colliding with Earth, caused the end-Cretaceous extinction.[111]

Illustration 24: The Chicxulub asteroid hitting Earth 65 mya. Source NASA.

These researchers also found many small droplets of basalt rock, called spherules, in the KT boundary layer, evidence that rock from Earth's crust had been melted and flung into the air by a violent impact. Add the presence of shocked quartz, tiny grains of quartz that form in high pressure impacts, to the materials found in the boundary layer, and the evidence seemed clear —an asteroid killed the dinosaurs.[112] Recent research suggests that the impact site may have been in the Yucatan Peninsula of Mexico.[113] This led scientists to name the collision the Chicxulub event, after an ancient Mayan village found close by.

There is still some controversy surrounding the Alvarez Theory. Competing theories put forth other explanations for the demise of the dinosaurs, but no matter what the cause, the Age of Reptiles had come to an end. The Age of Mammals was about to begin.

The Cenozoic Era

The Cenozoic Era began 65 million years ago with the extinction of the dinosaurs and continues into the present. The extinction of the dinosaurs at the end of the Mesozoic Era allowed early mammals and birds to adapt, occupying new habitats and ecological niches. It was as though life on Earth had started over again—for the third time.

Before examining the Cenozoic Era, our time-scale needs to be expanded again. We will be breaking Periods into smaller time spans called Epochs. The Cenozoic is broken into two Periods, the Paleogene and the Neogene, which have three and four Epochs, respectively. The names of these divisions and their dates are shown in Table 6. Note how, as the time periods approach the present day, the time spanned by each Epoch becomes smaller.

The Period names come from the Greek; *gene* means "born," and the prefixes *paleo* and *neo*, meaning "ancient" and "new." So Paleogene and Neogene mean "ancient-born" and "new-born." The Cenozoic Era is sometimes partitioned into different sub-periods, the *Tertiary* and *Quaternary* Periods, using an older naming scheme. The Tertiary spanned all of the Cenozoic, except the time of the Holocene Epoch, which was covered by the Quaternary.

Period	Epoch	Dates (mya)
Neogene	Holocene	0.0115-0
	Pleistocene	1.806-0.0115
	Pliocene	5.332-1.806
	Miocene	23.03-5.332
Paleogene	Oligocene	33.9-23.03
	Eocene	55.8-33.9
	Paleocene	65-55.8

Table 6 Periods and Epochs of the Cenozoic Era. Source ICS.[114]

This naming scheme has recently been abandoned by the International Commission on Stratigraphy (ICS), but the older Period names still appear in scientific papers. We mention this because the end-Cretaceous extinction event is often referred to as the KT extinction. The "K" is for Cretaceous, used to avoid confusion with the earlier Cambrian and Carboniferous Eras, and the "T" is for Tertiary.

The Paleogene Period

The Paleogene Period consists of the Paleocene, Eocene and Oligocene Epochs. During the Paleogene, which lasted about 42 million years, mammals evolved from relatively small, simple animals into many diverse, larger forms. These large animals would come to dominate the land, taking the place of the extinct dinosaurs. Mammals also returned to the sea, adapting and evolving into whales, seals and other present day marine mammals.

Earth's climate cooled throughout the Paleogene, eventually leading to the great Pleistocene Ice Age in the mid-Neogene Period. But before the world cooled down, something unexpected happened; an event that still confounds scientists—the Paleocene-Eocene Thermal Maximum (PETM).

This dramatic event occurred around 55 mya, at the start of the Eocene, and was one of the most rapid and extreme periods of global warming in geologic history. Sea surface temperatures rose between 10°F and 15°F (5°C and 8°C) over a period of a few thousand years. In the Arctic regions, sea surface temperatures rose to a sub-tropical 73°F (23°C).[115] Not only did the surface of the Arctic ocean heat up about 10 degrees, but the ocean depths warmed as well.[116] The temperature of shallow waters off the North American coast rose to a balmy 92°F (33°C).[117]

Rapid warming in the higher latitudes led to significant changes in animal life both on land and in the oceans. Oxygen levels in the oceans dropped sharply and 40% of marine life perished. Around the world, water flowing in ocean currents reversed directions. Scientists now think as much as 4.5 trillion tons of CO_2 were added to the atmosphere,[118] raising the level of CO_2 to ten times current levels. This was real, uncontrollable global warming.

What caused the PETM? As with many events in science, there are a number of competing theories. Some think it was a volcanic outbreak, others a massive release of methane from the deep sea floor.[119] No one knows for sure. Better known is the effect the PETM had on Earth's wildlife.

It is no surprise that such a dramatic shift in climate would have an equally dramatic effect on the living creatures of the time. Still, the effect was rather surprising—animals became significantly smaller. Paleontologists discovered this downsizing from the size of teeth found in the fossil record. During the period of hottest temperatures, mammals were half the size of those that came before them.[120] This warm interval lasted 150,000 years and marked fundamentally different climatic conditions than were present at any other time in the Cenozoic.

Aside from the anomalous PETM, the Eocene Epoch's climate was the most homogeneous of the Cenozoic, with an average temperature difference between the equator and the poles only half what it is today. The polar regions were significantly warmer than today, with temperate forests extending north and south, as far as land allowed. The deep ocean waters remained exceptionally warm, moderating temperatures worldwide. Rainy tropical climates extended as far north as 45 degrees latitude. Only in the tropics was the climate similar to today. Earth's climate during the Eocene was as close to a Garden of Eden as could be imagined.[121,122]Though the ancestors of modern animals were emerging during the Eocene, they were vastly different than animals of today. There were strange creatures around, like *Mesonyx,* a carnivorous wolf-like animal that had hooves at the ends of its stubby toes. Mesonyx belonged to an extinct order of animals called *mesonychids,* who were the only known group of ungulates (hoofed animals) to become carnivorous.

Illustration 25 Mesonyx, a wolf-like mammal from the Eocene Epoch.

After the PETM, larger animals began to reappear. Another member of the mesonychids, *Andrewsarchus,* was found in Mongolia and parts of Asia. "Andrews' Beast," named after paleontologist Roy Chapman Andrews, is the largest carnivorous mammal ever to walk on Earth. Standing more than 12 feet (3.5 meters) at the shoulder and weighing more than 3000 lbs, Andrewsarchus was twice the size of a modern grizzly bear.[123]

At the end of the Eocene, yet another extinction event occurred called the *Grande Coupure* ("Great Break"). It severely affected European and Asian mammals.[124] The cause of the extinction could have been cooling climate, large swings in sea level, or the impact of large asteroids in what is now Russia[125] and the Chesapeake Bay.[126] Whatever the cause, it brought the Eocene to a close.

During the Oligocene Epoch, roughly 34 – 23 mya, Earth continued its slow transformation into the world we know today. The first elephants with trunks appeared, along with early horses. Modern grasses made their debut, plants that would produce vast grasslands during the following Miocene Epoch. The climate still supported vast tracts of tropical greenery, but there was a distinct hint of drier, cooler times to come for much of the world.

The world's climate stabilized. Fragments of the supercontinent Gondwana had fully broken off, with Australia and India moving northwards while Antarctica remained isolated in the extreme south. Once Australia moved far enough north to allow ocean waters to flow unhindered around Antarctica, the circumpolar current formed. This isolated the south polar region and accelerated that continent's descent into frigidity.

Illustration 26 Mesohippus bairdi, a three-toed Oligocene horse. Painting by Heinrich Harder.

Temperatures cooled throughout the Oligocene, with an icecap forming for the first time over Antarctica. Global temperature dropped as much as 15°F (8.2°C).[127] There is evidence of glacial episodes occurring at 29.16, 27.91, and 26.76 mya, causing sea level fluctuations of 160–210 ft. (50–65 m).[128] More dramatic changes were in store as the Neogene Period began.

The Neogene Period

The Neogene started about 23 mya and continues on to the present day; it is our final geologic Period. As we approach modern times, our knowledge of Earth's past climate grows more detailed because there has been less time for age to destroy the evidence. During the Neogene, the warm and relatively stable climatic conditions, present for so much of the Mesozoic and Cenozoic eras, came to an end. This stable period lasted 200 million years—since the world recovered from the Permian-Triassic Extinction.

Through the Paleogene Period, the world was still a warm, humid, and heavily forested place, unlike the drier, colder and much harsher conditions of today. During the Neogene, continental forests were replaced by grasslands and broad savannas. In North Africa and Central Asia, large deserts formed.

The Neogene world looked much like our own. Continents began to take on the familiar shapes recognized today. The continents were changing, as they still are today, creating mountains by slow motion collisions. Throughout the Neogene, new mountain ranges grew. India collided with the underside of Asia and formed the Himalayas, Spain pushed into southern France raising the Pyrenees, Italy moved north into Europe and created the Alps. In what are now the Americas, the Rockies and Andes formed.

Also during the Neogene, northward-moving South America finally caught up with North America. Their collision created the Isthmus of Panama, one of several land bridges formed between continents during this period. The Isthmus of Panama allowed North and South America to exchange animals. Dogs, cats, bears, and horses migrated south while North America gained armadillos, porcupines, and opossums.

The Miocene Epoch, 23–5 mya, makes up most of the Neogene. During this time, the transition from forests to grasslands forced existing animals to adapt or die. Many forest dwellers became extinct and new animals developed that could live on grass. Horses, bison, sheep, giraffes and camels appeared on the plains.

With clear vistas, good eyesight and long legs, these new grazing animals proved harder for predators to catch, causing an escalation in the age-old arms race between predator and prey. Wolves and big cats emerged as the new apex predators. Able to run fast to catch prey, and equipped with powerful jaws and teeth to take them down, these new animals became the dominant predators on the Miocene grasslands.

Around six million years ago, colliding plates in Europe and Africa blocked off the Mediterranean basin, causing the Mediterranean Sea to dry up and disappear. In his book, *The Mediterranean Was a Desert,* geologist Ken Hsu paints a dramatic picture of a "deep, dry, hot hellhole," some 1.8 miles (3 km) below sea level. Geological evidence indicates that this happened not just once, but ten times or more over a million-year span.[129]

The evaporation of the Mediterranean Sea brought about the Messinian Salinity Crisis.[130] Aside from having rather drastic impact on the creatures trapped in the Mediterranean, water from the sea was redistributed to the world's oceans. This caused sea levels to rise and ocean salinity to drop, affecting life worldwide. Eventually, the barrier at the Strait of Gibraltar broke permanently, re-flooding the basin and temporarily creating a spectacular set of water falls bigger than Niagara and higher than today's Angel Falls,[131] over 3,200 ft (979 m).

Ice House World

For reasons that are still not fully understood, 14 million years ago the Antarctic ice sheets stabilized and became permanent. Around seven million years ago, a small temporary ice sheet formed on Greenland for the first time. Still, the early Pliocene was about 4°F (2°C) warmer than today. Some scientists think that the world was undergoing perennial *El Niño* conditions.[132]

El Niño initially referred to a weak, warm current appearing annually at the end of December along the coast of Ecuador and Peru. It usually lasted only a few weeks to a month. Every three to seven years, an El Niño event would last for many months, having significant economic and atmospheric consequences worldwide. During the past forty years, ten of these major El Niño events have been recorded, the worst of which occurred in 1997-1998.

Midway through the Pliocene Epoch, the great Pleistocene Ice Age began in earnest. This period of extreme glaciation was originally thought to have started during the next epoch, the Pleistocene, hence the name of the ice age. The boundary layer that marks the transition from the Pliocene to the Pleistocene has been dated between 1.8 and 1.6 mya, and shows a minor extinction event that affected populations of shallow water clams. Even though we now know this ice age started in the Pliocene, the name remains unchanged—it is called the Pleistocene Ice Age.

One possible explanation for the onset of global cooling is that ocean surface temperatures began dropping when winds started bringing cold, deep ocean waters to the surface in low latitudes. Other theories range from long-term cyclical variation of solar energy to massive volcanic eruptions. Whatever the cause, Earth's last, great, warm period was over.

About three million years ago, large ice sheets began forming over North America and Europe. These ice sheets were to become bigger than the Antarctic ice sheet of the present day. In North America, ice came as far south as Ohio and New York, while glaciers covered most of northern Europe. Ocean sediments from the northern Pacific show ice-rafted debris abruptly began appearing ~2.7 mya,[133] evidence of drifting icebergs that came from glaciers, not pack ice. Only ice that scrapes across land picks up embedded dirt and rocks. Earth had become an ice house world.

The ice age lasted the remainder of the Pliocene and throughout the Pleistocene, with alternating episodes of glacial advance and retreat. During periods of heavy glaciation, sea levels dropped by as much as 400 ft (120 m), establishing land bridges between continents that allowed animals, including man, to migrate to new areas.

The Holocene Epoch

Ice core records indicate that Earth started warming up about 18,000 years ago, though it proceeded in a number of fits and starts. The Pleistocene Epoch ended with the termination of the last glacial period of the Pleistocene Ice Age, about 12,000 years ago. The Holocene Epoch ("new whole") begins with the retreating ice at the beginning of the current interglacial.

Scientists do not know if the Pleistocene Ice Age is truly over, or if we will return to another glacial period, but most seem to think we are going back into the deep freeze.[134] There seems to be a cycle of 20,000 year interglacials separating 100,000 year glacial periods. This is possibly caused by changes in Earth's orbit and the wobbling of its axis. If this is true, then the start of the next glacial period could be right around the corner or 10,000 years in the future.

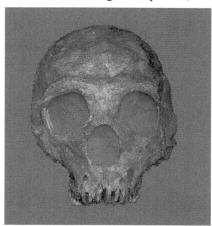

Illustration 27 Neanderthal man, an extinct cousin.

We are creatures of the Ice Age. Many scientists believe that, without the extreme challenges brought on by the advancing ice sheets, mankind would never have developed the survival skills that have brought us to planetary dominance. Great intelligence, ability to collaborate using language, and facile use of tools, has enabled *Homo sapiens sapiens* ("wise wise man") to spread from nearly pole to pole.

Most of our closest relatives were not as adaptable. They were also not as fortunate. Earth is littered with the bones of other failed members of the genus Homo; *Homo habilis, Homo ergaster, Homo erectus* and *Homo neanderthalensis.* Anatomically, modern humans generally have a lighter physical build compared to our earlier cousins. Physically, we were weaker, yet our species was the one that survived.

But we shouldn't be cocky, mankind hasn't been around very long. The oldest fossil evidence for anatomically modern humans is about 130,000 years old[135]— a mere instant in geologic terms. The entire recorded history of our species is contained within the Holocene, the rest of Earth's long past belongs to pre-history and to other species.

Lessons from the Past

So what lessons can we learn from Earth's long history? For one thing, we can say that life is tenacious. Everywhere you look on Earth you will find life, from the Arctic to the Tropics, from the highest mountains to the bottom of the deepest abyssal ocean trench. Living organisms can be found in pitch dark caves, boiling hot volcanic ocean vents, and underneath mile thick glaciers. There are microorganisms, related to ancient archaea, that live in underground oil deposits and have been found in oil wells.[136] There are even

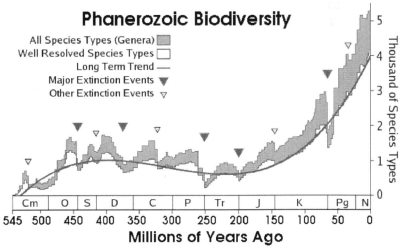

Illustration 28 Biodiversity in terms of number of species types (genera) during the Phanerozoic Eon. After R. Rohde, Global Warming Art.

69

"radiation eating," single-cell fungi that have been found thriving inside the collapsed nuclear reactor at Chernobyl, Ukraine[137]—when humans make a mess, life finds a new habitat.

We are constantly warned that "Mother Earth is in trouble," that we are losing species right and left, that diversity is vanishing. The notion that nature is fragile is puzzling. As we have seen, after every catastrophe, every mass extinction, life springs back. Illustration 28 shows a graph of species diversity throughout the Phanerozoic Eon. Note how after each extinction event diversity increases. Life rises to a challenge.

Far from being the fragile, breakable thing described by some ecologists, life is tougher than ice ages, tougher than asteroids, tougher than anything the Universe has been able to throw at it. A life is fragile, Life is tougher than rock.

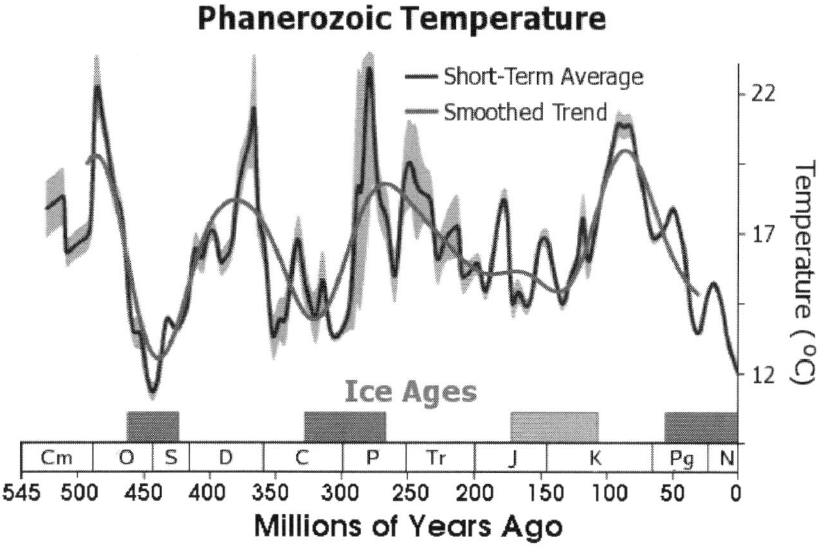

Phanerozoic Temperature

Illustration 29: Temperature over the Phanerozoic from oxygen-18 proxy data. After R. Rohde, Global Warming Art.

Another trend to observe is temperature. Illustration 29 shows estimated average temperature during the Phanerozoic derived from oxygen isotope proxy data. How these temperatures and other environmental factors from the past are measured will be discussed in Chapter 13, Experimental Data and Error. The main observation that can be drawn from these data is that Earth's modern day climate is the anomaly. For most of Earth's past, temperatures have been significantly higher than they are today. Only during previous ice ages are temperatures similar to today found. Comparing

70

modern CO_2 and temperature levels with the past, it becomes clear why scientists say we are still in an active ice age.

The final important observation is the variation of carbon dioxide levels during the Phanerozoic Eon, shown in Illustration 30. CO_2 levels at the beginning of the Cambrian were 25 times the average level during the Holocene. They steadily declined through the coal bed forming Carboniferous, but began to rise again during the Permian. Throughout the Age of the Dinosaurs, CO_2 levels were 5 to 10 times modern levels.

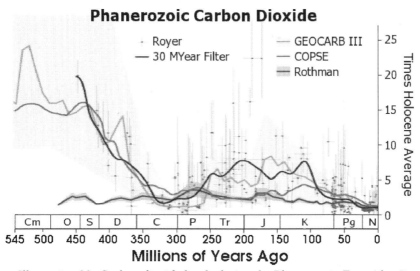

Illustration 30: Carbon dioxide levels during the Phanerozoic Eon. After R. Rohde, Global Warming Art.

The general trend of CO_2 reduction is not surprising, given that there are numerous biological mechanisms that work to store carbon within Earth's strata. The number of ways biology has conspired with geology, both on land and in the oceans, to remove CO_2 from the atmosphere, is astounding. These natural processes dwarf humanity's release of greenhouse gases. The sequestration of carbon will be discussed in greater detail in Chapter 7, Changing Atmospheric Gases. It will be shown that even if all the fossil fuels buried in Earth's crust were burned, it would not make an appreciable bump in the overall carbon cycle. The CO_2 graph would show hardly a wiggle, at least on the geologic timescale.

This is not to say that releasing that much CO_2 would have no effect on Earth's climate. It undoubtedly would. The real question is, would this be a catastrophe? Earth's ecosystem would change and adjust, like it always has, and humanity might not like all of the changes. But nature is always changing, as shown in the graphs. Species change, temperatures change and CO_2 levels change. One of the persistent mysteries of the global warming debate is how those who claim to love nature most, seem to understand it least.

Atmospheric carbon dioxide did not reach modern levels until the onset of the Pleistocene Ice Age. When placed in historical context, the current level of CO_2 in the atmosphere is near a historic low point. Clearly, life has thrived during periods when CO_2 levels were significantly higher than today. This makes the IPCC's obsession with CO_2 levels all the more puzzling. To summarize observations from this review of Earth's past:

- Earth's climate is constantly changing. There is no one "normal" climate pattern.

- More often than not, there have been no polar icecaps on Earth.

- The temperature on Earth has been much hotter than it is now and has been for long stretches of time (more than 100 million years in the Mesozoic).

- CO_2 levels in Earth's atmosphere have been much higher than they are now and were higher for most of the history of life.

- There have been a number of times when Earth has been very much colder than it is right now. There have been several Ice Ages.

- After each ice age, Earth's climate has warmed back up to modern temperature levels or higher.

All of these variations in climate, the warm periods and the ice ages, have come and gone without the presence of mankind. Earth has experienced climate shifts far more dramatic than the recent global warming. Conclusive fact? Man could not have caused them.

As we have seen, the claim that Earth's climate is undergoing "unprecedented" change is totally inaccurate. In fact, when viewed from a long-term historical perspective, the recent warming trend isn't even noteworthy. If anything, our current cool climate conditions are abnormal. We have also seen that elevated CO_2 levels have neither ended nor prevented ice ages in the past, implying that CO_2 is not the primary driver of climate.

What, then, does cause Earth's climate to change? We will examine the mechanisms of climate change that science has discovered in the next several chapters. We will start with a closer look at the causes of radical climate change in the past—ice ages.

Chapter 5 Ice Ages

"The snow doesn't give a soft white damn whom it touches."

— *E. E. Cummings*

People are fascinated by the thought of an Ice Age—a frozen world populated by club wielding cavemen and animals like mammoths and saber-toothed cats. We know that our ancestors actually lived during such an ice age because they left evidence of their presence behind.

Some of the earliest examples of human art date from the last glacial period, stone age cave paintings. Paintings dating back as far as 30,000 years ago have been found in France, China, Tanzania and Australia. This indicates that humans were widely spread across the world by that time.

Other than a few scattered cave paintings, our ancient ancestors left no record of the ice age behind. How did we find out about ice ages? When were they first discovered by Science? How many have there been, how long ago and how often? In this chapter, we will examine the geologic evidence for ancient ice ages predating our ice age, the Pleistocene. Ice ages that may have frozen the planet from pole to pole. But first we will start with the story of a scientific theory and the clues that led scientists to believe that our planet was once covered in ice.

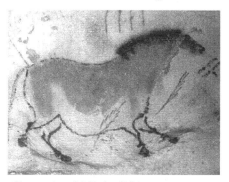

Illustration 31 Paleolithic cave painting of a horse, Lascaux France.

The Father of Glaciology

Louis Agassiz, a nineteenth century naturalist, is considered the "Father of Glaciology." The son of a minister, Agassiz was born in the village of Montier, in the French-speaking part of Switzerland, on May 28, 1807. A physician by training he was also one of the leading naturalists of his day, establishing his reputation with his meticulous studies of fish fossils.

An inquisitive man, with a keen eye for observation, it is not surprising that he would have noticed the signs of glacial activity in his native Switzerland. After all, he had grown up in the shadow of the Alps surrounded by mountain glaciers. For many years, early geologists had been debating the source of a number of unexplained natural phenomenon; scoured rocks, strange mounds of debris called *moraines*, and *erratic boulders*.

Erratic boulders, large rocks found in the oddest places, having nothing in common with the surrounding rock on which they rested had, for years, puzzled the young science of geology. In the Alps, large boulders had been found perched high up on valley walls, well out of the path of any river or stream. Many, at that time, thought that these boulders had been carried to their strange locations by icebergs during the biblical flood. Agassiz examined the evidence he saw all around him, and thought the answer clear —at one time, glaciers must have covered most of Europe.

He first publicly voiced his opinions in 1837 and later, in 1840, published his theory in a book, *Étude sur les glaciers*. It was not well-received by the scientists of the day. One reason for this chilly reception was that geologists and other scientists were still highly influenced by the history of Earth as told in the Bible. Not only did the Bible fail to mention a time when Earth was covered in ice, the time frames needed for Agassiz's glacial theory to work were far longer than what was theologically acceptable.

Illustration 32 Jean Louis Agassiz, 1807-1873.

Add to this theological opposition, the bruised egos and trampled reputations of the geological establishment, and the new theory of an age of ice was roundly met with ridicule. The debate over Agassiz's theory raged for decades. As is often the case in science, great advances can require a fresh point of view, the point of view of an outsider.[138] Agassiz, a physician who had made his reputation as a zoologist, was certainly that.

He was also a scientific genius and in the end, his ideas won out. But, it is not surprising that the geological community regarded the ideas of this outsider with hostility, or that it took decades to displace the older, inaccurate, but accepted ideas. As we will see in the later chapter on how science works, this is an all too frequent story in the history of science.

Agassiz came to the United States and, in 1848, he accepted a professorship at Harvard. He went on to help found the Museum of Comparative Zoology, the National Academy of Sciences and served as a regent of the Smithsonian Institution.[139] Today, he is rightly regarded as a great scientist and visionary.

Agassiz's original concept of an ice age was a period in the distant past when major parts of the northern hemisphere were covered by thick glacial ice.

74

Today, we know that there have been many such periods in the past, and that these periods are not single episodes of glacial conditions. Ice ages consist of many alternating periods of glacial advance and retreat, called *glacials* and *interglacials*. When modern scientists refer to an ice age, they mean the entire climatic episode, including both the warm and the cold periods.

Thanks to Louis Agassiz, and those who followed him, we now know that there have been many dramatic temperature swings in our planet's past. Geologic evidence suggests that Earth has experienced several intervals of intense, global glaciation during Precambrian times[140] and many times since.

Ancient Ice Ages

The true history of ice ages goes back much further than the stone age. In fact, scientists now think that there have been ice ages dating back all the way to the middle of the Archean Eon, around 2.8 billion years ago. We have evidence of this from layers of sediment found in rock formations known to belong to that period.

Glaciers are massive, weighing billions of tons, but they are not static. They move, flowing over the underlying rock of the lands they rest on. The constant grinding of ice against rock, under tremendous pressure, creates a type of sediment, called *glacial drift*. This finely ground material is carried away in melt water, often causing the water in glacial streams and rivers to be milky white.

Geologists have been able to find evidence of glacial drift lying on top of the grooved bedrock of ancient periods. Drift (also called till), turned to rock is called *tillite* and is recognized by geologists as proof of glacial activity. As early as 1891, glacial sediments dating to the Precambrian were found in Scotland and Norway.

Over time, many more sediment layers have been found and identified; the Gowganda tillite in North America and the Bigganjarga tillite in Finland are among the oldest. Presently, there is evidence for an ice age in the late Archean, and five or six in the Proterozoic. Within the Phanerozoic there is evidence for glaciations in the Cambrian, the Ordovician/Silurian, the Permian/Carboniferous, the Devonian, and the late Cenozoic.[141]

Snowball Earth

Eight hundred million years ago, during the Neoproterozoic Era, Earth underwent a monstrous ice age. There is evidence of glacial ice in tropical latitudes, only 15° to 30° north of the equator. In our world, this would mean glaciers as far south as Miami, Florida. Earth would have looked like a different planet, with almost no open ocean and few areas of exposed rock. Only ice and snow, a world of almost pure white.

At that time, most of the land belonged to the super-continent of Rodinia, which formed around 1,100 mya. Rodinia, contained the land that makes up the modern continents today, but not in a configuration we would recognize. North America was in the middle. South America, Australia and Antarctica were packed around North America. Rodinia straddled the tropics, leaving a single vast ocean sweeping across the other side of the globe. There was no land at either pole.

In 1992, Joseph L. Kirschvink, of the California Institute of Technology in Pasadena, put forward a theory that our planet had almost completely frozen from pole to pole, with the only open ocean choked with pack ice. He named this condition "Snowball Earth." Other researchers have calculated that some of the glacial periods during this time had lasted as long as 10 million years. During these periods, the ocean may have frozen over completely, blocking all sunlight and killing most ocean life.[142]

When Rodinia began to break up, 760 mya, it created a new ocean near the equator that was not covered in ice. This allowed a cyanobacterial bloom which depleted atmospheric CO_2. This, in turn, reduced the greenhouse effect, a change that, when combined with the Sun's lower output at the time, created conditions for a permanent deep freeze.

Scientists think that over time volcanoes replenished the atmosphere's CO_2. In fact, they think that the CO_2 levels in the atmosphere rose to 350 times current levels, leading to intense greenhouse warming. The ice sheets rapidly melted, and another giant algae bloom occurred, with similar results to the previous time. CO_2 dropped and the planet reentered its frozen state. This cycle repeated at least four times.[143] Even with the occasional thaw, this time was so pervasively cold, it was named the Cryogenian Period.

Eventually, the growing strength of the Sun, perhaps with a bit of help from emerging forms of life, managed to break the manic cycle of freezing and thawing. When the ice retreated for the last time, life exploded across the planet and the Cambrian Era began.

Researchers now believe that there have been ice ages dating back as far as 2.8 billion years ago. On occasion, these episodes lasted several hundred million years, and may have rivaled the ice age during the Neoproterozoic in intensity. There may have been several Snowball Earth periods in our planet's past.[144]

It is thought that a period of global glaciation ending 2.4 billion years ago may have been the cause of the Oxygen Crisis. Melting of the oceanic ice could have induced a cyanobacterial bloom, leading to an oxygen spike and radically changing Earth's atmosphere. It seems that ice ages, and the development of life on Earth, have been tightly linked from the earliest times.

Phanerozoic Ice Ages

Since the Precambrian, ~550 mya, ice ages have occurred at widely spaced intervals of geologic time, approximately every 200 million years. Varying in intensity and lasting for millions, or even tens of millions of years, these ice ages have had significant impact on the development of life on Earth.

Evidence has been found in South America, South Africa and Morocco for an ice age spanning the late-Ordovician / early-Silurian.[145,146] This ice age occurred from 460 to 430 mya and has been linked to the Ordovician Extinction that occurred 450 to 440 mya.[147] This extinction was the second most devastating extinction for marine life in Earth's history. At this time, much of the world's dry land was concentrated in the super continent Gondwana. One possible cause for this ice age was Gondwana's passing over the North Pole, which resulted in global climatic cooling and widespread glaciation.

A major ice age, from the end of the Carboniferous well into the Permian (350 to 260 mya), is known from South African strata. There is strong evidence for extensive polar ice caps during this period, which is known as the Karoo Ice Age.[148] The name derives from the Karoo region of South Africa where the first clear evidence of this ice age was uncovered. From the sediments found, there appears to have been major ice advance–retreat cycles lasting between 9 and 11 million years. Also found were shorter term ice fluctuations that can be linked to Earth's orbital variations.

There was a dip in temperature towards the end of the Triassic Period. Evidence has been found indicating there was limited glaciation on some continents at the time,[149] but temperatures did not sink to the level of an ice age.

During the late-Jurassic / early-Cretaceous, there is evidence of a rise in global sea level and a climatic shift from warm humid conditions to arid cold climates in the higher latitudes. Studies imply the existence of stable climatic belts during this time with significant temperature differences between the tropics and temperate zones.[150] While not a full ice age, the global climate experienced moderate "ice house" conditions.[151] Finally, an ice age started in the middle of the Cenozoic Era and has continued right up to the present day.

Our Ice Age, the Pleistocene

Technically, we are still in the midst of the most recent ice age, the Pleistocene. During the peak of the Pleistocene Ice Age, vast glaciers, thousands of feet thick, covered most of the northern hemisphere. At their greatest extent the ice sheets covered a third of all land on Earth. Though the ice has retreated significantly during the current warm period, the glaciers on Antarctica and Greenland still cover a tenth of Earth's surface.[152]

As ice ages go, the Pleistocene has not been the longest or coldest Earth has ever seen. Still, a return to glacial conditions would be very disruptive for human civilization. Many entire countries would disappear under the advancing ice. Twenty thousand years ago, the regions now occupied by Chicago, St. Paul, Toronto, London, Copenhagen, and Moscow were all covered by sheets of ice more than a half mile (~1 kilometer) thick.

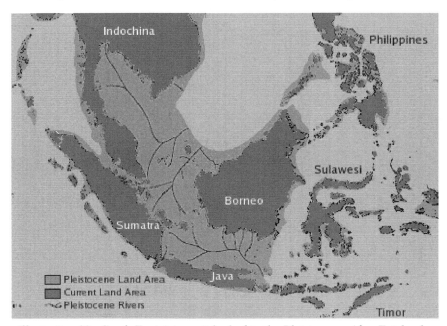

Illustration 33: South East Asia as it looked in the Pleistocene. After Fairbanks.

During an ice age, some areas lose land and others gain. Sea levels drop significantly changing the familiar shapes of coastlines. An example of this can be seen in the shallow seas of Southeast Asia, where there are prominent submerged river valleys, much like the Hudson Canyon off New York.[153,154]

As shown in Illustration 33, Borneo, Java, and Sumatra were once joined with the Asian continent, forming a great peninsula. In places where there are shallow seas today, there used to be broad, fertile plains crossed by mighty rivers.

During the last glacial period, shifting coastlines and merging landmasses caused some plant and animal species to go extinct due to loss of habitat. Others found new areas open to them as land bridges between islands and continents appeared. Make no mistake, the impact of a renewed ice age would be much more dramatic and disruptive than that predicted for global warming.

Will the ice sheets come again? That depends on a number of factors that help drive Earth's climate. As with the causes of global warming, there are many competing theories that attempt to explain why our planet should suddenly ice over. We will cover the major ones in the next section.

But before we talk about why ice ages occur, we will examine what the northern hemisphere looked like at the peak of the last glacial period. Illustration 34 shows the view looking down on the North Pole during the last glacial peak, 18,000 years ago, and how it looks today.

Quoting Doug Macdougall, Emeritus Professor of Earth Sciences at the Scripps Institution of Oceanography:

> "It is worth reiterating ... the Earth is still in an ice age. We are in a warm period, one of the many interglacial intervals that have occurred throughout the Pleistocene Ice Age, but even so, there are significant amounts of permanent ice in the polar regions. It is easy to forget that this may be just a short respite before another glacial interval begins."[155]

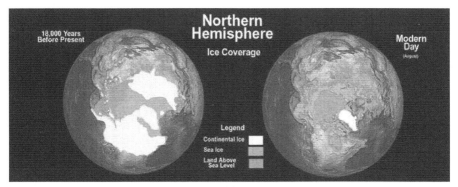

Illustration 34 Earth's North Polar region at the peak of the last glacial period and today. If the glaciers return Earth may look like this again. Source NOAA.

As strange as it seems, the globe in the future could again look like it did 18,000 years ago.

The Causes of Ice Ages

There are a number of factors that could cause Earth to slip into an ice age. Among these are changes in the energy output of the Sun, variations in Earth's orbit, depletion of greenhouse gases in the atmosphere, rising particulates in the atmosphere, and the influence of cosmic rays on low level cloud formation. Some factors are tied to natural cycles; changes in Earth's orbit, the sunspot cycle, the solar system's journey around the galaxy. Others factors are more random in their timing.

Even more interesting might be what caused the various ice ages to come to an end. Covering large portions of Earth's surface with ice and snow would make it reflect more sunlight. Scientists would say that glaciers raise Earth's *albedo*, the measure of how reflective the atmosphere and surface are. A small change in albedo can have a big impact on Earth's climate. A drop of as little as 1% in Earth's albedo would have a major warming influence on climate—roughly equal to the effect of doubling the amount of carbon dioxide in the atmosphere.[156]

Today, Earth's average albedo is approximately 30% meaning it reflects one third of the sunlight hitting it. If Earth was covered in ice, like a giant snowball, its albedo would be about 84%, causing it to reflect most of the Sun's rays. Without some external influence, a frozen Earth would tend to remain frozen.

In most discussions about CO_2 and the greenhouse effect, volcanoes are mentioned as a rather minor source of gas emissions (page 120). Yet scientists have credited volcanic CO_2 with ending the drastic snowball Earth ice ages of the Precambrian. Though biologic activity controls the short-term carbon balance, in the long-term volcanoes provide the majority of atmospheric carbon.[157] This can be shown with some simple math.

If we assume a constant rate of ½ Gt (gigatons) of CO_2 from volcanoes each year, as stated in Chapter 7, then the total emissions over a million years would be 500,000 Gt. This is an amount greater than all the carbon in all of the biologic sinks in the biosphere. If all of this gas remained in the atmosphere, the CO_2 level would be more than 680 times modern levels. We know that there are multiple processes removing CO_2 from the atmosphere, but scientists suspect that these processes might not always be sufficient to keep the CO_2 buildup in check. There are two reasons for this.

First, the amount of volcanic activity was much higher on primitive Earth. The geologic record contains proof of many periods with extensive eruptions.[158] Traditional paintings of dinosaurs, with smoking volcanoes as a backdrop, are geologically accurate.

Second, the presence of huge ice sheets, covering most of the planet would have drastically curtailed biological activity; the very activity that works to remove CO_2 from the atmosphere. Without large stretches of open ocean, CO_2 would continue to accumulate in the air. The buildup that ended the Cryogenian Ice Age is thought to have had CO_2 concentrations 350 times present levels. Atmospheric CO_2 levels would have exceeded 10%, a concentration that would asphyxiate human beings.

The time of the Permian-Triassic extinction, 251 million years ago, is marked by massive floods of volcanic rock in Siberia that cover an area 1.6 million square kilometers, roughly the size of western Europe.[159,160] This area, known as the Siberian Traps, contains rock deposits a mile thick in places.

The volcanic activity in Siberia did not resemble the spectacular, explosive eruptions of recent volcanoes like Mount St. Helens or Mount Pinatubo. Unlike cone-shaped mountains spewing out smoke and molten rock, the eruptions of the Siberian Traps were characterized by large amounts of lava oozing out of cracks in the ground. The volume of rock contained in these lava flows, called *flood basalts,* was so great it could have paved the entire Earth with a layer 20 ft (6 m) thick.[161]

Illustration 35: Lava at Kamaomao, Hawaii, photo by R. Hoblitt, USGS.

During this period of near constant eruptions, lasting a million years, Siberian volcanoes are thought to have pumped out 10,000 gigatons of carbon dioxide into the atmosphere, 14 times the amount present today.[162] This injection of CO_2 may have been the trigger that ended the Karoo Ice Age and there are those who think that volcanoes, not an asteroid collision or nearby supernova, were the cause of the Permian-Triassic extinction.[163]

These are examples of CO_2 causing significant global warming events, but the rise in atmospheric levels involved are hundreds of times the amount generated by human sources today. Even more telling are findings that during the latest ice age, increased CO_2 levels followed rising temperature. This implies that CO_2 doesn't drive temperature, it follows temperature.

Further evidence of the nature of CO_2 levels comes from a study of fossil birch tree leaves from the very beginning of the Holocene.[164] About three centuries after the initiation of Holocene warming, a 150 year cooling period took place, called the Preboreal Oscillation. During this temperature dip, the reconstructed CO_2 levels ranged from ~0.030% to ~0.034%. These levels contradict previous assumptions that Holocene CO_2 levels were in the 0.027% to 0.028% range until the beginning of the Industrial Revolution. Atmospheric CO_2 may reinforce change in global temperature, but it is not the reason the temperature changes, except under the most extreme circumstances.

A Species Shaped by the Ice Age

Modern humans first appeared on Earth between 130 and 150 thousand years ago. Our earliest ancestors probably began spreading out of Africa during the Eemian interglacial, but the worldwide human diaspora didn't occur until the Pleistocene Ice Age reasserted itself and the most recent glacial period began. Greatly lowered sea levels and barren wide open plains allowed humans to walk to Europe, Asia and North America, eventually reaching Australia and the farthest tip of South America.

Bands of hunter-gatherers spread across the face of the planet, collecting food during the shortened summers and living in caves for the nine month long winters. It was during this time that scientists think Homo sapiens acquired their full cognitive abilities.[165] For 100,000 years, humankind was tested and molded by the ice age climate. More than 200 generations, most of our species' existence, was spent under these conditions. We are truly a species shaped by the Ice Age.

The intelligence, cunning and skill at making weapons that allowed our species to survive the harsh climate, would serve our kind well when the glacial period ended and the Holocene warming ensued. The warming climate caused a shift in human society, a change from a nomadic existence to farming and, eventually, civilization. The rest is, quite literally, history.

Just as climate helped shape our species, there is no question that humans have altered the climate. Agriculture, irrigation, and deforestation have had a significant impact on Earth's environment from ancient times. With the advent of industrialization, the impact has increased. It is understandable that the most successful species on Earth has had a dramatic impact on its environment.

In altering the environment to better suit our needs, we have also changed the climate. But at the same time, we may have lost some of the adaptability that our ice age ancestors depended on to survive. Archaeologist Brian Fagan has theorized that, while civilization has made people less susceptible to small variations in the environment, our technology has amplified the impact of the infrequent, but inevitable, large catastrophes. He believes that our environmental vulnerability is greater than we imagine.[166] We will revisit these concerns in Chapter 16, The Worst That Could Happen.

Ice ages are among the most radical climate changes that occur on Earth. The fact our planet has transitioned several times from very warm conditions to very cold conditions, is an indication that there are large forces at work. We can ignore the Snowball Earth ice ages that occurred before Earth gained its complex biosphere, and concentrate on the more recent ice ages. We have shown that climate variation during ice ages is not driven by CO_2 levels. In fact, temperature does not follow CO_2 levels, CO_2 levels follow temperature. So, what does cause ice ages and, of more immediate concern, what causes the alternating glacial and interglacial periods of the Pleistocene Ice Age?

Most scientists think that it is the periodic variation in Earth's orbit around the Sun. According to Orrin H. Pilkey, Professor Emeritus of Geology at Duke University, and Linda Pilkey-Jarvis, a geologist with the Washington State Department of Ecology:

"The fundamental causes of the ice ages and their huge sea-level changes are various Earth orbital changes, such as the tilt of the axis of spin and the eccentricity of the orbit around the sun. These changes are responsible for changes in the location and intensity of solar radiation on the surface, which in turn determines global climate."[167]

The orbital changes referred to are called the Milankovitch Cycles, and we will examine them in detail in Chapter 9. But, as we have said, Earth's climate is very complex—there are other contributing factors.

Other drivers of ice ages, and by inference, climate, have been proposed. These include: changes in Earth's atmosphere, shifting continents, geologic uplift of mountain ranges, variation in the energy received from the Sun, and possibly radiation from exploding stars in the form of cosmic rays. We will discuss the prominent scientific theories for each of these topics in separate chapters. Before we investigate the more periodic sources of change, we will examine mass extinctions and the seemingly random events that cause them.

Chapter 6 Ancient Extinctions

"What a book a devil's chaplain might write on the clumsy, wasteful, blundering low and horribly cruel works of nature!"

— *Charles Darwin*

When looking at the history of life on Earth, the most dramatic events are the great mass extinctions. These ancient ecological incidents may seem to be terrible, wasteful, pointless catastrophes, but consider the following statement from the Smithsonian Institution Department of Paleo-biology:

"Extinction is the complete demise of a species. It takes place when all individuals of a species die out. Extinction has occurred throughout the history of life on Earth. It is the ultimate fate of all species. In fact, it has been estimated that 99.9% of all species that have ever lived on Earth are now extinct."[168]

A sobering thought, that the ultimate fate of all species is extinction. Since the threat of mass extinction is often cited as a consequence of global warming we will investigate the causes of previous extinction events. When did science first come to suspect that some species have vanished from Earth forever? That is an interesting story involving a pair of French naturalists, an American president, and some ancient bones found by the Ohio river.

The Ohio Animal

Thomas Jefferson, revered as the author of the American Declaration of Independence, and third president of the United States, was also an avid amateur scientist and noted naturalist. A man of the Age of Reason and a product of the Enlightenment, Jefferson was considered an expert in architecture, civil engineering, physics, mechanics, meteorology, anatomy, and botany.

He could read and write Greek, Latin, French, Spanish and Italian. A mathematician and astronomer, he made suggestions for improving almanacs and accurately calculated the eclipse of 1778. He was recognized as a pioneer in ethnology, geography, anthropology and, as our story will show, paleontology.[169]

Illustration 36: Thomas Jefferson, painted by C.W. Peale, 1791.

As John F. Kennedy said in 1962, to a White House gathering of 49 Nobel Prize recipients, "I think this is the most extraordinary collection of talent, of human knowledge, that has ever been gathered at the White House—with the possible exception of when Thomas Jefferson dined alone."

During his presidency, Jefferson commissioned the Corps of Discovery, led by Meriwether Lewis and William Clark, to explore the American interior. They were instructed to collect and send back samples of plants, animals, and rocks along the way. Jefferson filled rooms in the White House, and later his home, Monticello, with the specimens they sent back.

Jefferson had such an interest in paleontology that he commissioned, and personally paid for, William Clark to explore the famous fossil deposits at Big Bone Lick, Kentucky.[170] Because of his efforts to establish scientific stratigraphy in the US some have dubbed him the "Father of American Paleontology," but Jefferson never claimed to be more than an enthusiastic amateur.

When the influential French naturalist Georges-Louis Leclerc de Buffon spoke disparagingly of new world animal life in his 36 volume treatise, *Histoire Naturelle,* Jefferson was less than pleased. Jefferson assessed Buffon's pronouncements this way:

"The opinion advanced by the Count de Buffon, [Buffon, xviii. 100, 156.] is 1. That the animals common both to the old and new world, are smaller in the latter. 2. That those peculiar to the new, are on a smaller scale. 3. That those which have been domesticated in both, have degenerated in America: and 4. That on the whole it exhibits fewer species. And the reason he thinks is, that the heats of America are less; that more waters are spread over its surface by nature, and fewer of these drained off by the hand of man. In other words, that heat is friendly, and moisture adverse to the production and development of large quadrupeds."[171]

In *Notes on the State of Virginia,* written in 1781, Thomas Jefferson vigorously defended the wildlife of North America against these insults. Buffon claimed, wrote Jefferson, "that nature is less active, less energetic on one side of the globe than she is on the other ... as if both sides were not warmed by the same genial sun." Jefferson's analysis of Buffon's hypothesis continued:

"I will not meet this hypothesis on its first doubtful ground, whether the climate of America be comparatively more humid? Because we are not furnished with observations sufficient to decide this question. And though, till it be decided, we are as free to deny, as others are to affirm the fact, yet for a moment let it be supposed. The hypothesis, after this supposition, proceeds to another; that moisture is unfriendly to animal growth. The truth of this is inscrutable to us by reasonings a priori. Nature has hidden from us her modus agendi. Our only appeal on such questions is to experience; and I think that experience is against the supposition."[172]

Years of observation had convinced Jefferson that animals were actually larger in America than in Europe. When an expedition to the Ohio river valley returned with bones from what appeared to be a gigantic elephant, he viewed the remains as conclusive proof that Buffon was in error. Judging from the size of bones and teeth, this animal was five or six times larger than an elephant. Though nobody had ever seen such a beast roaming the American countryside, Jefferson was sure it existed somewhere in the vast continental interior.

Illustration 37: American Mammoth, source U.S. National Parks Service.

"Our quadrupeds have been mostly described by Linnaeus and Mons. de Buffon. Of these the Mammoth, or big buffalo, as called by the Indians, must certainly have been the largest. Their tradition is, that he was carnivorous, and still exists in the northern parts of America."[173]

Whether any of these "Mammoths" were found alive or not, here was proof that America was home to Earth's largest animal. In 1797, Jefferson arranged for some specimens to be shipped to Europe to bolster his assertions.

Georges-Louis Leclerc, Comte de Buffon, was the most influential naturalist of his day. He originated Buffon's Law, which states that despite similar environments, different regions have distinct plants and animals. These observations led him to conclude that species must have both "improved" and "degenerated" after dispersing away from a center of creation, a foreshadowing of Darwin's theory of evolution.

One hundred years before Darwin, in his *Histoire Naturelle,* Buffon noted the similarities of humans and apes and even talked about a common ancestry of apes and Man. When Buffon published *Les Epoques de la Nature,* in 1788, he openly suggested that Earth was much older than the

6,000 years commonly accepted and proclaimed by the church. Based on the cooling rate of iron, he calculated that the age of the planet was 75,000 years. This proclamation was condemned by the Catholic Church in France, who burned Buffon's books. A man ahead of his time, *Les Epoques* discussed concepts very similar to Charles Lyell's "uniformitarianism," which will enter our tale 40 years later.

A prolific and flamboyant writer, Buffon was criticized by many of his contemporaries. Voltaire did not approve of his writing style, and d'Alembert[‡] called him "the great phrasemonger." Unfazed, Buffon said of his critics' attacks, "I shall keep absolute silence ... and let their attacks fall upon themselves."

Illustration 38: Georges-Louis Leclerc, Comte de Buffon.

Though Buffon set the stage for the discovery of extinction he was not the naturalist who first offered proof of the transient nature of terrestrial species. But, it was through the spat between Jefferson and Buffon that another French naturalist, Georges Cuvier, would come to know of the "Ohio Animal." It was to Cuvier that Jefferson shipped his samples of Mammoth bones in 1797.[174]

George Cuvier, was born on August 23, 1769, at Montbéliard, a French-speaking community in the Jura Mountains that was not under French jurisdiction at the time. This area was ruled by a German nobleman, the Duke of Württemberg.[175] The son of a retired army officer, his family had emigrated to avoid religious persecution. After a strict Lutheran upbringing, the young Cuvier studied at the Carolinian Academy in Stuttgart, a school which the Duke had founded. During his studies he became familiar with the works of Linnaeus[†] and Buffon, sparking an interest in nature and science.

Eventually Cuvier moved to Paris, where he became a professor of animal anatomy at the newly reformed Musée National d'Histoire Naturelle. In his

---

‡   Jean d'Alembert, 1717-1783, was a French mathematician and pioneer in the use of differential equations and their application to physics.

†   Carolus Linnaeus, 1707-1778, Swedish botanist who laid the foundations for modern taxonomy. His system for naming and classifying organisms is still in use.

first important paper, read publicly in 1796, *Mémoires sur les espèces d'éléphants vivants et fossiles,* he analyzed skeletal remains of Indian and African elephants as well as mammoth fossils found in Siberia. Also included in his study was the fossil remains of the *animal de l'Ohio,* the "Ohio animal."

Illustration 39: George Cuvier, painted by F.A. Vincent.

Cuvier's analysis proved that African and Indian elephants were, in fact, different species, and that mammoths were not members of either species. He further stated that the "Ohio animal" represented another distinct species that displayed even greater differences from living elephants than mammoths. He went on to hypothesize that both mammoths and the "Ohio animal" represented species that must be extinct.

Those first fossils Cuvier examined turned out to not be the same species as the later bones sent to him by Thomas Jefferson. Jefferson's creature was identified as an American Mammoth, one of three species that lived in what is now the mainland United States. In honor of Jefferson, that particular species is now known as *Mammuthus Jeffersonii,* Jefferson's Mammoth. Years later, in 1806, Cuvier would return to the original "animal de l'Ohio" in another paper and give it the name Mastodon.

Ever since fossils had been discovered, scientists assumed that they were remains of living species. This belief was reinforced by religious teachings of the time, which said that no species had ever gone extinct. This was based on the assumption that anything God had designed was perfect and couldn't die out. By the start of the 19th century, this ideology began to come under increasing scrutiny from scientists, most notably Cuvier.

In his second paper, Cuvier analyzed the fossil remains of a giant ground sloth, named *Megatherium,* that had been found in South America. Megatherium was one of the largest mammals to walk on Earth. Weighing as much as an African bull elephant and, when standing on its hind legs, about twice as tall, it was unlike anything ever seen alive. Comparing the fossil bones with the skeletons of modern sloths, he found that Megatherium represented a distinct species, and again the evidence indicated that this species was extinct.

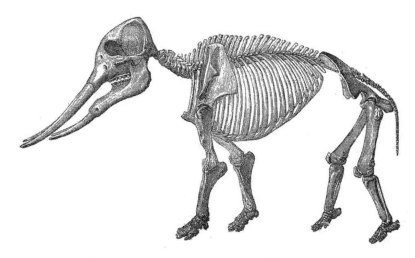

Illustration 40: The skeleton of an American Mastodon, source Kameno Doba by Jovan Zujovic, pub. 1893.

Cuvier now firmly believed that animal species went extinct. In his 1796 paper on elephants, he wrote, "All of these facts, consistent among themselves, and not opposed by any report, seem to me to prove the existence of a world previous to ours, destroyed by some kind of catastrophe." A belief that history was filled with natural catastrophes, which caused many species to become extinct, formed the basis of the geological school of thought called *catastrophism.*

Appointed Inspector-General of public education and State Councilor by Napoléon Bonaparte, Curvier outlasted the Napoleonic Empire. He continued as a state councilor under three successive Kings of France after the Bourbon Restoration. Showing an amazing talent for survival, he managed to serve under three different, opposing French governments; the revolutionary French First Republic, the Napoleonic French Empire, and the restored monarchy. In 1826, he was made grand officer of the *Legion of Honour,* and in 1831, he was elevated by King Louis Philippe to the rank of Peer of France.

Through all this, he helped establish vertebrate paleontology as a scientific discipline and created the method of comparative anatomy. For a brief period of time, starting in November of 1831, Louis Agassiz, whose work we discussed in Chapter 5, was Cuvier's student and colleague. Cuvier died in April 1832, and though their relationship had lasted only six months, Agassiz credited Cuvier with opening his eyes to the workings of nature. For the rest of his life, Agassiz promoted and de-fended Cuvier's geological catastrophism and classification of animals.

After Cuvier's death, the catastrophic school of geological thought lost ground to *uniformitarianism*. This school of thought claimed that the geological features of Earth were best explained by observable forces, such as erosion and volcanism, acting gradually over an extended period of time. Uniformitarianists, or gradualists, were among the first to argue that Earth must be several billion years old.

Championed by Scottish geologist Charles Lyell (1797-1875), who had been influenced by fellow Scotsman, James Hutton, this theory of slow, gradual change became the dominant mode of thinking in geology and

Illustration 41: Megatherium, subject of Cuvier's 2nd paper.

paleontology. In his widely read, three volume *Principles of Geology*, Lyell showed that Earth must be very old based on the supposition that it was subjected to the same sort of natural processes in the past that shape the land in the present.

So strong was the influence of uniformitarianism, that it was impossible to deviate from its tenets up until a few decades ago. Uniformitarianism is credited with much of the resistance to the Alvarez Theory that an asteroid strike caused the KT extinction. The strength of that theory, and its eventual widespread acceptance, has led to a re-evaluation of catastrophism and the work of Cuvier. Nowadays, it is recognized that portions of both schools of thought are correct. Earth and living creatures do undergo gradual long-term changes, but sometimes catastrophes occur causing more abrupt shifts.

The Tree of Life

When scientists discuss extinctions, they rate the severity of an event based on how many species were extinguished. But they also consider the impact on groups of species that are closely related. There are several complicated classification schemes for living things. Among the most recent is the three domain system introduced by Carl Woese in 1990.

The *phylogenetic tree,* shown in Illustration 42, reflects the very top of this classification scheme. This type of organized classification is called a *taxonomy*, and the entries are called *taxa* (singular *taxon*). A taxon name can designate an organism or group of organisms. A taxon is assigned a rank and is placed at a particular level in the systematic hierarchy reflecting evolutionary relationships.

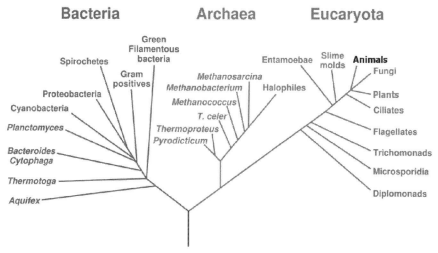

Illustration 42: A phylogenetic tree of living things, based on RNA data and proposed by Carl Woese, showing the separation of bacteria, archaea, and eukaryotes. Wikipedia.org.

Aristotle divided the living world between animals and plants. This system was followed over 2,000 years later by the first hierarchical classification of Carolus Linnaeus, who we mentioned earlier in this chapter. Originally, scientists assigned taxa based on the shape or appearance of organisms, their *morphology*. More recently, a school of thought has arisen where taxa are classified only by their evolutionary relationships. Called *Cladistics*, from the Greek word *klados*, meaning branch or twig, this approach has caused much reorganization of the Tree of Life. To add to the confusion, Cladists sometimes refer to taxa as *clades*.

The taxon hierarchy, from the largest units to the smallest, looks like this: Domain, Kingdom, Phylum (or Division), Class, Order, Family, Genus and Species. There is a mnemonic to help remember this scheme, "Distinguished Kings Play Chess On Fine Green Silk." These classes can be further subdivided into Sub-Classes, Sub-Orders, Sub-Families, etc.

All members of a species are, at least in theory, able to interbreed and produce fertile offspring. Several species may belong to a genus but, if they interbreed, members of different species within a genus are unable to produce fertile offspring. An example of this is when horses and donkeys are bred to produce sterile mules. Our genus, Homo, has one surviving species, sapiens. The other members of our genus, twenty-two species at last count, all died out in prehistoric times.[176] But, for a time, modern humans cohabitated with other human species, most notably *Homo neanderthalensis*. In recent times, there has been much speculation that modern humans interacted, and may have interbred, with Neanderthals.[177] Though there is no proof[178] of this, there have been a number of science fiction novels based on this premise.

Several *genera* (plural of genus) can belong to the same family, several families to an order, and so on up the hierarchy. Eventually, kingdoms are placed into one of the three domains; Bacteria, Archaea, or Eucaryota. Humans belong to the Eucaryota domain, in the kingdom *Animalia* (i.e. The animal kingdom), phylum *Chordata,* class *Mammalia*, order *Primates,* family *Hominidea*, genus *Homo*, species *Sapiens*. Even for scientists, this is too long a naming scheme for daily use. Instead, the *binomial* naming scheme is commonly used. Binomial designations use both the genus and species names, with the genus capitalized. In this scheme, we are Homo sapiens. There is also a *trinomial* naming scheme that adds a subspecies designation. Under this scheme, we are Homo sapiens sapiens, Latin for "wise wise man."

Membership in classes, orders and the other higher taxa, are decided on structural or anatomical similarities among life-forms. For example, both modern birds and dinosaurs belong to the sub-class *Diapsida*. Diapsids (Latin for "two arches") are a group of animals that developed two holes in each side of their skulls (*temporal fenestra*).

These creatures first appeared about 300 million years ago during the late Carboniferous period. Living diapsids are extremely diverse, including all birds, and all reptiles except turtles. Over time, lizards lost one skull hole and snakes both holes, but they are still classified as diapsids based on their ancestry. Modern bird skulls also diverge significantly from the skulls of primitive diapsids, but they are also classified as Diapsids.

Illustration 43: Skull of a Diapsid.

The reason we mention all of this is because paleontologists will often express the impact of an extinction in terms of lost genera or families, instead of species. A genus going extinct is worse than a single species going extinct, and a family going extinct is even more noteworthy. When the dinosaurs were wiped out, their entire order, *Dionsauria,* disappeared. But birds, who belong to the same sub-class, survive to the present.

Major Phanerozoic Extinction Events

Since the advent of complex life on Earth there have been five major mass extinctions. Recently, evidence has been found for another extinction during the early-Cambrian, 512 mya. This event is so far in the past that not much is known about its causes, but it is an indication that major extinctions have been happening for half a billion years.

The most famous extinction is also the most recent, the KT or end-Cretaceous Extinction, 65 million years ago. The subject of many TV shows, most people know the story of the asteroid that killed the dinosaurs. What most people don't know is that, along with the dinosaurs, 85% of all species on Earth vanished during that time. Because it was the most recent extinction event, scientists know more about the KT event than the other great extinctions. Here is a summary of the six major extinctions:

- Early-Cambrian (512 mya): earliest recognized mass extinction eliminated 50% of all marine species.

- End-Ordovician (439 mya): 85% of marine species disappeared, including many trilobites. Second largest marine extinction with 60% of marine genera and 26% of marine families.

- Late-Devonian (365 mya): 70-80% of animal species went extinct, including many corals, brachiopods, and some single-celled organisms. Accounting for 57% of marine genera and 22% of marine families.

- Permian-Triassic (251 mya): extinction of 96% of marine species, including all trilobites and many terrestrial animals—Earth's biggest extinction.

- End-Triassic (199 mya): extinction of 76% of species including many sponges, gastropods, bivalves, cephalopods, brachiopods, insects, and vertebrates. Mostly affecting ocean life killing 57% of marine genera and 23% of marine families.

- KT or end-Cretaceous Extinction (65 mya): 80% of all species on Earth vanished, most notably the dinosaurs. Eliminated 47% of marine genera and 16% of families.

The causes of these natural catastrophes are varied. Ice ages may have played a major role in several. There is evidence that fluctuations in sea level was the primary cause of the end-Ordovician extinction, which mostly affected marine life This is unsurprising since, during the Ordovician, the majority of life was found in the seas. Joseph Sepkoski called this extinction one of the two or three worst extinctions of the Phanerozoic, noting that generic diversity dropped to about the level of the pre-early-Ordovician proliferation.[179]

The late-Devonian mass extinction was a prolonged marine crisis spread over 20-25 million years and punctuated by 8-10 extinction events.[180] Because it most severely affected warm water marine species, leading many paleontologists to attribute the Devonian extinction to an episode of global cooling, similar to the event which caused the late-Ordovician mass extinction. Glacial deposit evidence, found in what is now Brazil, points to a glaciation event on the supercontinent Gondwana during this period. Much like the late-Ordovician crisis, global cooling and widespread drop in sea level may have triggered the late-Devonian crisis. There are suggestions that a meteor impact could be the cause, but the evidence is inconclusive.[181]

The Permian-Triassic Extinction is widely considered the worst of all the major extinction events, killing off an estimated 95% of terrestrial life. From rock layers in Texas and Utah comes evidence that this extinction came in two parts, called extinction pulses, separated by about 10 million years. Either of the two events alone was worse than the KT Extinction that killed off the dinosaurs. Between the two events, 82% of marine genera and 50% of all marine families were extinguished. In earlier chapters, we mentioned the extent of damage this extinction inflicted on Earth's creatures, but the most telling feature was the length of time needed for life to recover. Well into the Triassic, as many as 20 million years later, the effects were still felt. Geologists and paleontologists consider this extinction a major turning point in the history of life on Earth.[182]

Causes put forward for the biggest of all extinctions pretty much cover the entire range; climate change, sudden release of CO_2 or methane, volcanoes and an asteroid strike have all been suggested. Douglas Erwin, in his excellent book "Extinction," covers all the theories in detail, and finds no single explanation fully satisfying. He has suggested what he calls the "murder on the orient express" theory, a combination of several or even all of the causes listed above.

The end-Triassic Extinction doesn't get much press, coming on the heels of the worst ever extinction, and before the dramatic meteorite impact that extinguished the dinosaurs. At least two impact craters have been found

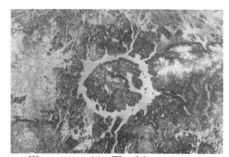

Illustration 44: The Manicouagan impact structure seen from space, Quebec, Canada. Source NASA/JPL.

from around the time of this extinction. One is in Western Australia, where scientists have discovered the faint remains of a 75 mile (120 km) wide crater. The other is a 212 million year old crater in Quebec, Canada, forming part of the Manicouagan Reservoir. The Manicouagan impact structure is one of the largest impact craters still visible on the Earth's surface, with an original rim diameter of approximately 62 miles (100 km).

Others have suggested that a sudden, gigantic overturning of ocean water created anoxic conditions causing the massive die-off of marine species. About 23% of terrestrial families also died out,[183] so there is doubt that such an aquatic event could account for all of the vanished species. There is recent evidence that the end-Triassic experienced an extended period of massive volcanism called the Central Atlantic Magmatic Province (CAMP). This event is associated with the breakup of the super-continent Pangaea and the appearance of a giant rift that eventually formed the basin of the Atlantic Ocean. Carbon isotope anomalies at the Triassic–Jurassic boundary reflect the effects of volcanically derived CO_2, possibly combined with methane release from gas hydrates due to global warming.[184] This makes volcanoes the current favorite trigger for this extinction.

The final major extinction in our list is the KT or end-Cretaceous Extinction. We have already mentioned, in Chapter , how Luis and Walter Alvarez, having found an unexpected spike in iridium content in sediment from the Cretaceous-Tertiary boundary, hypothesized that this extinction was due to an impact with an extraterrestrial body. They later found evidence of that impact in the Yucatan peninsula of Mexico. That evidence came in the form of a crater between 105 and 185 miles (170 to 300 km) in diameter.

Illustration 45: Impact point of the Chicxulub asteroid. Source Tufts.

Impact models estimate the object was between 6 and 10 miles (10-15 km) in diameter, and would have ejected 25,000 cubic miles (100,000 km³) of rock and debris. Such an impact would have directly killed

96

every-thing for thousands of miles, and also triggered earthquakes and tsunamis. As bad as these effects were, the real killing was caused by ash and vaporized rock that filled the atmosphere, which formed dust clouds that blocked the Sun. These clouds are thought to have lasted many months, stopping plant growth and chilling the planet.

A summary of the major extinctions is shown in Illustration 46, depicting diversity in terms of marine families. Starting with the newly discovered early-Cambrian event, the extinctions are numbered from zero. This is to avoid changing the normal numbering of the "big five" extinctions, as they are widely called in the literature.

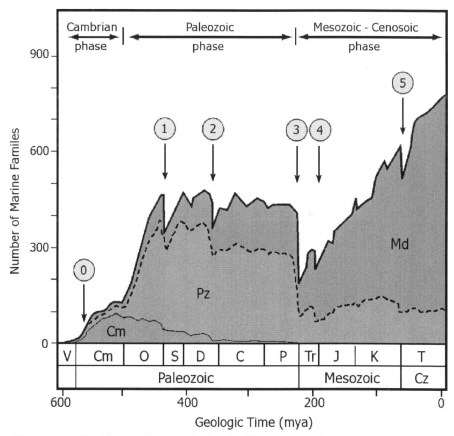

Illustration 46: Marine diversity during the Phanerozoic showing major extinction events. Modified from a figure by Jack Sepkoski (1984).

The three differently shaded areas of the graph represent the numbers of families belonging to the Cambrian, Paleozoic, and Mesozoic periods. As you can see, the early Cambrian life forms started being superseded by the intermediate forms of the Paleozoic prior to the end-Ordovician Extinction and had almost vanished by the late-Devonian. The life-forms characteristic of the Paleozoic undergo a major decline until the Permian-Triassic Extinction, and some related species linger to this day. Even so, the double blow of the Permian-Triassic followed by the end-Triassic cleared the way for the rise of the dinosaurs and, eventually, mammals.

The Causes of Extinction

As we have seen, there is no single answer to the question of what causes mass extinctions. Even the cause of the KT extinction is in question. Though there is no doubt that there was a major asteroid impact at the Cretaceous Tertiary boundary, there is good evidence that the dinosaurs had been declining for millions of years. Some speculate that only a few species were left in the late Cretaceous when the Chicxulub object collided with Earth.[185]

Much like the controversy surrounding the Permian Triassic extinction, there were other events taking place at the time of the Chicxulub event. Prior to the KT boundary, a massive series of volcanic eruptions had taken place in what is now western India and Pakistan. Called the Deccan Traps or lavas, these eruptions produced over 10,000 cubic kilometers of lava during continuous eruptions lasting a million years. The eruptions peeked a half million years before the K/T boundary and continued for some time beyond it. These eruptions could have changed the atmosphere, the ocean and the global climate, causing the extinction of many species.

At this same time there was a noted decline in sea levels, exposing wide swaths of previously covered continental shelf. This change may have been responsible for the extinction of families of ichthyosaurs and plesiosaurs, aquatic dinosaurs which were in decline long before the impact. Some tiny forms of marine live, such as coccoliths, vanished while diatoms and benthic foraminifera survived unscathed. The KT, Permian Triassic and other extinctions may have been caused by the coincidence of several factors, not a single earth shaking event. Looking back 65 million years or more, extinctions taking place over a million years can appear to have occurred suddenly. The Chicxulub impact may have just been the *coup de grâce* that finished what nature had already started. As with most things in science, the arguments continue.

In recent years astrophysicists have proposed links to other extraterrestrial phenomena, including nearby supernovae[186] and the solar system's course in and out of the plane of the galaxy.[187] Here is a list of causes that have been proposed for one or more mass extinction:

- Extraterrestrial Impacts — asteroids or comets striking Earth.

- Massive Volcanoes — in particular the effect on climate.

- Moving Continents — destruction of habitat due to continental drift.

- Ice Ages — glaciation, global cooling, lowered sea levels.

- Disappearing Oxygen — deep water overturn or methane ice.

- Cosmic Peril — impact of cosmic rays and supernovas.

- Coincident Causes — the "murder on the orient express" model.

Our planet's past is filled with extinctions, some large, some small, some solitary. All the ages in the fossil record chronicle the departure of species from this Earth. The sweep of geologic time, comprising more than 90 recognized time periods, is partitioned by changes in the fossil record. The most complete records come from the seas and oceans where the departed settle to the bottom in a silent rain of death. The vast ocean sediment beds accept all comers.

Though life on land has been similarly affected by changing climate, drifting continents, and the occasional rogue asteroid, the fossil record there is much poorer. This is because it is much less likely for a land animal or plant to die under circumstances conducive to preservation. Sometimes, a tree toppled into a peat swamp where the oxygen-poor water prevented decay until rising oceans buried the swamp with sediments. Hundreds of millions of years later, we find the imprint of a leaf or twig embedded in coal. On other occasions, animals were buried by volcanic ash or blundered into a tar pit. But, despite the astonishing number of extinct creatures that have been found, scientists believe many times more have vanished than we have uncovered. Of these creatures, we will never know. For us, the history of life is written in rock by the remains of the extinct.

Just as climate experts have been unable to find a model that would explain the sudden onset of ice ages, scientists are now looking to more sudden events for the cause of mass extinctions. Just as ice ages seem to need a "jump start" to begin or end, extinction events seem to need a trigger. These sudden catastrophes underscore our planet's often violent past.

Invasive Species

As we have seen, ice ages sometimes contribute to extinctions. They are suspected as the major cause of the Ordovician and Devonian extinctions and may have participated in many other minor extinction events. But sometimes the occurrence of an ice age in juxtaposition with a species' demise is just coincidence.

We know from evidence that both mammoths and mastodons survived the worst of the Pleistocene Ice Age in North America. What made them go extinct is not certain but many think it was related to the ice age. During the last glacial period of the Pleistocene, ocean levels were as much as 400 ft lower than today. This would have created numerous bridges between land masses, including one connecting Asia to North America. In the waning days of the ice age, a new, super-predator arrived in the Americas—Homo sapiens.

The Clovis people, migrants from Asia who entered North America approximately 12,000 years ago, are thought to have hunted mastodons, mammoths, and many other large mammal species to extinction. The Clovis people were primitive hunter-gatherers that subsisted on wild animals and plants. The height of their technology is embodied in the well-formed flint spear points they left behind (Illustration 47).

From 12,000 to 8,000 years ago, the Great Plains of North America were populated by small bands of hunters. These groups developed the weaponry and the expertise necessary to successfully hunt large animals. For a band of primitive hunters, wandering from place to place, living off what they could kill, a mammoth would have been a treasure on legs. They would have valued not just the meat, but the woolly hide, the sinew and bone, and the ivory in the animal's tusks. Not that this treasure was easily taken. Attacking a creature several times the size of a modern elephant, with only stone-tipped spears, would have been as hazardous to the hunters as the hunted.

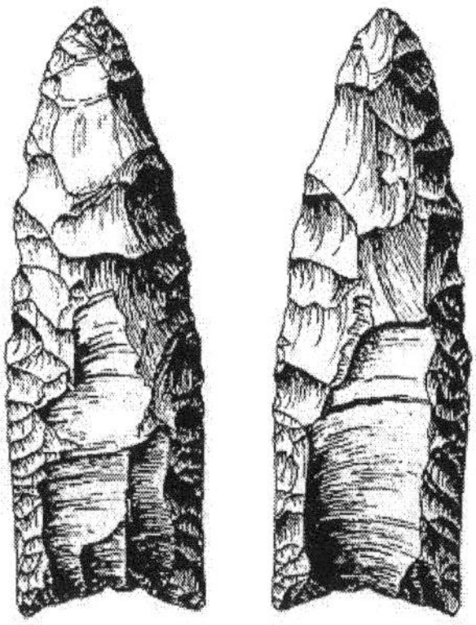

Illustration 47: Early Paleoindian Clovis points. Drawings by Richard McReynolds, from Chandler and Kumpe (1997).

Over time, these ancient people developed more effective hunting methods. In a number of locations in North America, scientists have found evidence that ancient hunters lured or herded their prey into dead falls. Mammoths became extinct on the Plains not long after the arrival of humans and, although climate conditions were worsening, many researchers think hunting by *Paleoindians* hastened the mammoth's demise.[188] With the disappearance of the mammoth, the main target of the hunters shifted to other large plains animals, *Bison antiquus* and *Bison occidentalis*, sometimes (incorrectly) called buffalo.

Though modern day bison are large animals, the bison of the late stone age were even larger. From skulls found at various sites, it has been estimated that these creatures were twice the size of their modern cousins. They were alert, powerful wild animals who were not easily killed. Noted anthropologist George Frison, of the University of Wyoming, described Paleoindian hunting this way:

"Large mammals (mammoths, mastodons, and bison) usually are portrayed mired in bogs with the hunters throwing spears and rocks, dogs barking, and crippled and dead hunters being dragged from the scene. Hunters wave blankets at a small group of bison and they jump off a cliff to their deaths. In reality, bison are extremely agile and, to force them over precipices, a large herd must be stampeded in the proper direction so that the ones in the rear can push the leaders over the edge."[189]

While stampeding a herd of bison off a cliff may not seem very sporting, these hunters were worried about survival, not ethical questions. This is primitive Man in his natural state as alpha predator. No different than crocodiles schooling at the one river crossing where all wildebeest must transit during their annual migration, or a pack of hungry wolves harassing and driving a caribou herd, killing the old, the young and the weak. This is why humans survived the ice age, humans are the most fearsome predators on the planet.

Contributions of human hunting to mammoth and mastodon extinction are still debated.[190] Sudden climate change, a nearby supernova,[191] and even trees have been blamed for their demise.[192] Still, many researchers believe that human over-hunting directly caused the extinction of the mastodon and mammoth. People like to think that early Man existed in harmony with nature, all creatures living in idyllic balance. This is a static and naive view. Often, as Earth and climate change, species previously separated come into contact for the first time. Then, nature arrives at a new balance. Unfortunately for some species, this means their time on Earth is over.

Species invasions have occurred innumerable times in the past from natural causes, but the actions of humans have also been responsible. Regardless of the source, the effect of introducing foreign species to new habitats can be devastating.

Examples of these unintended invaders are legion: Formosan termites imported in the wood of shipping pallets, tropical fish species flushed down toilets, zebra mussels carried in ships' bilge water, the introduction of kudzu and water hyacinths, and on and on.

Australia has been suffering a plague of rabbits for over 100 years. Introduced by the first European settlers in 1788, rabbits were released into the wild. By 1890 they had reached plague-like proportions, causing widespread eco-logical damage. They compete with native animals for habitat and food, such as the now-endangered Greater Bilby. Also known as the Rabbit-eared Bandicoot, the Bilby is an iconic marsupial once found across wide areas of Australia.

By consuming seeds and seedlings, rabbits are driving native Australian plants to extinction. They also cut away strips of bark around a tree's circumference, called ring-barking, which kills the tree. Rabbits cost Australian agriculture $600 million a year in lost production.[193]

Sometimes, invasive species are introduced unintentionally. Around the time of World War II, the brown tree snake was accidentally transported from its native range in the South Pacific to Guam, probably as a stowaway in a cargo ship. With no natural predators, it has wreaked havoc on the native ecology and threatened the island's human inhabitants.[194] Good climbers, that hunt primarily at night, these aggressive, venomous pests grow up to 8 ft (2.5 m) long. There are reports of these snakes entering houses during the night, attacking pets and small children.

In one incident, a brown tree snake bit a human baby, and attempted to eat the child, starting with its hand and arm. Discovered in time by the child's parents, the snake was dispatched and the baby recovered after a stay in hospital.[195] The brown tree snake has since migrated to other Pacific islands and now threatens Hawaii.

Illustration 48: A dodo bird, now extinct.

These stories have been repeated all over the world, sometimes by design, and sometimes by accident. The introduction of domestic animals, dogs and pigs, to the islands of Mauritius is thought to have caused the extinction of one of the most famous recently extinct animals, the *Dodo*. The Dodo bird is famous for two things—being dumb and being dead. The phrases "dead as a Dodo" and "gone the way of the Dodo" are synonymous with extinction. For the most part, the dodo was described as a lazy, rather dumb animal, with foul-tasting flesh.[196] As American humorist Will Cuppy said, "The Dodo never had a chance. He seems to have been invented for the sole purpose of becoming extinct and that was all he was good for."

The Sixth Extinction

Despite the impressive advances of science we are unable to bring back a single vanished species. When a species becomes extinct all the knowledge we might have gained from studying it, all the pleasure of observing it, is also gone forever. The genetic code of each life-form, its *genome*, contains a treasure house of information. New drugs, cures for disease, and unique evolutionary insights can all be lost. We should be very careful when dealing with such living treasure.

Some ecologists have cited Man's impact on nature, accelerating the extinction of numerous species, as the sixth major Phanerozoic extinction event.[197,198] Others have stated that it will take humans another 10,000 years to join the extinction big league. Either way, we are the only species that actually has a choice about which species we drive to extinction. Regardless of the outcome of the global warming debate, we humans need to pay closer attention to the effect our actions have on the environment. Sometimes technological advances can have a positive effect on nature. Man had hunted whales to the brink of extinction when the discovery of oil, in 1859, granted them a reprieve. Of course, oil is part of the problem today.

But even without Man's actions species will go extinct, and not all can or should be saved. Remember, every ecosystem—tropical rain forests teaming with life, magnificent redwood forests towering above the ground, northern tundra that springs to life during the brief Arctic summer, sun baked deserts that wait years for a rain shower to bloom—all these environments will be destroyed over time. Any species that overspecializes, whether plant, animal or bacterium, has signed on for early extinction. All species go extinct—that is nature's way.

So we see that there are numerous causes for extinctions, with climate change a contributing factor, but not the leading cause. Past climate fluctuations much more dramatic than the current, slight warming trend have failed to have a significant impact on Earth's *biota* (living things). Polar bears, penguins, and coral reefs have all lived through interglacial periods before. The ancestors of most modern species lived through the Paleocene-Eocene Thermal Maximum, a period of global warming that was an order of magnitude more intense that our current "crisis." In truth, Man is more likely to cause extinctions by the accidental introduction of foreign species than by warming the planet.

Man's recent impact aside, scientists have come to suspect that many of the events that have shaped life and climate on Earth have extraterrestrial sources. These events are not limited to the odd, random asteroid impact. Recent research has identified cyclic changes in Earth's climate on scales as short as tens of years to as long as hundreds of millions. But these suggestions come from outside the climatological community, and have been resisted or summarily dismissed. Others point to long-term terrestrial processes as the root cause of climate change. Next, we will examine these theories, starting with earthbound processes that can cause climate to change and species to die out. Because greenhouse warming caused by CO_2 emissions is central to the IPCC's case, we will start there.

Chapter 7 Changing Atmospheric Gases

"A theory's validity depends on whether or not it can be verified, it is constantly tested against the facts; wherever it can no longer explain the latter, it shows its limitations and unsuitability. It must then be rethought."

— *Pope John Paul II*

Life requires energy to exist, and the energy, for almost all life on Earth, comes from the Sun. Without the Sun, our planet would be a frozen, dark ball drifting through space, lifeless and alone. Energy is transferred from the Sun to Earth in the form of *electromagnetic radiation*. Part of this radiation appears to us as visible light, but there are other frequencies that we cannot see: infrared, ultraviolet, and x-rays are also forms of light.

When radiation from the Sun strikes Earth, part is reflected by the atmosphere and clouds, part is reflected by the surface, and part is absorbed by the land and oceans. It has long been known that some gases in the atmosphere trap heat radiating from Earth's surface, preventing it from escaping into space. This has an effect similar to a greenhouse, trapping heat and warming the planet. For comparison, look at Earth's sister planets, Venus and Mars.

Venus, named after the Roman goddess of love, is almost Earth's twin in terms of size. Unfortunately, Venus is so close to the Sun that its temperatures are far too hot for life to exist. The Venusian atmosphere is a thick blanket composed chiefly of carbon dioxide with swirling clouds of sulfuric acid. This massive blanket of carbon dioxide generates a surface pressure 90 times greater than that on Earth. This causes a runaway greenhouse effect that heats the planet's surface to an average temperature of 872°F (467°C)—hot enough to melt lead.

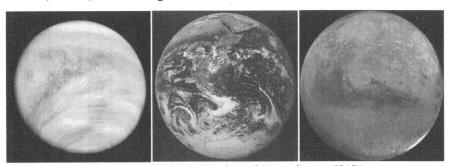

Illustration 49 Venus, Earth, and Mars. Source NASA.

Mars, on the other hand, is farther away from the warming Sun. As a result, it is a cold, forbidding world. The Martian atmosphere, like Venus's, is made mostly of CO_2. Unlike Venus, the surface pressure on Mars is only about 0.7% of the average sea level pressure on Earth, with an average surface temperature around -63°F (-53°C). The polar icecap, visible in Illustration 49, is frozen CO_2, not water. Mar's atmosphere is frozen, thin and unbreathable; perhaps befitting a planet named after the Roman god of war.

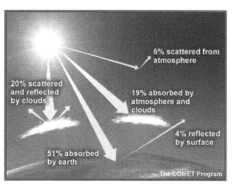

Illustration 50: Scattering and absorption of solar radiation. Source the COMET Project.

When compared to its sisters, Earth is often called the "Goldilocks" planet: Venus is too hot and Mars is too cold, but Earth is just right. Earth is, indeed, the "Miracle Planet" for we currently know of no other world where life exists. It is still possible that one of the moons of the outer planets—Jupiter, Saturn, Uranus, or Neptune—will harbor life. If they don't we will have to look to other star systems for company.

Earth isn't quite close enough to the Sun for temperatures to rise above freezing without help from the atmosphere. If it were not for the greenhouse effect, our planet would be about 60°F (33°C) colder than it is—far too cold for life. Earth's climate is the result of a balance mediated by gases in the atmosphere. Therefore, since life regulates atmospheric gases through a number of *feedback mechanisms,* it is fair to say that life itself creates livable conditions on our planet.

Discovering the Greenhouse Effect

Sir Frederick William Herschel (1738-1822) was a well-known English musician and astronomer. Born in Hanover, Germany, he moved to England in 1757 with his sister Caroline. Together, they constructed telescopes and surveyed the night sky. Their work resulted in several catalogs of double stars and nebulae. As an astronomer, Herschel is most famous for his discovery of the planet Uranus in 1781, the first new planet found since antiquity.[199]

In 1800, Herschel made a dramatic discovery. While observing sunlight through colored filters, he noted that filters of different colors seemed to pass different amounts of heat. To determine if the colors themselves were of varying temperatures, he devised a clever experiment to investigate his hypothesis.

Herschel directed sunlight through a glass prism to create a rainbow spectrum and then measured the temperature of each color. To measure the temperatures, he used three thermometers, their bulbs blackened to better absorb the heat. He placed one bulb in the visible light of a single color, while the other two were placed out of the light for comparison. As he measured the individual temperatures, he noticed that all the colors had temperatures higher than the unilluminated thermometers. Moreover, he found that the color's temperatures increased from the violet to the red part of the spectrum.

After noticing this pattern, Herschel decided to measure the temperature just beyond the red portion of the spectrum in a region where no sunlight was visible. To his surprise, he found that this region had the highest temperature of all. Herschel had discovered a form of light beyond red, invisible to human eyes. This is a classic example of how the unexpected often leads to important scientific discoveries. Herschel had not been looking for a new form of light, he was just trying to measure the heat from the visible spectrum.

In 1816, Herschel was made a Knight of the Royal Guelphic Order by the Prince Regent, granting him the title "Sir." Sir William helped found the Astronomical Society of London in 1820, which later became the Royal Astronomical Society. On 25 August 1822, Herschel died at Observatory House, Slough, and was buried at nearby St Laurence's Church.[200] It would fall to one of his contemporaries, Joseph Fourier, to make the connection between the heat carried by invisible light and the warming of Earth's atmosphere.

Jean Baptiste Joseph Fourier was a French mathematician and physicist who is best known for the mathematical tool called a *Fourier Series* and its application to the problem of heat flow. He also invented the eponymous *Fourier Transform,* which figures prominently in modern infrared spectroscopy and astronomy.

In 1827, Fourier published his theory that gases in the atmosphere might increase the surface temperature of Earth. He attributed

Illustration 51: Jean Baptiste Joseph Fourier (1768-1830).

this to interactions we now call the greenhouse effect. In his paper, he proposed the idea of planetary energy balance—that some sources warm a planet while heat is radiated away into space. Fourier recognized that Earth primarily gets energy from solar radiation, to which the atmosphere is transparent, and that geothermal heat contributes little to the energy balance. He theorized that a balance is reached between heat gain and heat loss, and that the atmosphere shifts the balance toward the higher temperatures by slowing the heat loss.

Fourier stated that planets lost energy due to what he called "chaleur obscure" ("dark heat"). Fourier understood that the emission of dark heat increased with temperature, but the exact relationship was not discovered for fifty years. Today, Fourier's dark heat is called infrared (beyond red) radiation. The next link in the chain of discovery belongs to an Irishman.

Illustration 52: John Tyndall (1820-1893), photo R. MacDonald.

John Tyndall was born in Ireland, in 1820, the son of a local constable. Tyndall attended a common primary school and joined the Irish Ordnance Survey in 1839. Later, he did survey work in England and railway construction during the boom of the 1840s. In 1847, he taught mathematics at Queenwood College Hampshire. In 1848, Tyndall began studies in Germany and became one of the first British subjects to receive the new PhD at Marburg. Though he contributed to many fields, Tyndall's major scientific work was in atmospheric gases.[201]

An accomplished mountaineer, Tyndall was fascinated by Louis Agassiz's daring proposal of ice ages, in which glaciers once covered enormous parts of the world (page 73). Looking for mechanisms to explain climate change, he established the absorptive power of *clear aqueous vapour*—water vapor. Tyndall's experiments showed that, in addition to water vapor, carbonic acid[†] (H_2CO_3) can absorb a great deal of heat energy.[202] He suggested this phenomenon was linked to changes in climate—changes that caused glaciers to advance and retreat.

By this time, scientists knew that the atmosphere consisted of a number of distinct gases. A few, primarily nitrogen and oxygen, were plentiful, while

---

† Carbonic acid (H_2CO_3) is a form of carbon dioxide dissolved in water.

others, notably water vapor and carbon dioxide, were present only in small amounts. The components of Earth's atmosphere are listed in Table 7.

Constant components *(proportions remain fixed over time and location)*	
Nitrogen (N_2)	78.08%
Oxygen (O_2)	20.95%
Argon (Ar)	0.93%
Neon, Helium, Krypton	0.001%
Variable components *(amounts vary over time and location)*	
Carbon dioxide (CO_2)	0.035%
Water vapor (H_2O)	0-4%
Methane (CH_4)	*trace*
Sulfur dioxide (SO_2)	*trace*
Ozone (O_3)	*trace*
Nitrogen oxides (NO, NO_2)	*trace*

Table 7: Components of Earth's atmosphere.

Correctly identifying water vapor as the strongest absorber of radiant heat, Tyndall marveled at the ability of transparent, colorless gases to trap heat. In his own words, he stressed the importance of water vapor in the atmosphere.

> "Aqueous vapour is a blanket more necessary to the vegetable life of England than clothing is to man. Remove for a single summer-night the aqueous vapour from the air which overspreads this country, and you would assuredly destroy every plant capable of being destroyed by a freezing temperature. The warmth of our fields and gardens would pour itself unrequited into space, and the sun would rise upon an island held fast in the iron grip of frost."[203]

Where others speculated and theorized, Tyndall carefully planned and executed laboratory experiments, using equipment of his own design and construction. To take accurate measurements of the absorptive properties of various gases, he constructed the first spectrophotometer, shown in Illustration 53.

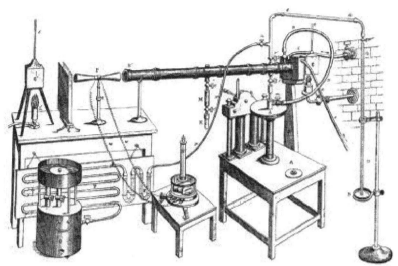

Illustration 53: Tyndall's experimental apparatus, the first ratio spectrophotometer.

Tyndall's experiments clearly demonstrated that trace atmospheric constituents were active absorbers of heat radiation, at least in the infrared. His meteorological and climatological speculations kept alive what was called the "hot-house theory," and suggested to Svante Arrhenius (page 13) and others, that Earth's heat budget may be controlled by changes in the trace constituents of the atmosphere. Today, we know this is true.

The Greenhouse Today

The portion of solar radiation that is absorbed by Earth's surface, about 51%, is re-radiated at longer, infrared wave lengths. The so-called greenhouse gases, mainly water vapor (H_2O), carbon dioxide (CO_2), methane (CH_4), and nitrous oxide (N_2O), trap this infrared radiation, preventing it from leaving Earth quite so rapidly, as shown in Illustration 54. This is the source of the greenhouse effect.

The case for anthropogenic global warming is based on human generated greenhouse gases. These gases build up in the atmosphere, increasing greenhouse warming. The primary culprit in this scenario is CO_2, the second most plentiful greenhouse gas in the atmosphere after water vapor.

Because it is well-know that mankind pumps billions of tons of CO_2 into the air each year, it is easy to lay the blame for global warming squarely on human activity. As a result of the tremendous world-wide consumption of fossil fuels, the amount of CO_2 in the atmosphere has increased over the past century and continues to rise at a rate of about 1 part per million (ppm) per year.

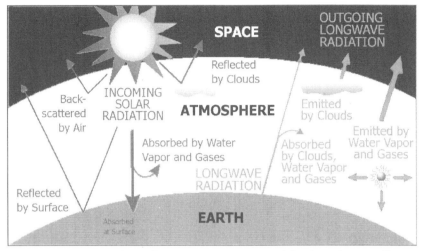

Illustration 54: Solar radiation striking Earth and the Greenhouse Effect, source USGS.

Most anthropogenic CO_2, some 79%, is produced from burning fossil fuels; coal, oil, and natural gas. Electricity generation accounts for 41% and transportation accounts for 22%.[204] There are other minor industrial sources for CO_2 and other greenhouse gases as well, but fossil fuels account for the lion's share of human greenhouse gas emissions (see Illustration 55).

The contributions of other greenhouse gases to global warming are usually glossed over because they complicate the story (i.e. CO_2 causes global

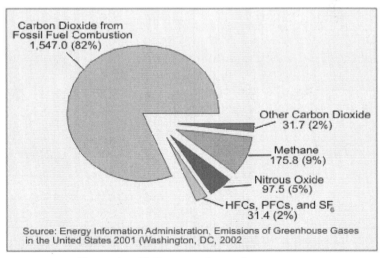

Illustration 55 Sources of greenhouse gases.

111

warming, Man causes CO_2). When we talk about the causes of climate change in detail, the effects of the other greenhouse gases will be analyzed. For the purpose of discussion, we will continue under the assumption that CO_2 is the main source of greenhouse warming even though it makes up only 0.038% of the atmosphere.

Predicted world CO_2 emissions through the year 2030 are listed in the table in Illustration 56. Notice that the majority of future increases will come from developing countries (Non-OECD) which have an average annual change of 3.0%, while the developed countries will only increase by 1.1%. This is one of the main problems with trying to freeze global CO_2 emissions. Developing nations are still building their economies and have not yet stabilized their emissions levels.

World Carbon Dioxide Emissions by Region, 1990-2030
(Billion Metric Tons)

Region	History		Projections					Average Annual Percent Change	
	1990	2004	2010	2015	2020	2025	2030	1990-2004	2004-2030
OECD	11.4	13.5	14.1	14.7	15.2	15.9	16.7	1.2%	0.8%
North America	5.8	6.9	7.3	7.8	8.2	8.8	9.4	1.3%	1.2%
Europe	4.1	4.4	4.5	4.6	4.6	4.6	4.7	0.5%	0.3%
Asia	1.5	2.2	2.3	2.4	2.4	2.5	2.6	2.5%	0.6%
Non-OECD	9.8	13.5	16.8	19.2	21.6	23.9	26.2	2.3%	2.6%
Europe and Eurasia	4.2	2.8	3.1	3.3	3.5	3.7	3.9	-2.8%	1.2%
Asia	3.6	7.4	9.7	11.4	13.1	14.8	16.5	5.2%	3.1%
Middle East	0.7	1.3	1.6	1.8	2.0	2.1	2.3	4.4%	2.3%
Africa	0.6	0.9	1.1	1.3	1.4	1.5	1.7	2.5%	2.3%
Central and South America	0.7	1.0	1.2	1.4	1.4	1.6	1.7	3.1%	2.3%
Total World	21.2	26.9	30.9	33.9	36.9	39.8	42.9	1.7%	1.8%

Sources: **1990 and 2004:** Energy Information Administration (EIA), *International Energy Annual 2004* (May-July 2006), web site www.eia.doe.gov/iea. **2010-2030:** EIA, System for the Analysis of Global Energy Markets (2007).

Illustration 56: Predicted world CO_2 emissions. Source DOE.

The United States and other developed countries show a lesser increase in CO_2 emission rates than developing countries. Even though most of the projected future increase come from developing nations, the Kyoto Protocol mostly exempts developing countries from emissions reduction obligations. The treaty requires developed countries, and *only* developed countries, to return their greenhouse gas emissions to 1990 levels.

Developing countries have surpassed the developed countries in tons of CO_2 emissions. Still, the Kyoto Protocol Treaty wants the United States and other industrialized countries to return to emissions levels of 3,000 million tons, while developing countries, like India and China get a free pass to raise their emissions to 4,000 million tons. Quoting from *Harvard Magazine* regarding Kyoto, "Harvard scientists and economists who study climate change express almost universal criticism of the accord, which they fault as economically inefficient, unobjective, inequitable, and—worst of all—

ineffective."[205] Is it so strange the US and Australia did not sign the Kyoto Treaty?

Note that Illustration 56 shows emissions by total weight, in metric tons of CO_2. IPCC figures are given in carbon equivalent units (CEUs). Carbon equivalent units are based on the carbon content of a gas, 27% in the case of CO_2. Using CEUs makes it easier when comparing carbon emissions in the form of other gases, like methane, and the amounts of carbon stored in plants and rock. To convert the Department Of Energy (DOE) numbers to the IPCC numbers, multiply by 0.27.

A graphical view of the same emissions trend, using carbon equivalent data, can be seen in Illustration 57. Again, notice the rise of emissions from developing countries, and the projected displacement of the developed nations as the world's largest emitters. A review of this data raises the question: Does pumping 6,500 million tons of carbon in the form of CO_2, into the atmosphere every year, create a problem? It sounds like it could, but is 6,500 million tons a large amount when all factors are considered?

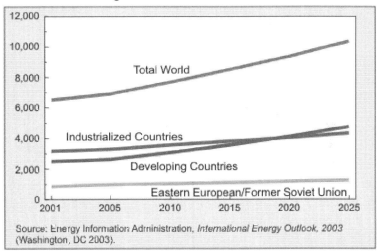

Illustration 57 Predicted CO_2 emissions until 2025 in millions of tons.

In order to understand the magnitude of the problem we need to know how much carbon is in circulation, and to do that, we need to understand more about carbon dioxide and the *carbon cycle*.

Carbon Dioxide

Carbon dioxide, CO_2, is one of the gases found naturally in Earth's atmosphere. It was first identified by Joseph Black, a Scottish chemist and

113

physician, in 1750. A simple molecule, CO_2, consists of one atom of carbon and two atoms of oxygen joined together with double bonds.

Under normal terrestrial temperatures and pressures, CO_2 is a slightly toxic, odorless, colorless gas with a slightly acidic taste. When present in concentrations greater than 5% by volume, it can be dangerous to human health. Air, with a carbon dioxide content of more than 10%, will extinguish an open flame and can be life-threatening by causing asphyxiation. Such high concentrations may build up in silos, wells, and sewers, so this is not an idle warning.

Illustration 58: A CO_2 molecule. The carbon atom is in the center.

Fortunately, CO_2 is uniformly distributed throughout the atmosphere at a concentration of only 0.038% or 380 ppm (parts per million). Though this sounds like a very small portion of the atmosphere, and it is, CO_2 is still the most abundant of the greenhouse gases, with the exception of highly variable water vapor (H_2O).

Water vapor is present in the atmosphere at an average concentration of 1%, but that amount varies widely, from 0-4%. If you have ever been in the American South during the summer, you have probably felt the greenhouse effect first hand. In the summer, even well inland, the heat of a southern day lingers well into the night, clinging like a hot, moist towel to the land. This is water vapor in action, retaining the heat that would otherwise escape into space.

Contrast this with nights in the desert southwest. Death Valley is one of the hottest places on Earth,[†] with daytime temperatures routinely reaching 122°F (50°C) during the summer. Located in the state of California near the Nevada border, Death Valley is also one of the driest places on Earth, receiving only 2 inches (5 cm) of rain each year[206]. From the rock-splitting heat of the day, temperatures drop below freezing at night under clear cloudless desert skies.

Water vapor also affects the climate by participating in energy transfer between the ocean and the atmosphere. It affects energy absorption from the Sun by forming clouds which trap heat under some conditions and reflect

---

† The hottest temperature measured in Death Valley was 134°F (57°C) recorded on July 10, 1913, which held the record for the hottest place on earth until 1922 when a temperature of 136°F (58°C) was recorded at Azizia, Libya in the Sahara Desert.

radiation under others. Combine water vapor's complex environmental interactions with the fact that water covers 70% of Earth's surface, and it seems implausible to identify H_2O as the cause of the current dilemma. Because of these complexities, CO_2 is usually treated as the most abundant greenhouse gas and prime suspect in the global warming crisis. H_2O is just too complicated.

CO_2 finds commercial uses in beverage carbonation, fire extinguishers, as a refrigerant, and many other applications.[207] Dry ice is CO_2 in its solid form, which only exists at temperatures below -108°F (-78°C). Dry ice is often used to keep perishable items frozen during shipment, like mail order steaks from Omaha. In the United States, 10.89 billion pounds of carbon dioxide were produced by the chemical industry in 1995, ranking it 22nd on the list of top chemicals produced.

Because of the low concentration of carbon dioxide in the atmosphere, it is not practical to obtain the gas by extracting it from air. Most commercial carbon dioxide is produced as a by-product of other processes, such as manufacturing ammonia or the production of ethanol by fermentation. CO_2 can also be made as a primary product by burning coke or other carbon-containing fuels.

In addition to being a component of the atmosphere, carbon dioxide also dissolves in the water of the oceans. At room temperature, the solubility of carbon dioxide is about 55 in^3 per quart (900 cm^3 per liter) of water. When absorbed by water, carbon dioxide exists in many forms but most of it remains as dissolved gas.

The Carbon Cycle

Carbon is the fourth most abundant element in the universe, accounting for ~4.6% of all normal matter. But it is not nearly as common on Earth where the *lithosphere* (Earth's crust) is only 0.032% carbon by weight. In comparison, oxygen and silicon make up 45.2% and 29.4% of Earth's surface rocks, respectively.[208] Given the relative scarcity of carbon on Earth, it may seem surprising that carbon is so abundant in living things. But there are good chemical reasons our planet is full of carbon-based life-forms.

Carbon is a very versatile element. An atom of carbon is capable of forming links, called *chemical bonds,* with up to four other atoms at a time. Forming bonds with itself or other elements, carbon can create an incredible number of different molecules. Compared with bonds formed by other elements, carbon bonds are not that hard to break. This allows carbon to react quickly with other elements. Carbon makes a good building block for life because it can form multiple bonds and do so quickly. But, why carbon, to the exclusion of all other elements?

One row below carbon, in the periodic table of elements, lies silicon— carbon's heavier and more plentiful twin. Silicon also forms four bonds, but they are stronger than carbon bonds. As a result, silicon compounds tend to be less reactive than carbon compounds. But sometimes, when silicon is substituted for carbon, the resulting molecules are unstable.[209]

Silicon is an essential element in biology, though only tiny traces of it are required by animals. In 1893, James Emerson Reynolds, a British chemist, speculated that the heat stability of silicon compounds might allow life to exist at very high temperature.[210] Silicon life-forms have been a staple of science fiction ever since. On Earth, there are no silicon-based life-forms—according to NASA's Astrobiology Institute, "silicon simply doesn't have the moves."[211]

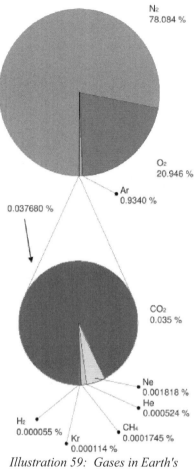

Illustration 59: Gases in Earth's Atmosphere. Source Wikipedia.com.

Silicon's real problem is that it bonds too tightly with oxygen. When we breath, carbon is oxidized forming the gas carbon dioxide. When silicon combines with oxygen it forms a solid, called silicate (SiO_4^{-4}), in which each silicon atom is surrounded by four oxygen atoms. Silicate *anions* connect to each other by sharing oxygen atoms, forming crystals. A gas is easy to exhale, a solid is much harder to dispose of. For this and other reasons, life chose carbon and, by extension, CO_2.

Life extracts carbon from the environment in order to build *organic molecules* (molecules based on carbon). Organic molecules can assume a bewildering number of shapes: rings, long chains, multi-ring chains, and folded sheets to name a few. Carbohydrates (starches and sugars), lipids (fats), nucleic acids (DNA and RNA), and proteins are all based on carbon. Carbon is the stuff of life.

There are two large pools of terrestrial carbon; geologic carbon stored in rock and fossil fuel deposits, and biologic carbon stored in living things or "in play" in the atmosphere and oceans. The source of both types of stored carbon is life. Over billions of years, uncountable billions of living things have collected carbon; growing, eating, breathing, reproducing and finally, dying. Most of this carbon came from Earth's primitive atmosphere in the form of carbon dioxide, which has steadily decreased over time.

Through a number of different living processes, vast amounts of carbon are now trapped in Earth's crust in the form of sedimentary rocks; limestone, dolomite, and chalk. This type of storage, or *sink*, accounts for the majority of carbon on Earth, 66,000,000 to 100,000,000 billion metric tons[†] (gigatons or Gt).

As seen in Table 8, fossil fuel deposits, the other major type of geologic carbon, accounts for a paltry 4,000 Gt. The important difference between geologic and biologic carbon sinks is that geologic carbon is out of short term circulation. It is only released by slow processes, such as volcanism,[212] degassing,[213] and rock weathering, which can take millions of years. At least that was true before Man started digging up fossil fuels and burning them.

The other carbon sinks shown in Table 8, the oceans, soil, atmosphere, and plants, all participate in what is called the carbon cycle of life. This carbon is involved with life over the short term.

Sink	Amount in Billions of Metric Tons (Gt)
Ocean	38,000 to 40,000
Soil Organic Matter	1500 to 1600
Atmosphere	578 (as of 1700) - 766 (as of 1999)
Terrestrial Plants	540 to 610
Marine Sediments and Sedimentary Rocks	66,000,000 to 100,000,000
Fossil Fuel Deposits	4000

Table 8: Amount of carbon stored in sinks. Source Dr. Michael Pidwirny[214].

---

† A metric ton is 1000 kilograms, about 2,200 lb in English units. A gigaton is 1,000,000,000 tons.

Carbon is constantly cycling between the atmosphere, the oceans, and living matter. Plants take in atmospheric CO_2 in order to grow. When the plants die and decay, the carbon returns to the biosphere. When animals eat plants, some of the carbon becomes incorporated into the living tissues of the consumer. Part is released as gases, byproducts of digestion, and the rest becomes organic waste. Other organisms break down the organic waste as it becomes mixed with the soil.

Eventually all animals die and, regardless of whether they are eaten or decompose, the carbon in their bodies eventually gets returned to the biosphere. In this way, carbon passes through Earth's biota ultimately ending up in either the ocean or the atmosphere. The major exchange paths, along with the amounts exchanged in gigatons, are shown in Illustration 60.

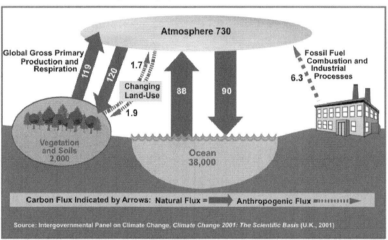

Illustration 60: The carbon cycle of life. Source IPCC.

Looking at the numbers in Illustration 60, the first thing to notice is that the atmosphere is the smallest reservoir of carbon shown (soil and plant matter are combined here). By far, the largest carbon reserve is the ocean, holding a staggering 38,000 Gt, mostly in the form of dissolved CO_2. CO_2 doesn't dissolve easily in seawater, but Earth's oceans are very big, so even a small amount of dissolved gas per cubic foot ends up being a sizable total amount. Even Earth's plant matter, organic material in the soil and growing plants hold almost three times as much carbon as is freely available from the atmosphere. Still, 730 Gt of CO_2 in the atmosphere is a respectable amount.

Another thing to notice from Illustration 60 is the amount of carbon being exchanged. Living matter gives off 119 Gt while absorbing 120 Gt. The oceans give off 88 Gt while taking in 90 Gt. The other exchanges shown are the results of human activity, changing land use gives off 1.7 Gt and consumes 1.9 Gt while fossil fuels and industrial activity pumps 6.3 Gt into the atmosphere. If everything in the diagram is added up, natural mechanisms are running a carbon deficit of around 3 Gt, almost half of the human-caused emissions. This means that the CO_2 level in the atmosphere is not growing as fast as human emissions rates would indicate.

What is not shown in the carbon cycle diagram are the geologic carbon sinks that store carbon for long periods of time. Carbon in these sinks take millions of years to cycle back into the biosphere. As mentioned, there is a tremendous amount of CO_2 dissolved in the oceans. Though much of the CO_2 in seawater remains as dissolved gas, a portion is converted into other chemical compounds. Among the compounds that are formed are carbonate (CO_3) and bicarbonate (HCO_3). Many forms of sea life (labeled *Aquatic Biomass*) have the ability to modify bicarbonate by adding calcium (Ca), producing calcium carbonate $(CaCO_3)$. Calcium carbonate is used by these organisms to build shells and other body parts. Illustration 61 shows how carbon cycles through the oceans and sedimentary rock deposits.

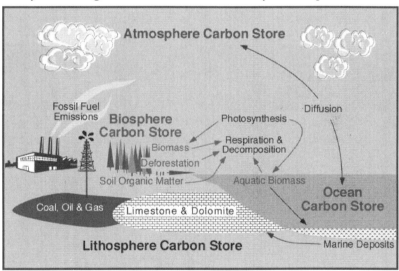

Illustration 61: Carbon cycle with marine sediments. Source PhysicalGeography.net.

In this way corals build reefs, while shellfish, such as clams and oysters, make shells for themselves. Some smaller organisms, protozoa and algae, also build skeletons out of carbonate.[215] When these organisms die, their shells and body parts sink to the ocean floor where they form marine sediments.

Marine sediments can be classified as either shallow water or deep water deposits, depending on the environments that formed them.[216] Shallow water deposits tend to be made up from the disintegrating skeletons of coral, mollusks, sea urchins, and the like. There also may be the remains of algae and inorganic material as well. Shallow water carbonate-rich sediments are found mostly in the tropics and subtropics. These deposits can form

Illustration 62: Coquina. Source Florida Department of Environmental Protection.

aggregates called *coquina*, consisting of whole and fragmented mollusk shells in a matrix of sand cemented by calcite. Coquina is found in large quantities in Florida and the Caribbean where it is used as a building material. It was used historically to build forts, where the soft nature of the stone allowed it to absorb cannon balls without shattering. Deposits of coquina, harder coquinite, and coquinoid limestones are found around the world.

Deep water deposits, on the other hand, tend to be made primarily from the skeletons of pelagic (Greek for "open sea") organisms. These tiny creatures die and their remains begin the long fall to the deep ocean floor where they accumulate as carbonate-rich deposits. After long periods of time, these deposits are physically and chemically altered into sedimentary rocks. The result of this process can be seen in the large beds of limestone and dolomite present in Earth's crust today.

Volcanoes generate CO_2 when rock containing carbon, such as limestone, is melted by magma underneath an active volcanic vent (see Illustration 63). Release of CO_2 from volcanoes is a minor contribution to the carbon cycle on a yearly basis. Mount St. Helens, during its 1980 eruption, had a maximum daily emission rate of 22,000 metric tons. The total amount of gas released during the non-eruptive period from the beginning of July to the end of October was 910,000 tons.[217] All volcanoes combined produce about 500,000,000 tons (½ a gigaton) of CO_2 per year, about 2.8% of man's

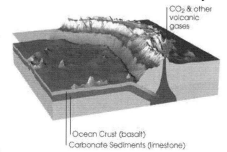

Illustration 63: Volcanic release of CO_2. Source NASA.

yearly emissions. But, over long periods of time, the contribution of volcanoes can be significant. There have been periods in Earth's past that had higher levels of volcanic activity.

After examining all these complex interactions among air, water, rock and life, scientists have a mystery on their hands. If human CO_2 emissions are removed, the world should be in balance—carbon in should equal carbon out. It seems that the numbers do not all add up and each year about 2.9 Gt of carbon goes missing. To quote Richard Houghton, Senior Carbon Research Scientist at the Woods Hole Research Center:

> "When considered with the other terms in the global carbon equation (the atmosphere, fossil fuels, and the oceans), there is an apparent imbalance in the global accounting, and considerable effort has gone into explaining and finding this residual sink, or missing sink, of carbon."[218]

The Missing Sink

For more than three decades, the attention of biologists and ecologists studying the global carbon cycle has focused on an apparent imbalance in the carbon budget. The so-called "missing sink" is a result of the following equation:

Atmospheric CO_2 Increase = Human Emissions + Land Use − Ocean Uptake

This equation is simple enough: the amount of carbon produced by humans plus the carbon produced by other living things, less the amount absorbed by the oceans, must end up as atmospheric CO_2. But, if actual numbers are used, the equation does not balance.

The average annual emissions of 8.5 Gt during the 1990s, 6.3 Gt from fossil fuels and 2.2 Gt from land use, are greater than the sum of the annual buildup of carbon in the atmosphere (3.2 Gt) and the annual uptake by the oceans (2.4 Gt). Here, land use includes carbon from decaying dead vegetation, soil organic matter, and wood products less the uptake by regrowing ecosystems. An additional sink of 2.9 Gt is required to balance the carbon budget. Though this is a small amount, over time, it adds up, 115 Gt of missing carbon over the period 1850-2000. The amounts involved over time can be seen in Illustration 64.

Despite the best efforts of scientists to account for the "missing" carbon, no good answer has been found. During the 1990s, the world's ecosystems are calculated to have been a net sink of 0.7 Gt of carbon per year to the atmosphere, causing speculation that plants are absorbing the missing

Flux of Carbon (Gt per Year)

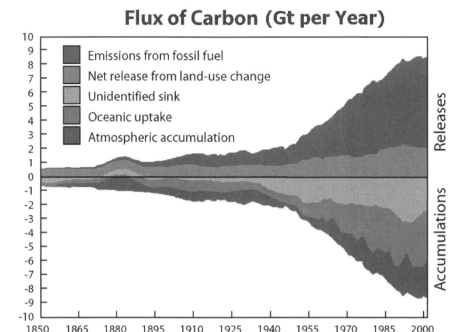

Illustration 64: Carbon flux showing sources and sinks. Source Woods Hole Research Center.

carbon. The most popular theories revolve around an observed greening of North America, Europe and Russia. To quote again from Dr. Houghton:

> "In the last few years several independent analyses based on geochemical data (data from the atmosphere and oceans) and a series of carbon budgets based on data from forest inventories have shown that carbon is accumulating in northern mid-latitude terrestrial ecosystems, although estimates of the magnitude and location of the accumulation vary among the analyses."[219]

Plant life in the US has thrived over the last 100 years, and the increased vegetation growth has absorbed more atmospheric CO_2. This growth is due to the recovery of ecosystems from development in the 1800s and 1900s, when prairies and forests were turned into farmland.

As farming became mechanized, the shrinking number of draft animals greatly reduced the amount of land needed for farm production. This reduction, combined with fertilizers and modern farming techniques, has reduced the amount of land needed to feed each citizen. As a result, since 1950, US forest cover has increased.[220] Similar trends have been noticed in developed countries around the world.

Some think an increase in CO_2 spurs plant growth[221,222] while others say wetter weather and extended growing seasons are the cause. NASA scientists credit an 8% increase in precipitation from 1950 to 1993, combined with higher humidity, with an overall 14% increase in plant growth in the United States.[223]

Scientists' best guess is that carbon is being absorbed by undisturbed or resurgent ecosystems, but they don't know exactly where these are. Since the results of this accumulation remain unidentified the carbon remains missing and the "missing sink" remains unfound. It would seem that there are things scientists still don't understand about the carbon cycle, CO_2, and Earth's climate.

The Bottom Line On CO_2

With this understanding of CO_2 at hand, we can begin to place human-caused global warming in some perspective. Human activity only adds around ½-1% to the amount of carbon dioxide in the atmosphere each year. Even so, that amount is growing and, as Dr. Lomborg has said, "the important question is not *whether* the climate is affected by human CO_2, but *how much*."[224]

Having seen how the carbon cycle works, with both short term and long term carbon sinks, we are faced with a number of new questions. How do Earth's plants respond to the higher levels of CO_2?[225] Why doesn't the warming caused by increasing CO_2 levels result in even more CO_2 release from the oceans?

We know that the ocean gives off CO_2 when it heats up. Why this happens can be demonstrated with a bottle of carbonated water. Soda water and other carbonated beverages contain dissolved CO_2, just like ocean water, but under greater pressure. If you open a warm bottle of soda water there will be a hiss of escaping gas and the liquid will bubble up, possibly overflowing. If the bottle is placed in a refrigerator to cool prior to opening, uncapping the bottle results in a slight psst, few bubbles and no overflow. The reason for the difference is that cold water can hold much more dissolved gas than warm water.

The same is true of ocean water. As the oceans grow warmer, huge volumes of CO_2 are released—not quite as dramatically as a foaming bottle of warm soda, but for precisely the same reason. Scientists have come to believe that the oceans may have released large quantities of CO_2 in ages past, causing periods of intense global warming.[226] This is particularly troubling if CO_2 is the primary driver of earthly temperatures.

If mankind's release of extra CO_2 into the atmosphere causes the oceans to warm, more CO_2 will be released. This will cause further warming, which will cause the release of even more CO_2 and so on, in an ever-increasing spiral of rising temperatures and CO_2 levels. This is an example of what engineers call a *positive feedback loop,* where a process feeds on itself, spinning out of control. But this doesn't seem to be happening, temperatures have not spiraled wildly upwards.

A clue as to why Earth doesn't have a runaway greenhouse effect might be hidden in the missing carbon sink that has been confounding scientists for the past 30 years. The missing CO_2 hints at other mechanisms actively helping to control Earth's temperature. Whatever the reason, it is fortunate that other forces are at work beside CO_2.

We do know that the planet is warming around 1.8°F (1°C) per century. The obvious question is how much of that is due to added CO_2, and how much might be due to other factors? Emissions from ruminants (cud chewing animals) have been recently cited by the European Parliament as "the greatest threat to the planet."[227] These emissions, making up 18% of all greenhouse gases, are mostly in the form of methane and nitrous dioxide. More recently, a study of the pervasive, low-level particulate pollution, found over much of Asia, has revealed that the ubiquitous "brown clouds" are having an unsuspected warming effect. Based on air samples taken by drone aircraft, researchers have concluded that man-made haze is warming the lower atmosphere by as much as 1.4°F (0.8°C).[228] Given these latest findings, perhaps the case for CO_2 being the major driver of Earth's climate needs to be re-evaluated.

Much of the carbon cycle involves interaction between land, sea and air— and the huge geologic carbon sink that is directly related to both ocean life and Earth's active geology. We will next examine those areas in detail. One of the factors that sets Earth apart from its sister planets is Earth's ever-changing face, which we examine in the next chapter. The changes are caused, not by meteor impacts, but by the moving continents.

Chapter 8 Moving Continents & Ocean Currents

"Rocks crumble, make new forms, oceans move the continents, mountains rise up and down like ghosts yet all is natural, all is change."

— *Anne Sexton*

Continental drift has dramatically reshaped the face of Earth during the Phanerozoic, causing significant impact on climate by affecting ocean and atmospheric currents. As the continents move, warm ocean currents can be rerouted or blocked altogether, keeping tropical waters from warming colder regions. This can cause changes in precipitation, desertification and weathering patterns. In turn, these changes may affect the release of CO_2 into the atmosphere. It is probably not a coincidence that during the periods of coldest climate, all of the land masses were gathered into single giant super-continents; Rodinia during the Cryogenian Period (850-635 mya, snowball Earth) and Pangaea during the Permo-Carboniferous Ice age (350-260 mya).

Collisions between continental plates raise mountain ranges, which in turn affect air currents and precipitation patterns. High mountains make snow fall in temperate regions year round and cause glaciers to form, providing islands of cold in otherwise tropical latitudes. The bright white of snow and glaciers reflect more sunlight causing Earth to cool. Mt. Kilimanjaro has kept a permanent snow cap throughout recorded history despite being only 120 miles (300 km) from the equator. The snows of Kilimanjaro have been visibly retreating over the past 100 years, and are expected to disappear by 2020.[229] The melting of mountain glaciers and snowfields is normal during an interglacial period.

Though continents seem rock solid, they do move. Continents are made primarily of granitic rock that is lighter than the basaltic rock that forms the ocean floors and the molten mantle. Though it is hard to think of granite as light, it is in a comparative sense. Continents float like corks on top of heavier, molten rock. *Plate Tectonics,* the geological theory that describes the movement of Earth's crust and uppermost solid material, attributes continental motion to the mechanism of sea floor spreading. In Greek, *tectonics* means "to build," and, in geological terms, a *plate* is a large, rigid slab of solid rock. According to this theory, new rock is created by volcanism at mid-ocean ridges and returned to Earth's mantle by *subduction* at ocean trenches. Earth's crustal plates are in constant, slow motion as new ocean floor is created at submarine ridges. As this happens, the continents go along for the ride.

How was continental drift discovered? Francis Bacon, Benjamin Franklin, Alexander von Humboldt and many others had noted that the shapes of continents on either side of the Atlantic Ocean, Africa and South America, seem to fit together like pieces of a puzzle. But they could not explain why this should be so.

Drifting Continents

Friedrich Wilhelm Heinrich Alexander Freiherr von Humboldt (1769-1859) was a Prussian naturalist and explorer. The son of a Prussian Army officer belonging to a prominent Pomeranian family, he was well-educated but had a sickly childhood. In 1789, he produced his first scientific treatise, *Mineralogische Beobachtungen über einige Basalte am Rhein* (Mineralogical Observations over some Basalts on the Rhine), after an excursion up the Rhine river.

As a result of his experience on the Rhine, Humboldt decided to pursue a career as a scientific explorer. He set himself to learning astronomy, geology, anatomy and botanical science. His early travels were mostly confined to Europe but, after the death of his mother in 1792, he began to explore more distant realms.

The name of the Jurassic Period derives from Humboldt's use of the term *"Jura Kalstein"* for carbonate deposits he found in the Jura region of the Swiss Alps in 1799. Later that year, after a false start on a world-circling expedition, Humboldt traveled to the Americas.

Illustration 65: Alexander von Humboldt, painted by Georg Weitsch.

During this trip, which was to last until 1804, Humboldt recorded observations on geography and wildlife. He commented on the Leonid meteor shower, the transit of Mercury, electric eels and searched for the source of the Amazon river. By the end of his journey he was well-positioned to make future contributions to meteorology, geography and oceanography.

In 1817, he developed the concept of "isothermal lines," providing a way of comparing the climatic conditions of various countries. He was the first to investigate the decrease in temperature with increasing altitude above sea level. By investigating the origin of tropical storms, Humboldt laid the groundwork for the future discovery of more complicated laws governing atmospheric disturbances in higher latitudes.

One of his many contributions to geology was his observation that volcanoes seemed to fall into linear groups. Based on his study of the New World volcanoes, he suggested that these bands of volcanoes were a reflection of extensive subterranean fissures, thousands of miles long. Humboldt had discovered an important clue that would eventually help resolve the mystery of continental drift, but he pursued the idea no further.

Though Humboldt investigated and wrote about many different aspects of nature, for the purposes of this chapter we will note only one more of his discoveries—the Humboldt Current. This cold ocean current, consisting of low salinity, nutrient-rich water, extends along the West Coast of South America, from Northern Peru to the southern tip of Chile. The waters of the Humboldt Current flow in the direction of the Equator and extend 1,000 kilometers offshore.

The Humboldt Current is one of the major upwelling systems of the world, supporting an extraordinary abundance of marine life. We will return to the discussion of ocean currents and their effect on climate after continuing with our exploration of the discovery of continental drift.

Alfred Wegener (1880-1930), a German physicist and meteorologist, was the first to use the phrase "die Verschiebung der Kontinent," German for "continental drift." In 1912, he formally published the hypothesis that the continents were once all together in one large land mass and had somehow drifted apart. Unfortunately, he couldn't convincingly explain what caused them to drift. His best explanation was that the continents had been pulled apart by centrifugal force resulting from Earth's rotation. This force caused the continents to "plow" through the sea floor to new locations. Wegener's explanation was considered unrealistic by the scientific community.

Illustration 66: Alfred Wegener. Source USGS.

Wegener had attended the University of Berlin, where he earned a PhD in astronomy in 1904. As is often the case in science, he was soon distracted by pursuits in other areas. A long-time fascination with climate soon led him to the developing fields of meteorology, climatology, and geophysics. He made several key contributions to meteorology and wrote a textbook that became standard throughout Germany. An avid and experienced balloonist, Wegener pioneered the use of balloons to study the atmosphere. He also became a noted Arctic explorer, joining a 1906 expedition to Greenland to study polar air circulation. It was his passion for the Arctic that would eventually lead to his untimely death.

In 1911, Wegener became a lecturer at the University of Marburg. While teaching there, he pursued his own research into geophysics and climate. He noticed, as so many others before him, that the continents on the globe fit together like the pieces of a jigsaw puzzle. But Wegener looked deeper into the matter, comparing the geological strata found at the matching continental edges. He was struck by the presence of identical fossils in rock strata that were on opposite sides of the ocean. At the time, the accepted explanation for the similarities in fossil records was that land bridges during the Ice Age allowed creatures to migrate among continents. Wegener did not find this explanation credible.

He came to believe that the modern-day continents had all broken away from a single massive supercontinent some time in the prehistoric past. Judging from the fossil evidence, this supercontinent must have broken apart more than 200 million years ago. Using geological features, fossil records, and climate as evidence, he built an overwhelming case to support his continental drift hypothesis. Among the evidence: coal fields in North America and Europe matched, mountain ranges in Africa and South America lined up, and fossils from the same species of dinosaurs were found in places that are now separated by oceans. Finding it implausible that these beasts could have swum across oceans, Wegener was convinced that these reptiles had once lived on a single continent.

Wegener introduced his theory at a meeting of the German Geological Association in 1912. His paper was published later that year and he published a more detailed explanation of his theory in a book in 1915. His ideas about moving continents proved quite controversial.

At the outbreak of World War I, in 1914, he was drafted into the German army. After being wounded in combat, he served out the war as a weatherman in the Army weather forecasting service. Wegener returned to Marburg after the Armistice, but soon became frustrated with his slow advancement. In 1924 he accepted a professorship in meteorology and geophysics at the University of Graz, in Austria.

Between the World Wars, the idea of continental drift caused sharp disagreement among geologists. In 1921, the Berlin Geological Society held a symposium on the theory. By 1922, Wegener's book had been translated into English, introducing his ideas to a wider audience. In 1923, the theory was discussed at conferences by the Geological Society of France, the Geological Section of the British Association for the Advancement of Science, and the Royal Geological Society. In 1926, the American Association of Petroleum Geologists (AAPG) held a symposium at which the continental drift hypothesis was vigorously debated. The resulting papers were published in 1928, under the title *Theory of Continental Drift*, with Wegener contributing a paper to the volume.

Even though he had amassed copious volumes of supporting data, geologists mostly ignored Wegener's theory. The theory's main problem was the lack of a satisfactory explanation of how continents moved through the solid rock of the ocean basins. Most geologists did not believe that this could be possible. Instead of investigating possible sources for Wegener's anomalies, the geological community rejected the theory out of hand, preferring their erroneous but familiar existing beliefs. As Wegener himself said in his "The Origins of Continents and Oceans:"

> "Scientists still do not appear to understand sufficiently that all earth sciences must contribute evidence toward unveiling the state of our planet in earlier times, and that the truth of the matter can only be reached by combing all this evidence... It is only by combing the information furnished by all the earth sciences that we can hope to determine 'truth' here, that is to say, to find the picture that sets out all the known facts in the best arrangement and that therefore has the highest degree of probability. Further, we have to be prepared always for the possibility that each new discovery, no matter what science furnishes it, may modify the conclusions we draw."

Wegener made what was to be his last expedition to Greenland in November, 1930. On that ill-fated journey, Alfred Wegener and a companion went missing. Wegener's body was eventually found in May, 1931. The suspected cause of death was heart failure through overexertion. His theory of continental drift had not found acceptance during his life.

Continental Movement in the Past

Four major scientific developments spurred the formulation of the plate-tectonics theory: discovery of the youth of the ocean floor; confirmation of repeated reversals of Earth magnetic field in the geologic past; proof of seafloor-spreading and the recycling of oceanic crust; and the realization that

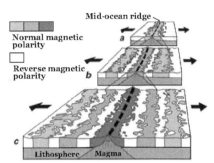

Illustration 67: Ocean floor Magnetic striping, Source: USGS.

the world's earthquake and volcanic activity is concentrated along oceanic trenches and submarine mountain ranges.[230]

Harry Hammond Hess, a Professor of Geology at Princeton University, helped set the stage for the emerging Plate Tectonics theory. While serving in the US Navy during World War II, Hess was able to conduct echo-sounding surveys in the Pacific while cruising from one battle to the next. From these data, he formulated a theory called *seafloor spreading,* stating the ideas that new crust was added at mid-ocean ridges and consumed in deep trenches. In 1962, his ideas were published in a paper titled "History of Ocean Basins."[231] The paper would prove to be one of the most important contributions in the development of plate tectonics.

In the 1950s, scientists using *magnetometers,* instruments that detect changes in magnetic fields, began noticing odd magnetic variations across the ocean floor. These unexpected variations occur because the iron-rich, volcanic basalt rock that makes up the ocean floor contains *magnetite.* Magnetite is a strongly magnetic mineral that can locally distort compass readings.

As molten rock emerging at an ocean ridge solidifies, it takes on the same magnetic orientation as Earth's magnetic field. From time to time, Earth's magnetic field reverses its orientation, swapping north and south poles. This leads to a striped magnetic pattern in the rock expanding away from ocean ridges: a record of both seafloor spreading and the reversal of Earth's magnetic field over time. This phenomenon showed that the seafloor becomes older with increasing distance from ocean ridge crests, helping confirm Hess' theory. Unlike Wegener, Hess, who died in 1969, lived to see his seafloor-spreading hypothesis become accepted during his lifetime.[232]

Over geologic time, the continents have shifted radically. By comparing the layers of rock, geologists try to piece the parts of the puzzle back together. Unfortunately, over long periods of geologic time, the rock layers become jumbled and new rock is created, confusing the matter. It has taken years of scientific investigation to produce a detailed history of Earth's changing continents. Geologists' records now extend back more than a thousand million years.

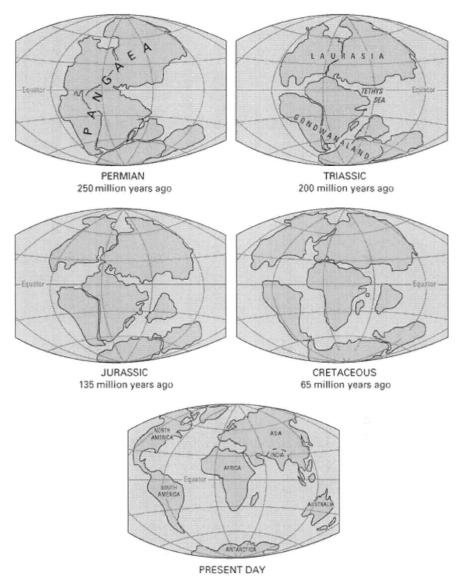

PERMIAN
250 million years ago

TRIASSIC
200 million years ago

JURASSIC
135 million years ago

CRETACEOUS
65 million years ago

PRESENT DAY

Illustration 68 The drifting continents since the Permian Era. Source NOAA.

Illustration 68 shows a series of views of Earth since the end of the Permian Era, 250 million years ago. Notice how most of the land was concentrated in a single continent around the time of the Permian-Triassic extinction, 250 mya. Here is Wegener's single supercontinent, where dinosaurs once roamed freely.

Plate Tectonics

It was not until the 1960s that geologists came to accept Earth's continents did move. This realization came after geologists began to figure out the structure of Earth's interior. Earth's surface solidified permanently after the

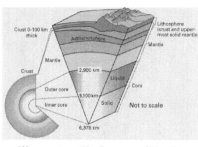

Illustration 69: Layers of Earth's interior. Source USGS.

Great Bombardment ended over 4 billion years ago (page 48). The top layer of Earth's surface is called the lithosphere, and it consists of two parts, the crust and the upper mantle. The lithosphere is only a thin layer on top of many that make up Earth's interior, as can be seen in Illustration 69.

The crust is Earth's outermost layer, forming the familiar continents and ocean floor. The crust varies in thickness from 3 to 60 miles (5 to 100 km) and though it is solid, it is not a single, continuous surface.

Earth's crust is broken into a number of pieces, called plates, that ride on top of the more pliable layers of the *mantle*. Oceanic crust is a thin layer of dense, basaltic rock averaging about 3 miles (5 km) thick. The continental crust is less dense, consisting of lighter-colored granitic rock, and varies from 18 to 60 miles (30 to 100 km) in thickness. Most of Earth's seismic activity takes place at the boundaries where plates interact. Like the shell of an egg, Earth's crust is brittle and can break. Oceanic trenches, mountain ranges, volcanoes and earthquakes are all the result of collisions and partings among these plates.

The continents ride around on top of the crustal plates, which extend under the world's oceans. The various different types of plates interaction are depicted in Illustration 70. If two plates are sliding past each other, heading in different directions, a transform plate boundary occurs. An example of this type of boundary is the San Andreas Fault that runs the length of California.

Where plates are pulling apart from each other, a *rift* is formed. At rifts, new crust is formed by upwelling magma. When this happens under the ocean, an oceanic ridge is formed, like the mid-Atlantic ridge. But rifts can also occur on land, as along the African Rift Valley.

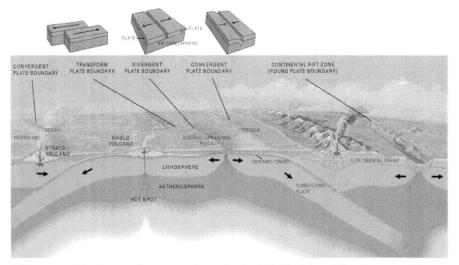

Illustration 70: Types of tectonic plates, by José F. Vigil from This Dynamic Planet, U.S. Geological Survey.

When two plates collide, moving toward each other, a *subduction zone* is formed. Subduction zones are places where two lithospheric plates come together, one riding over the other. Usually, when two oceanic plates collide, the younger of the two plates will ride over the edge of the older plate. This is because oceanic plates grow more dense as they cool and move further away from the Mid-Ocean Ridge. This leads to the younger plates being lighter than older ones. Subduction zones are often highly active seismically, meaning they are prone to earthquakes.

Because the less dense continental material cannot sink, a descending oceanic plate carrying a continent will dive into a trench behind the leading oceanic crust until it collides with the overriding plate. This causes the continent's leading edge to fold up, forming a mountain range. This can also result in some of the oceanic crust of the overriding plate to be deposited on top of the continent. Pressure builds up until the trench flips, and the previously overriding oceanic plate dives under the continental crust. This explains why most ocean trenches are found along the edges of continents.[233]

If two plates collide at subduction zone, and both plates carry a continent, a collision of land masses is unavoidable. When this happens, subduction stops along the collision zone and the trench disappears. The continents pile up, resulting in the birth of a new mountain range. A good example of this is the collision between the Indian plate and the Eurasian plate, shown in Illustration 71. Their ongoing engagement continues to raise the Himalayas and the Tibetan Plateau to new heights.

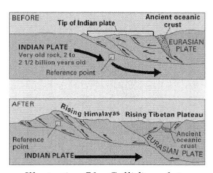

Illustration 71: Colliding plates raising the Himalayas. Source USGS.

Colliding plates create the bands of volcanoes first noticed by Humboldt. A map showing earthquake locations around the world, for the decade starting in 1990, can be seen in Illustration 72. These areas also show high levels of volcanic activity. Today, the geologically active area around the rim of the Pacific Ocean are called the *Ring of Fire*.

Plates have been in motion since Earth's crust formed, but a plate may not exist forever. If a plate's leading edge is being consumed in a subduction trench faster than new crust is being added to its trailing edge, the entire plate can disappear. When this happens, the ridge slowly moves toward the trench and the whole plate is eventually drawn down into the mantle, swallowed by Earth. The disappearance of an entire plate causes a global realignment of the remaining plates and their borders.

One of the dire predictions made by global warming proponents is that the level of the ocean will rise, flooding coastlines and low-lying areas. What is usually not mentioned is that geology has as much to do with local sea levels changing, as do melting glaciers. Continental elevations are due to contributions from compositional buoyancy, thermal and geological forces.

Illustration 72: Earthquakes from 1990-2000. Source NOAA.

134

Researchers have calculated elevation adjustments for parts of North America ranging from -3600 feet (−1100 m) for the southern Rocky Mountains to +7500 feet (+2300 m) for the Gulf of California.[234] Adjusted elevations show clear trends, with an average 1.86 mile (3 km) difference between hot and cold crustal regions. The internal forces of the planet cause significant, ongoing shifts in elevation. When the shifted land is on a coast, sea levels change.

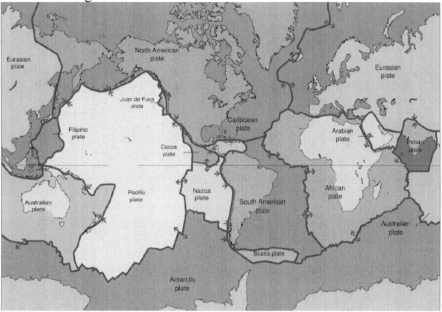

Illustration 73: Tectonic plates circa 2000. Source USGS.

Such tectonic effects can be seen on San Clemente Island, off the coast of California, which has been rising since the early Pleistocene at a rate of 8-16 inches (20-40 cm) per thousand years. Similar rising coasts can be found in places like New Guinea, Barbados, and the Palos Verdes Hills on the Californian mainland.[235]

Earth's Plates Today

Today, Earth's outermost layer is broken into seven large plates: the African, North American, South American, Eurasian, Australian, Antarctic, and Pacific plates. Several minor plates also exist, including the Arabian, Indian, Nazca, and Philippines plates.

The current configuration of Earth's tectonic plates is shown in Illustration 73. Areas where the arrowheads point in opposite directions are ocean rifts where new crust is being created. Prominent among these rift zones are the Mid-Atlantic Ridge and the East Pacific Rise. The Mid-Atlantic Ridge extends from 200 miles (330 km) south of the North Pole, to Bouvet Island in the far south Atlantic and the East Pacific Rise runs from near Antarctica north through the Gulf of California. Ocean ridges run around the world like the stitching of a baseball, forming a nearly continuous system 25,000 miles (40,000 km) long.

South America and Africa are moving apart at an average of 2.25 inches (5.7 cm) per year, about as fast as a human fingernail grows. This is due to sea floor spreading along the Mid-Atlantic Ridge, which has been happening since Pangaea broke apart during the CAMP (page 96). The fastest recorded sea floor spreading is along the East Pacific Rise at 6.7 in (17.2 cm) per year.

Trenches formed where plates are colliding include some of the deepest parts of the ocean. The Puerto Rico Trench, with maximum depth 28,232 ft (8,605 m), is the deepest point in the Atlantic. The Mariana Trench, with a depth of 35,798 ft (10,911 m) holds that distinction for the Pacific. Trenches are generally found parallel to volcanic island chains, with the trench about 120 miles (200 km) from the arc of land. An undersea view of the Puerto Rico Trench and islands of the Greater and Lesser Antilles demonstrates this arrangement in Illustration 74.

Trenches can also be found near rift boundaries. The South Sandwich Trench in the South Atlantic Ocean, 60 miles (100 km) to the east of the Sandwich Islands, is formed by the small Sandwich Plate being subducted under the South American plate. This is an anomaly, since the Sandwich Plate is one of the youngest plates on Earth. Geologists think that some force is actively pulling the Sandwich Plate down, even though it is the younger plate.[236]

The Great Ocean Conveyor Belt

Worldwide circulation of ocean water helps to redistribute heat around the globe. This causes some areas to be warmer than would be expected, given their latitude, and others to be colder. The great Humboldt Current lowers the temperature of the western coast of South America while the Gulf Stream helps to keep Western Europe temperate. Without the Gulf Stream, Europe would have the same climate as Siberia.

Over the ages, ocean circulation patterns changed as the continents shifted. We have already mentioned South America and Australia moving north, forming the circumpolar current that thermally isolates Antarctica today.

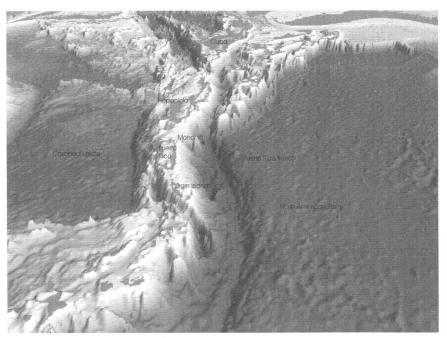

Illustration 74: The Puerto Rico Trench, the arc of the Greater and Lesser Antilles, and surrounding ocean basins. USGS.

Equatorial currents linked the Pacific and Atlantic oceans, until North and South America became joined by the Isthmus of Panama. Over time, other changes will occur in the world's circulatory systems, as the continents continue their slow but inexorable movement.

One of the major consequences of the way continental land masses partition the oceans is the meridional overturning circulation (MOC), more commonly called the great ocean conveyor belt. The stability of Earth's climate largely depends on oceanic and atmospheric currents transporting heat energy from low latitudes to high latitudes. Up to four petawatts[†] of thermal energy is redistributed by these currents.

The thermally driven MOC carries one-third of this heat, making it a major factor in Earth's climate system. This circulation pattern is also referred to as the thermohaline circulation because it is driven by differences in water density which, in turn, are determined by both temperature and salinity. Warm, less dense water flows to higher latitudes in surface currents, like the Gulf Stream. This moderates the climate in higher latitudes, but the flow of warm water must be balanced by a similar return volume of water.

---

† A petawatt is a billion megawatts or 1,000,000,000,000,000 watts.

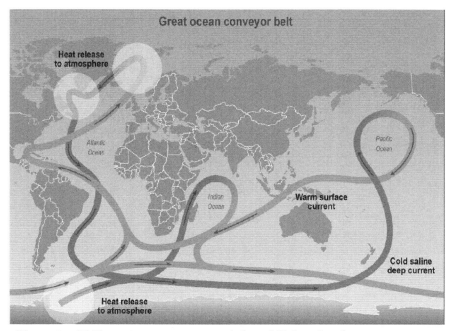

Illustration 75 The great ocean conveyor belt, or MOC, redistributes temperatures around the world. Source IPCC.

The return flow for the Gulf Stream comes from dense water flowing out of the Arctic basin, which flows south along the deep sea floor. The water in the Arctic basin is dense because it is both colder and saltier than average. The increased salinity is caused when water crystallizes to form sea ice—salt ions are excluded raising the concentration of salt in the remaining sea water. The salinity of water in the tropics can also be increased through surface evaporation. The interaction of surface and deep ocean currents, influenced by solar radiation, ice formation, fresh water runoff and evaporation, is highly complex and poorly understood. Observation of the MOC has been ongoing for years, but still, little is known about its fluctuations and their effect on global climate.[237]

We do know that this mechanism is not directly tied to the El Niño mechanism. The El Niño-Southern Oscillation (ENSO) creates surface water temperature anomalies in the equatorial Pacific because of changes in trade winds. It temporarily affects atmospheric circulation patterns over a sizable portion of the globe. The effects of the MOC are longer term and more substantial than an El Niño or the corresponding North Atlantic Oscillation (NOC).

It is known that temperature fluctuations in the Northern Pacific have a significant effect on marine ecosystems and the climate of North America. Observations over the past 50 years show that this overturning circulation has been slowing down since the 1970s, causing a rise in equatorial sea surface temperatures of about 1.5°F (0.8°C). Like many aspects of Earth's climate system, the physical mechanisms responsible for these fluctuations are also poorly understood.[238]

Scientists suspect that the conveyor belt has been disrupted in the past. Toward the end of the last glacial period, when the Wisconsin ice sheet was retreating northward into Canada, a large freshwater lake formed. The lake water's path to the sea was blocked by glacial ice. As the climate warmed, the ice continued to melt, adding to the water trapped in the lake.

This lake, named after glaciologist Louis Agassiz, was twice the size of the Caspian Sea, the largest body of freshwater on Earth today. At its greatest extent, Lake Agassiz stretched from Saskatchewan, east to Quebec, and from Minnesota, north to the Hudson Bay.[239] After collecting water from melting glaciers for 4,000 years, the ice dam broke and the lake suddenly emptied into the Arctic Ocean through Hudson Bay. The water flowed from the Arctic basin into the North Atlantic, and the great ocean conveyor belt stopped.

This incident, which occurred 12,700 years ago, is believed to have been the cause of a period of sharply colder temperatures called the Younger Dryas.† This period is named after *Dryas octopetala*, a flowering plant found in Arctic tundra regions, because great quantities of dryas pollen were found in ice core samples from that time. The Younger Dryas was the most significant rapid climate change event to occur during the onset of the Holocene warming. Recent studies show this interval began with a 12°F (7°C) drop in temperature, in the span of only 20 years.[240]

Work done by Wallace Broecker provided the link between the emptying of Lake Agassiz, the disruption of the ocean conveyor, and the sudden shift back to glacial conditions.[241] The

Illustration 76: Seed head of Dryas octopetala on the Burren, County Clare.

---

† The Younger Dryas is the most recent of the three Dryas periods, the others being the Older Dryas (14,000-13,700 BP) and Oldest Dryas (18,000-15,000 BP).

Allen Simmons & Doug L. Hoffman

Younger Dryas lasted about 1,300 years, then the Holocene warming trend reasserted itself. Theory suggests that this event is an example of Earth's climate shifting between different stable modes, and has led to warnings that current melting of glacial ice might trigger another sudden cold snap.[242,243]

Interestingly, the end of the Younger Dryas was marked by a period of rapid temperature increase—an estimated 27°F (15°C) rise over half a century. A visual comparison of these changes and more recent temperature variation can be seen in Illustration 77.

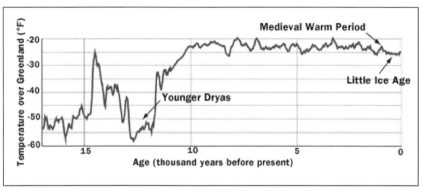

Illustration 77: Temperatures during the Younger Dryas and the present day. Source data from Richard B. Alley.

While some species may have gone extinct during this period, it was not a noticeable biotic calamity. This is a direct contradiction of claims that a temperature rise of a few degrees will lead to widespread extinctions. Climate stress or habitat destruction might have helped kill off the mastodon and the mammoth, but nature did not collapse, nor did our stone-age ancestors. As usual, Earth's resilient life adapted and thrived.

No wonder the IPCC's claim of "unprecedented" global warming is always qualified by "in the last 10,000 years." Go back a bit farther in time and there is a temperature drop of 12°F (7°C) and a subsequent rise of 27°F (15°C). The rate of temperature rise at the end of the Younger Dryas is 20 times the most widely quoted IPCC prediction of 1.5°C per century.

The Only Constant is Change

As we have seen, the ever shifting arrangement of Earth's tectonic plates is a source of slow but inexorable climate change. This change is due to modification of ocean currents and atmospheric currents that redistribute heat energy around the world. Though these changes assert themselves over

140

very long time periods, they are a reminder that there are much more powerful influences on Earth's climate than human activity. Other effects of ocean currents are visible on a more human time scale.

The global nature of ocean circulation is shown by a recent story about rubber bath toys, reported in England's *Daily Mail*. It seems that 15 years ago a shipment of 29,000 plastic yellow ducks, blue turtles and green frogs, traveling on a cargo vessel from China to America, were washed overboard. Most of the toys circled the northern Pacific once before being washed up on the shores of Alaska and the West coast of Canada and the US. Since then, some of the toys have traveled 17,000 miles, landing in Hawaii, spending years frozen in the Arctic pack ice, and floating over the site where the Titanic sank.[244]

At some point, the toy armada is expected to wash up on beaches in South West England. This slow journey from the Pacific to the Atlantic, by way of the Arctic, was predicted by British oceanographer, Curtis Ebbesmeyer. Having heard about the toys' escape at sea, Ebbesmeyer theorized that some of the floating objects would eventually end up in the Arctic Ocean and become embedded in the ice. Hitching a ride on the Arctic ice cap, which moves as much as a mile a day, the rubber ducks found their own Northwest Passage to the Atlantic.

This event illustrates how interconnected all Earth's oceans are, and that ocean circulation includes both water and ice. Scientists are now using the lost toys' journey to help understand ocean currents. In the future, as the continents continue to drift and ocean currents change, scientists may dump more rubber duckies overboard on purpose.

The circulation of the deep Atlantic Ocean during the height of the last glacial period appears to have been quite different from today. The conveyor belt didn't halt, as it did during the Younger Dryas, but changes in temperature and flow occurred.[245] The British Oceanographic Data Center, as part of the Natural Environment Research Council (NERC) Rapid Climate Change (RAPID) program, and the University of Miami's Rosenstiel School of Marine and Atmospheric Science (RSMAS) have moored buoys in the Atlantic to monitor the MOC in real time.[246] Perhaps they will be able to provide a warning if the MOC shuts down again. As Earth's climate continues to change, variation in ocean circulation should be expected. Whether these changes will cause sudden shifts in temperature, in some regions, cannot be foretold.

Even so, we need to be prepared. Man can no more control Earth's climate than he can stop the drifting continents. As we have seen from studying the causes of extinction, successful species are those that can adapt to Earth's ever changing environment. To quote physicist Stephen Hawking, "We've got to start adapting, we've got to start understanding how severe the changes are and that there's very little than we can do to stop them."

Chapter 9 Variations In Earth's Orbit

"At present the globe goes with a shattered constitution in its orbit.... No doubt the simple powers of nature, properly directed by man, would make it healthy and a paradise"

— *Henry David Thoreau*

The changing seasons are among the most familiar short term climate cycles. The change of summer into fall, or winter into spring illustrates the variability of temperature and the adaptability of Earth's living creatures and plants. The parade of seasons is caused by the tilt in Earth's axis of rotation, an imaginary line drawn through the north and south poles.

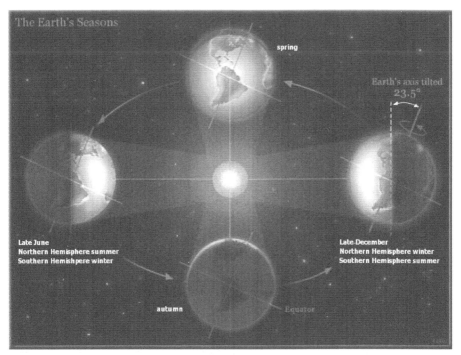

Illustration 78: Earth's seasons due to axial tilt. Source NOAA.

Earth spins like a top, completing a revolution every 24 hours, but it is slightly tilted from the plane of its orbit around the Sun. Today, this tilt is 23.5° but, as we shall see, this amount changes over time—referred to as variation in *axial obliquity*. Why Earth tilts like this is a mystery. Some think it might be a result of the collision that formed the moon.

143

As Earth orbits the Sun, the direction of the tilt remains constant, just like a spinning top. So, during the summer, the tilt in the northern hemisphere is toward the Sun and, during the winter, it points away. The longer summer days allow more sunlight to warm Earth, while shorter, winter days mean less warmth and colder weather. When the northern hemisphere is tilted toward the Sun, the southern hemisphere is tilted away. This causes the seasons to be reversed between the north and south. At the South Pole, winter starts in June.

People have long understood what causes the seasons and ancient observers of the night sky noticed other changes as well. Early astronomers noticed that the canopy of stars was slowly moving around Earth, causing the stars of the zodiac to slowly move through the seasons. This is due to a phenomenon known as *precession*. The discovery of precession is usually attributed to the Greek astronomer Hipparchus of Nicaea,[†] in the 2nd century BC. He was the first to develop accurate models of solar and lunar motion.

Reportedly, Hipparchus made use of the observations recorded by the Chaldeans for several centuries preceding his own observations. Virtually all Hipparchus' writings were lost, including his work on precession. We know of it today because Ptolemy mentioned the works of Hipparchus in his *Almagest*. Ptolemy explained precession as the rotation of the celestial sphere around a motionless Earth. It is reasonable to assume that Hipparchus, like Ptolemy, thought of precession in geocentric terms as a motion of the heavens (see page 218).

There has been speculation that other cultures discovered precession prior to Hipparchus. The Babylonians may have known about precession as early as 330 BC. According to al-Battani,[‡] Chaldean astronomers had measured the difference between the solar and sidereal year. The solar year is the length of time that the Sun takes to return to the same position along its path among the stars relative to one of the equinoxes. The sidereal year is the time it takes for the Sun to return to the same position with respect to the stars of the celestial sphere. This is the same length of time as the orbital period of Earth. Because precession causes the equinoxes to move backward along the ecliptic, a solar year is shorter than a sidereal year. The value of precession is the difference between the solar and sidereal years.

---

† Hipparchus (ca. 190 BC-120 BC) Greek astronomer, geographer, and mathematician.

‡ Muhammad ibn Jabir al-Harrani al-Battani (ca. 853-929) Arab astronomer, astrologer and mathematician.

Claims have been made that precession was known in Ancient Egypt. Some buildings in the Karnak temple complex were allegedly oriented toward the point on the horizon where certain stars rose or set at key times of the year. When precession made the orientations inaccurate, the temples would be torn down and rebuilt. Of course, observing precession's effects doesn't mean the Egyptians understood what caused the changes.

The first known reference to precession resulting from the motion of the Earth's axis appears in Copernicus's *De Revolutionibus Orbium Coelestium*, where he called precession the third motion of the Earth. Over a century later it was explained in Newton's *Philosophiae Naturalis Principia Mathematica* to be a consequence of gravitation. Newton's original precession equations contained errors and were later revised by d'Alembert.

Scientists came to suspect that these variations, linked to Earth's travels around the Sun, might have an impact on climate. Establishing the connection between variations in Earth's orbit, attitude and long term climate change took more than 150 years and was an on again, off again affair.

Cycles of Earth

Scientists in the mid 19th century had come to accept Louis Agassiz's Ice Age theory and were casting about for explanations as to why they would occur. In the ninth edition of his *Principles of Geology*, Charles Lyell speculated that variations in Earth's orbit might be the cause, but provided no proof for the conjecture. The first scientist to attempt to prove this link between Earth's orbit and ice ages was a self-educated Scot named James Croll.

James Croll (1821-1890) was one of those people who seem to have been prevalent in the Victorian British Empire, the self-taught amateur scientist who ended up a major contributor to a number of fields of study. Born to a poor rural family, Croll only attended school until 13 years of age, when he had to quit to help work the family farm. He had a varied career outside of science, working as a hotel manager, an insurance salesman, an itinerant industrial equipment repairman, and other odd jobs. Eventually, he found a position

Illustration 79: James Croll, 1821-1890.

as caretaker at Anderson College in Glasgow, which gave him time to pursue more scientific matters.

Though his first publication was on philosophy, after reading a book by the French scientist Joseph Adhémar on the influence of wobble in Earth's axial tilt and its eccentric orbit on ice ages, Croll became obsessed with climate change. Adhémar claimed that the combined effects of these two factors would result in alternating glacial periods in the northern and southern hemispheres. Some of Adhémar's other claims were a bit wacky, which diminished the impact of the good ideas contained in his book. But, Croll and several others took note and pushed the investigation further. For 25 years, Croll was to be captivated by what he termed "The Fundamental Problem of Geology."

Croll was a very methodical man, with a deep desire to understand the fundamental nature of the problems he studied. His approach to understanding the impact of Earth's orbital eccentricities involved arduous, intricate calculations. Though the elliptical nature of Earth's orbit had first been calculated by Laplace, in 1773, Croll's calculations were much more painstaking.

He published his first paper on the causes of climate change in 1864, and it rocketed him to prominence in geological circles. Ice ages and their causes were a topic of great interest at the time. No one had developed an accurate way to tell when the last glacial period had ended or how long it had lasted. Croll corresponded with Charles Lyell, sending his ideas of links between ice ages and variations in Earth's orbit.

In 1875, Croll published his major work on the linkage between Earth's orbital variations and climate, *Climate and Time in their geological Relations: A Theory of Secular Changes of the Earth's Climate*. In it, he estimated that the last glacial period had ended 80,000 years ago. He also reiterated Adhémar's conjecture that the glaciers would alternate between the poles in a 11,500 year cycle. Acceptance of Croll's theory was at its peak.

Geologists working in the field began to use glacial drift deposits and erosion to try and date the glacial episodes experimentally. As their data came in, discrepancies with Croll's carefully calculated, but theoretical dates began to arise. The alternation between northern and southern hemispheres was also shown to be erroneous. By the end of the 19th century, Croll's astronomical theory of climate change was discredited and discarded.

The rise and fall of Croll's theory is a good example of how scientific progress is made. Croll built on the works of others, including Agassiz and

Adhémar, both of whom he readily acknowledged. His work corrected and refined their theories while making new predictions of its own. Other scientists, trying to confirm or disprove Croll's predictions, showed the astronomical theory to be incorrect in several ways, leading to its fall from favor. But sometimes, the basic idea behind a theory is correct, even if its expression is flawed. In this case, the fundamental ideas were correct— Croll's theory would rise again.

The scientist who revived Croll's theory was Milutin Milankovitch, a Serb mathematician and engineer whose initial fame was due to concrete. Born in 1879, in the town of Dalj on the Danube, he was well-educated during his youth. He attended university in Vienna, eventually earning his PhD in engineering. His dissertation was on the uses of reinforced concrete in buildings, and provided a wealth of data about strengths and shapes used for construction.

With his detailed knowledge of concrete, Milankovitch quickly found work with a large engineering firm in Vienna, eventually becoming head engineer. He earned notoriety designing buildings all across Europe, and his career seemed on a steady course. But the political storm clouds, that would eventually lead to World War I, were starting to gather.

Though he loved his work and life in Vienna, in 1908, political tensions led him back to Serbia. He became a professor at the University of Belgrade, giving up his much higher paying job as an engineer. Milankovitch was a man driven to succeed at whatever he worked on, and he was soon looking around for problems where he could apply his considerable mathematical skills.

Illustration 80: Milutin Milankovitch

After reading Croll's work on orbital variation and climate, Milankovitch decided to investigate the causes of climate change.

A great deal of scientific progress had been made since Croll's work. Better measurements of planetary movements were available, along with better historical climate data. Milankovitch was not trying to explain ice ages, but all climate changes. He was seeking a mathematical theory that would give the temperature for any point on Earth's surface at any given time. In fact, he would eventually extend his calculations to the Moon and Mars.

Though his work was interrupted by WWI, during which he was briefly imprisoned, by 1920 he published *A Mathematical Theory of the Thermal Phenomenon Produced by Solar Radiation*. His book generated interest among meteorologists, but gained little notice from geologists of the time.

The German meteorologist Wladimir Köppen was one of the scientists who took note of Milankovitch's theoretical approach to climate. Köppen's daughter was married to geophysicist Alfred Wegener. Wegener is the scientist who first put forth a serious theory of continental drift, which we investigated in Chapter 8. Milankovitch began a collaboration with Köppen and Wegener that was to last their lifetimes. When Köppen and Wegener published *Die Klimate die Geologischen Vorzeit* (Climates of the Geological Past), in 1924, it helped to bolster Milankovitch's work.

Among the key contributions of Milankovitch's work were the ideas that different latitudes experienced the impact of orbital variation differently, and that the key to the onset of glaciation is cool summer weather, not colder winters as had previously been assumed. Understanding that there were a number of factors involved in regulating climate, Milankovitch didn't try to give absolute temperature values. Instead, he used "equivalent latitude." If a location's equivalent latitude decreased, effectively moving the climate south, temperatures were warmer. Similarly, an increase in equivalent latitude meant temperatures were colder.

Once again, Europe was preparing for war when Milankovitch decided to publish his definitive book on climate, *Cannon of Insolation and the Ice Age Problem*, printed in German in 1941. For a time, this resurgent theory of climate change found wide acceptance, but it was not to last.

New dating techniques and geological data began to uncover discrepancies in Milankovitch's predictions. Around the time of his death, in 1957, Milankovitch's theory was out of favor for much the same reasons as Croll's. History was repeating itself.

Milankovitch had managed to out-live the Nazis and the second World War, only to find himself trapped in the dismal reality of Communist Yugoslavia. Fortunately, his international reputation as a scientist garnered him some respect from the new regime, allowing him to live out his days in relative peace. But the story doesn't end here.

In the 1960s, several advances were made in geophysics that improved the collection of data about Earth's past climate. Analysis of radioactive isotopes of hydrogen and oxygen, along with the ability of scientists to collect sediment cores from deep ocean beds, greatly improved knowledge of ice age climate changes. In 1970, Wally Broecker and J. van Donk published a paper that detailed temperature changes going back 400,000 years.

In this paper, a number of the apparent discrepancies in Milankovitch's theory were resolved. Though he didn't live to see his theory vindicated, Milankovitch's astronomical theory of climate change is now recognized as the best explanation of the cycles of glacial-interglacial change. In his honor, these periodic changes in Earth's orbital orientation are called the Milankovitch Cycles.

The Croll-Milankovitch Cycles

Though scientists had long considered the variation in insolation warming from the Sun too weak to cause the waxing and waning of ice ages, the cycles found by Croll, and expanded on by Milankovitch, fit the climate data so well some form of link had to exist. Unlike atmospheric carbon dioxide, where several Earth-bound explanations exist for changing CO_2 levels, there is no way that terrestrial forces can cause the changes in Earth's orbit.

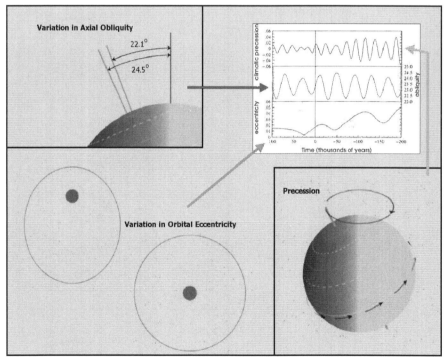

Illustration 81: Variation in Axial Obliquity, Orbital Eccentricity, and Polar Precession. Images from NOAA.

Today, we know variations in the intensity and timing of heat from the Sun are the most likely cause of glacial/interglacial cycles. This solar variability is partially driven by changes in the Sun's output, but is affected more strongly by variations in Earth's orbit.

There are three major components of Earth's orbit about the Sun that contribute to changes in our climate. These are, in order of longest to shortest cycle, *Orbital Eccentricity, Axial Obliquity,* and *Precession of the Equinoxes.* These three variations are shown in Illustration 81.

Earth's orbit goes from measurably elliptical to nearly circular in a cycle that takes around 100,000 years. Presently, Earth is in a period of low eccentricity, about 3%. This causes a seasonal change in solar energy of 7%. The difference between summer and winter is a 7% difference in the energy a hemisphere receives from the Sun.

When Earth's orbital eccentricity is at its peak (~9%), seasonal variation reaches 20-30%. Additionally, a more eccentric orbit will change the length of seasons in each hemisphere by changing the length of time between the vernal and autumnal equinoxes.

The variation in eccentricity doesn't change regularly over time, like a sine wave. This is because Earth's orbit is affected by the gravitational attraction of the other planets in the solar system. There are two major cycles; one every 100,000 years and a weaker one every 413,000 years.[247]

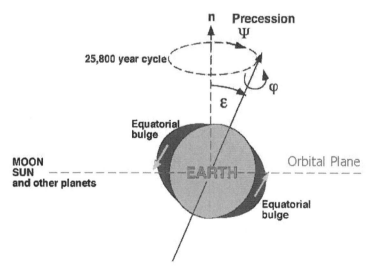

Illustration 82: Precession of Earth's axis of rotation.

The second Milankovitch cycle involves changes in obliquity, or tilt, of Earth's axis. Presently Earth's tilt is 23.5°, but the 41,000 year cycle varies from 22.1° to 24.5°. This tilt is depicted in the upper-left panel of Illustration 81. The smaller the tilt, the less seasonal variation there is between summer and winter at middle and high latitudes.

For small tilt angles, the winters tend to be milder and the summers cooler. Cool summer temperatures are thought more important than cold winters, for the growth of continental ice sheets. This implies that smaller tilt angles lead to more glaciation.

The third cycle is due to precession of the spin axis. As a result of a wobble in Earth's spin, the orientation of Earth in relation to its orbital position changes. This occurs because Earth, as it spins, bulges slightly at its equator. The equator is not in the same plane as the orbits of Earth and other objects in the solar system. This is shown in Illustration 82.

The gravitational attraction of the Sun and the Moon on Earth's equatorial bulge tries to pull Earth's spin axis into perpendicular alignment with Earth's orbital plane. Earth's rotation is counter-clockwise; gravitational forces make Earth's spin axis move clockwise in a circle around its orbit axis. This phenomenon is called precession of the equinoxes because, over time, this backward rotation causes the seasons to shift.

The full cycle of equinox precession takes 25,800 years to complete. Presently, Earth is closest to the Sun in January and farther away in July. Due to precession, the reverse will be true 12,900 years from now. The Northern Hemisphere will experience summer in December and winter in June. The North Star will no longer be Polaris because the axis of Earth's rotation will be pointing at the star Vega instead (see Illustration 83).

A consequence related to this phenomenon is that the Moon is slowly becoming more distant from Earth. The Moon is departing from us at the rate of 1.5 inches (3.8 cm) each year. The idea of the Moon retreating proposed over a century ago by English mathematician and geophysicist George Howard Darwin, Charles Darwin's son, has been confirmed by measuring the distance to the Moon with lasers.[248,249]

The tidal drag caused by the Moon's gravity slows Earth's rotation and accelerates the Moon. One of the counter intuitive things about physics is that, if you speed up an object in orbit, it takes longer to complete a full orbital revolution. In orbit, you slow down if you speed up, at least when viewed from Earth's surface. This means that both days and months are getting longer.

This mechanism has been working for 4.5 billion years, since Earth and Moon first formed. There is evidence in the geologic record that Earth rotated faster and that the Moon was closer to Earth in the distant past. We know from silt deposits that, 620 million years ago, a day was 21.9 hours long, there were 13 months/year and 400 solar days/year.[250]

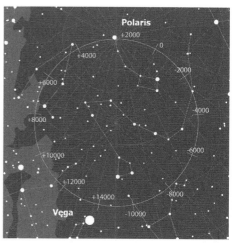

Tidal forces are also why the Moon always keeps the same side facing Earth. One day, Earth will reach a similar state—a day will be as long as a month. A month will be longer than it is today because the Moon

Illustration 83: Path of the north celestial pole among the stars due to precession. Original image by Tau'olunga.

will be farther away, and Earth will show only one face to the Moon. But this shouldn't be a major concern: There is good reason to believe that the Sun will expire, taking Earth and its Moon with it, long before this happens (see Chapter 10).

Cycles Summarized

Individually, each of the three cycles affect insolation patterns. When taken together, they can partially cancel or reinforce each other in complicated ways. Illustration 84 shows how the three cycles combine to affect solar forcing over the past 200,000 years. It is the complex pattern of insolation change created by the interaction of all three factors that caused so much confusion verifying Croll and Milankovitch's predictions.

Adhémar based his predictions on the mathematics of d'Alembert, who calculated precession in 1754. Alexander von Humboldt discredited this theory by pointing out that, though the seasonal insolation varied, the total energy received remained constant. Precession alone was not enough.

Croll used both eccentricity and precession, as well as the effects of glaciation and changing air currents. In 1875, he added tilt to his calculations. Even then, his interpretation of the factors was not totally correct. Not until Milankovitch produced integrated curves combining all the orbital elements were the cycles on firm footing. But the apparent disagreement with experimental data still caused the theory to fall into disrepute.

Only after work by Erickson, Broecker, et al,[251] in the 1950s, was the theory revisited. In 1978, Berger corrected the theory with more accurate formulae for calculating insolation variations.[252] Today, we use numerical integration to calculate the effects of all the bodies of the Solar System. This is difficult, requiring minimization of the error in present observational data and running long calculations on supercomputers.[253] Croll and Milankovitch had to do their calculations by hand. Science is often a long and tortuous process.

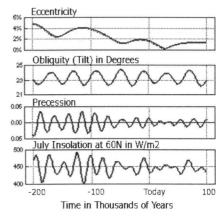

Illustration 84: Relationship of cycles with insolation. Source NOAA.

Today, it is widely accepted that Milankovitch cycles are the forcing that decides the timing of glacial/interglacial periods. Data from the glaciation record are in strong agreement with this theory. In particular, during the last 800,000 years, the dominant period of glacial-interglacial oscillation has been 100,000 years, which corresponds to changes in Earth's eccentricity and orbital inclination.

Glacial periods can be triggered when tilt is small, eccentricity is large, and perihelion, when Earth is closest to Sun, occurs during the Northern Hemisphere's winter. Perihelion during the Northern Hemisphere winter results in milder winters but cooler summers, conditions that keep snow from melting over the summer.

Deglaciation is triggered when perihelion occurs in Northern Hemisphere summer and Earth's tilt is near its maximum. There are other factors which act to enhance the forcing effects of the cycles. These include various feedback mechanisms such as snow and ice increasing Earth's albedo, changes in ocean circulation and enhanced greenhouse heating due to increased CO_2 and water vapor concentrations. Earth's current place in the three cycles are as follows:

- **Eccentricity**. Earth's current orbital eccentricity is 0.0167, which is relatively circular. Presently, Earth's distance from the Sun at perihelion, on January 3rd, is 95 million miles (153 million km). Earth's distance from the Sun at aphelion, on July 4th, is 98 million miles (158 million km). This difference between the aphelion and perihelion causes Earth to receive 7%

more solar radiation in January than in July. Currently, Earth's orbital eccentricity is close to the minimum of its cycle.

- **Obliquity.** Currently, axial tilt is approximately 23.45 degrees, reduced from 24.50 degrees just a thousand years ago. Even so, Earth's current tilt is almost at its maximum. This explains the contrast in Earth's seasons. A lower degree tilt would result in cooler summers and warmer winters, thus, moderating global temperatures. Some argue that this would cause growth in ice sheets in the high latitudes. Snow would accumulate over the winter and would be less prone to melting and recession during the summer.

- **Precession.** The variation in the direction of Earth's axial tilt is thought to be the most important influence of climate. Today, Earth is closest to the Sun during Northern Hemisphere winter and farthest away during Northern Hemisphere summer.

So currently, Earth's orbit meets only one of the three conditions which lead to the onset of glaciation, Perihelion during Northern Hemisphere winter. As stated above, glacial maximums have occurred roughly every 100,000 years for the past 800,000 years with the last glacial maximum occurring 18,000 years ago. After each glacial period there has been a period of rapid warming. These warming periods are followed by a relatively slow cooling trend leading to the next glacial maximum. Today, we are poised to enter another glacial period but, there are still uncertainties.

The European Project for Ice Coring in Antarctica (EPICA) team has noticed the interglacial period of 400,000 years ago closely matches our own because the shape of Earth's orbit was the same then as it is now. That warm spell lasted 28,000 years so we might not be as close to the next glacial episode as often thought.

Limits of Orbital Forcing

To sum up, scientists believe the Croll-Milankovitch cycles caused the onset of the Holocene interglacial period in the following way. At the beginning of the Holocene, around 15,000 years ago, variation in Earth's orbit and attitude caused a small increase in the solar radiation received from the Sun. Those changes also resulted in a redistribution of solar energy within Earth's atmosphere and ocean, which caused a slight warming, ending the glacial period. Retreating glaciers, melting snow cover, and diminished sea ice exposed larger areas of land and open ocean. The exposed areas absorbed more solar radiation, reinforcing the warming trend. This accelerated warming of the ocean, releasing of large amounts of CO_2 and further reinforcing the warming trend. Temperatures increased to modern levels and have stayed there since—though, the climate continues to undergo variations on century and decade long time scales. How long the warm period will last cannot be predicted.

Today, scientists believe that the principal cause of glaciations is the intensification of the hydrologic cycle caused by Earth's orbital cycles. Variation in insolation patterns cause tropical oceans to warm, increasing the equator-to-pole temperature gradient. This leads to the growth of land-based ice in high latitudes. In other words, increased heating of the oceans is needed to start a glacial period.

This argument was first made by John Tyndall, the Irish physicist, naturalist and educator, in the late 1800s.[254] Greater solar radiation in winter and spring at the expense of summer and autumn, leads to higher frequency of El Niño anomalies. From studying the start of the previous glacial period, similarities can be seen in current orbital changes.[255]

Although the current variations are less extreme, researchers have concluded, "association of recent positive seasonal anomalies of global mean temperature with El Niño events suggests that the ongoing global warming may have a significant, orbitally influenced natural component. The warming could continue even without an increase of greenhouse gases."[256]

Even if there is a connection to current global warming, the Croll-Milankovitch cycles do not explain short term, decadal variations, or the longer term changes that signal the beginning and end of ice ages. We need to look for other influences—in particular, the Sun.

Chapter 10 Varying Solar Radiation

*"The sun is an example. What it seems it is and, in such
seeming all things are"*
— *Wallace Stevens*

In our modern, technology-driven world, it is easy to take the Sun's light for granted. With electric power and indoor lighting, our civilization is no longer a slave to daylight. We can work and play around the clock, with illumination available at the flick of a switch. But, for most of human history this was not the case.

In earlier times, candles and oil lamps provided a poor substitute for the natural light of the Sun. And, in ancient times, there was often no substitute for sunlight available, aside from the flickering of a fire. Weak stuff compared with the golden radiance of our local star. It is no wonder that people around the world revered and worshiped the Sun as a deity.

To the ancient Greeks, the Sun was the chariot of the god Helios, driven across the heavens by four horses. For the ancient peoples of Meso-America, the Sun was a god, and they carefully observed and recorded the changing path that the Sun traveled across the sky throughout the year. The Inca, Maya and Aztecs created detailed calendars and astronomical tables going back thousands of years.

According to astronomer and stellar physicist David Dearborn, there is reason to believe that some Meso-American cultures recognized sunspots. There is no clear evidence that the Inca or the Maya noticed sunspots, but the Aztec myth of creation involves a Sun god with a pock-marked face. This strongly suggests that they had seen dark blemishes on the face of the Sun.[257] As early as 28 B.C., astronomers in ancient China recorded systematic observations of what looked like small, changing, dark patches on the surface of the Sun. These observations were recorded in the *Hanshu*, or *Book of Han*, an historical text chronicling the history of the early Han dynasty.

In western cultures, the visage of the Sun tended to be viewed as perfect and without blemish. There are some early references to sunspots in the writings of Greek philosophers, but Aristotle held that the Sun and the heavens were ideal geometric objects. The ancient Greeks, and the Europeans after them, were highly influenced by Aristotle. Even when faced with contradictory observations, it was hard to supplant his teachings. Not until Galileo pointed his telescope at the Sun did the opinions of Aristotle begin to lose their grip on the western imagination.

Galileo Galilei (1564-1642) was an Italian physicist, mathematician, and philosopher, and probably the most famous astronomer of all time. Often called the "Father of Astronomy," Galileo was an important figure in the development of modern science.

Galileo was born in Pisa, the eldest of six children. He was the son of Vincenzo Galilei, a famous musician of the day. As a young man, he seriously considered becoming a priest, but at his father's urging, he enrolled at the University of Pisa to study medicine. He never completed his medical degree, changing to mathematics instead.[258]

Illustration 85: Galileo Galilei,

In 1616, Galileo's support for the theories of Copernicus[†] led him into conflict with the Catholic Church. Seventy years earlier, Copernicus had advocated a heliocentric view of the solar system. He placed the Sun, not Earth, at the center of the Universe, with the planets orbiting around it. Among the first to use a refracting telescope, Galileo became an advocate of the Copernican view of the solar system after viewing the previously undiscovered moons of Jupiter in orbit around that planet. During a hearing before the Inquisition, Galileo was ordered not to "hold or defend" the idea that Earth moves or that the Sun stands still at the center of the universe. Not an openly-rebellious person, as often portrayed in historical accounts, he stayed well away from the controversy for the next several years.

After the election of Cardinal Barberini as Pope Urban VIII, in 1623, Galileo wrote a book presenting the evidence for and against a heliocentric solar system. Pope Urban, a friend and admirer of Galileo, had asked for the book to be published, but had warned Galileo to be careful with his pronouncments. The book, *Dialogo sopra i due massimi sistemi del mondo* (Dialog Concerning the Two Chief World Systems), was published in 1632, with encouragement from the Pope, and under a formal license from the Inquisition.

Although the book presented both sides of the issue, there was no question which side got the better of the argument. For overstepping his bounds by clearly advocating the Copernican view, Galileo was again hauled before the Inquisition. This time, he was forced to recant his views, his book was banned, and he was imprisoned. An apocryphal embellishment to this

---

† Nicolaus Copernicus (1473-1543) Polish astronomer who formulated the first heliocentric model of the solar system.

conflict is the claim that, after his forced recantation that Earth orbited the Sun, he muttered, "Nevertheless, it does move."[259] The imprisonment was later reduced to house arrest but, for the rest of his life, Galileo worked quietly out of the public eye. He eventually went blind and died from natural causes in January, 1642. But not before completing his final and greatest work.

The *Discorsi e dimostrazioni matematiche, intorno a due nuove scienze* (Discourses and Mathematical Demonstrations Relating to Two New Sciences) was published in 1638. Highly praised by both Isaac Newton and Albert Einstein, this book has resulted in Galileo being called "the father of modern physics—indeed of modern science."[260] Interestingly, Galileo was a practicing astrologer during most, if not all, of his career.[261] Practicing astrology was still a normal activity for a mathematician in the early 17th century but, nowadays, is considered anything but modern or scientific.

More than 300 years later, during the Pontificate of Pope John Paul II, there was a considerable effort for reconciliation between Church and science. In a speech given on 19 November 1979, for the centenary commemoration of the birth of Einstein, the Pope deplored the failure of the Church to perceive the legitimate autonomy of science. In 1992, a study by the Pontifical Council for Culture resulted in Pope John Paul issuing a statement expressing regret for the way the Church had responded to Galileo's theses.

The first person known to have suggested that the Sun is a star was the Greek philosopher Anaxagoras (circa 450 BC). Around 1590 AD, Giordano Bruno suggested the same thing. For that, and other heresies, he was burned at the stake by the Roman Inquisition in 1600. During the 16th and 17th centuries, through the work of Copernicus, Kepler[†] and Galileo, the nature of the solar system, and the Sun's place in it, slowly became apparent. Finally, in the 19th century, the distances to stars were accurately measured. After establishing how far away the stars were, their brightness could be fixed. This allowed scientists to finally prove that our Sun is a star—one of hundreds of billions in our galaxy.

Stellar Evolution

Stars come in a wide-range of sizes and colors; from massive super-giants to diminutive brown dwarfs, and from hot blue-white to cool deep-red. Their lifespans also vary widely, from a few tens of millions of years to several times longer than the age of the Universe. To a great extent, the color and

---

† Johannes Kepler (1571-1630) German mathematician and astronomer, best known for his laws of planetary motion.

lifespan of a star depends on its size. But, there are also exotic stars that do not fit the normal rules; novae, pulsars, X-ray binaries, and gamma-ray bursters, to name a few.

Most stars are what are called *main-sequence* stars, for reasons we will explain. There are three main phases in the evolution of a star: the pre-main-sequence phase, the main-sequence phase, and the post-main-sequence phase. In the pre-main-sequence phase, gravitational contraction provides most of a star's energy. During the main-sequence phase, energy is provided from thermonuclear fusion of hydrogen into helium, a process called *nucleosynthesis*. As a star's hydrogen runs out, it moves into the third post-main-sequence phase where hydrogen burning moves away from the star's core. This phase continues through the fusion of successively heavier elements until the star runs out of nuclear fuel.[262]

Stars are formed out of interstellar clouds of gas and dust, though scientists are not quite sure what causes a cloud to collapse into a new stellar system. Possible causes include radiation pressure from exploding supernovae and the shock waves that form at the leading edges of the galaxy's spiral arms.

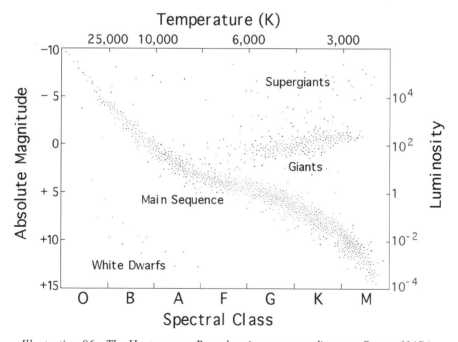

Illustration 86: The Hertzsprung-Russel main-sequence diagram. Source NASA.

Since the arms seem to be stellar nurseries, and also contain the large hot stars that tend to explode as supernovae, both mechanisms may play a part in stellar formation. Regardless of the cause, once the gravitational threshold is crossed, a potential new star enters the pre-main-sequence phase of its life, powered by the conversion of gravitational potential energy into heat.

If the object being formed has enough mass, its internal density and temperature will become great enough for nuclear reactions to begin. The needed amount of mass is about 0.08 times the mass of our Sun. Below this limit, the nuclear fires won't ignite. The giant planet Jupiter, the largest planet in our solar system, is considered a failed star by many astronomers. Above the limit, enough energy is produced from fusion to halt the star's gravitational collapse.[263]

When a star has stabilized, due to the nuclear burning of hydrogen, it has entered the main-sequence, shown in Illustration 86. How long a star remains on the main-sequence depends on how massive it is. A star like our Sun will be on the main-sequence for 10 billion years or more. Stars below 0.8 solar masses will last for as long as 1,000 billion years, a hundred times the current age of the Universe. At the other end of the weight scale, a star massing 30 times that of the Sun will burn up its hydrogen in about 5 million years. No stars heavier than 50 solar masses have been seen on the main-sequence.[264]

The mass of a star also affects the frequencies of light that it produces. Stars emit light in a temperature-dependent frequency distribution scientists call *black-body radiation*. This means that the electromagnetic radiation a star emits follows the *Plank curve*. We will explain more about the significance of this later in the chapter. Heavy stars are bluer, hotter and shorter-lived than our Sun.

The fate of a star when it leaves the main-sequence, after exhausting its supply of hydrogen, also depends partly on its mass. Once a star's hydrogen is spent, it will start burning helium in a nuclear reaction that forms carbon and oxygen. A side effect of this change in nuclear fuel is swelling and cooling of a star's outer layers. When this happens, the star becomes a *red giant*.

Once helium is exhausted, a large star will burn carbon and oxygen, proceeding with successively heavier elements until iron and nickel are produced. Stars massing less than eight solar masses are not big enough to burn carbon so their careers as red giants ends when their helium supply is gone. With no ongoing thermonuclear reaction to provide heat to support it, a star will collapse.

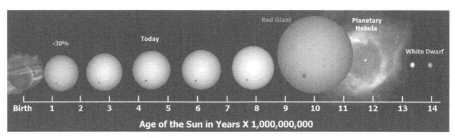

Illustration 87: Life Cycle of the Sun (images not to scale), Source: original images from NASA, composition by D. L. Hoffman.

When a star the size of the Sun runs out of fuel and collapses, the sudden contraction of matter causes one or more explosions. These explosions blow off the outer layers of the star, forming a planetary nebula. Many of the more spectacular photographs from the Hubble Space Telescope are of planetary nebulae. Over time, the matter in the nebula will disperse, leaving the burned-out remains of the star behind—a *white dwarf.*

Since a white dwarf is no longer supported against gravitational collapse by the heat of fusion, its further collapse must be resisted by other forces. Main-sequence stars, with medium or smaller masses, are supported by *electron degeneracy pressure.* Matter in this state is highly compressed, and therefore extremely dense. A star with mass comparable to the Sun ends up shrinking down to a volume the size of Earth. Becoming a white dwarf is thought to be the final state of over 97% of the stars in our galaxy.[265] This is the expected fate of our Sun, as shown in Illustration 87.

Electron degeneracy pressure is a quantum mechanical force, resulting from the *Pauli[†] exclusion principle,* which limits how tightly matter can be squeezed together (electrons don't like to be crowded). If a star weighs more than 1.4 solar masses, its collapse cannot be halted by forming electron-degenerate matter. This mass is called the *Chandrasekhar Limit,* named after Indian-American physicist, Subrahmanyan Chandrasekhar. It is the maximum non-rotating mass that can be supported against gravitational collapse by electron degeneracy pressure. More massive stars will continue to collapse because the degeneracy pressure provided by electrons is weaker than the inward pull of their gravity.

When a star heavier than the Sun contracts, it ejects its outer layers. If it ejects enough mass to slip under the Chandrasekhar Limit, its core will become a white dwarf. But, when the star remnant contains more than 1.4 times the mass of the Sun, it will continue to collapse. Its electrons will be

---

† Wolfgang Ernst Pauli (1900-1958) Austrian theoretical physicist and pioneer in quantum mechanics, most notably spin-statistics.

forced to combine with protons in the star's atoms, forming a strange form of matter called *neutron-degenerate matter.*[‡] The star becomes a *neutron star.*

Stars with 10-20 solar masses become unstable after they exhaust their helium and begin burning carbon. Such stars can blow off their outer layers in incredibly violent explosions called a supernova. Even more massive stars, after exploding as supernovae, can become *black holes.*[266]

Black holes are thought to form from heavy stars which start off with masses more than 20 or 25 times that of the Sun. How black holes form is still an area of active research so exact numbers are not available. When one of these stars expires in a supernova explosion and its core collapses, gravity wins out over any other force that might be able to hold the star up. Eventually, the star collapses to such an extent that it is contained within its *Schwarzschild radius,* or *event horizon,* from which even light cannot escape.

Albert Einstein's[†] 1915 theory of general relativity provides the theoretical foundation for black holes. Einstein showed that gravity can bend the path of light just as it bends the path of any other moving object. When a massive star collapses, its gravitational pull becomes strong enough to prevent light from escaping. Though black holes can be extremely massive, their size is relatively tiny; a black hole with the mass of the Sun would have an event horizon the size of a small asteroid.

It is now thought that all large galaxies contain gigantic black holes at their centers. These galactic black holes can be millions or even billions of times more massive than the Sun. Some black holes are among the most violent and energetic objects in the universe—forming *quasars,* quasi stellar objects, which shoot off jets of x-rays and gamma-rays as they suck in surrounding gas. Others, like the black hole at the center of the Milky Way, are more docile, quietly consuming surrounding stars.

This information about stellar evolution and supernovae is important to fully understand our solar system: where it came from and how it will end. When the Universe began, 14 billion years ago, almost all normal matter was in the form of hydrogen.[267] As mentioned, elements heavier than hydrogen, from helium through iron, are created as by-products of thermonuclear reactions

---

‡ *Neutronium* is a term originally used in science fiction and in popular literature to refer to the material present in the cores of neutron stars. It is rarely used in scientific literature; typically, neutron-degenerate matter is used instead.

† Albert Einstein (1879-1955) German-born theoretical physicist best known for his theory of relativity and mass-energy equivalence equation, $E=mc^2$.

—the ash of fusion. When stars explode, at the end of their main-sequence lifetimes, their matter is redistributed. Even heavier elements, up through uranium, are created by nucleosynthetic alchemy when massive stars erupt in supernovae explosions.

This is where the material that makes up Earth and the other planets comes from. The oxygen and silicon in Earth's crust, the iron in its core and all the other elements, so essential to life, are all the remnants of dead stars. The nearby supernova that threatened the start of life on Earth may have contributed to its ultimate emergence instead.

Our Star, the Sun

The Sun is a fairly average, middle-aged star of the type astronomers call a type G dwarf. It is often said that the Sun is an "ordinary" star, but there are many more smaller stars than larger ones. The median size of stars in our galaxy is probably less than half the mass of the Sun, with the Sun in the top 10% by mass. It is currently composed of approximately 72% hydrogen, 26% helium and 2% of all other chemical elements in gaseous form. The source of the Sun's brightness, and that of other stars, is due to thermonuclear reactions occurring in their innermost parts. The temperature inside the Sun can reach 27,000,000°F (15,000,000°C).

The Sun is 900,000 miles (1,400,000 km) in diameter and emits about 3.86×10^{26} watts of energy from its surface.[268] This translates to about 1366 watts per square meter at Earth's orbit. It is the source for 99.98% of the energy that drives Earth's ecosystem, the other 0.02% coming from the cooling of Earth's molten interior.

Prior to the discovery of the process of thermonuclear fusion, gradual gravitational contraction, commonly called *Helmholtz Contraction*, was thought the most likely way stars generated energy. Since the 1930s, however, astrophysicists have become convinced that thermonuclear fusion is responsible for lighting the stars. According to contemporary stellar models, the physical conditions within the core of a star make the fusion process unavoidable.

Each second, more than 4 million tons of matter are converted into energy within the Sun's core, producing neutrinos and solar radiation. So far, during its life, the Sun has converted around 100 Earth-masses of matter into energy. Unlike the perfect disk of many ancient myths, we know the Sun is actually a violent, roiling mass of gas driven by nuclear fire. Flares leap from the solar surface that dwarf Earth in size.

As a consequence of changes brought about by thermonuclear fusion, a slow increase in size and energy output is predicted over time. We stated earlier, when Earth was young, the Sun was 30% fainter than today. The Sun's estimated brightness four billion years ago would not have been enough to warm Earth's atmosphere to a comfortable level for life. Any liquid water exposed to the surface would have quickly frozen solid. But geological observations from that time, evidenced in sedimentary rock, required the presence of flowing liquid water. That evidence led to what astronomer Carl Sagan[†] called the "Faint Young Sun Paradox."

Scientists have offered a number of possible explanations for this apparent contradiction. Noting that before the advent of abundant life, atmospheric oxygen levels were much lower than today and there was also a significant amount of methane (CH_4) present. In today's atmosphere, oxygen interacts with methane, rapidly turning it into carbon dioxide. The absence of oxygen in the early atmosphere would have allowed methane concentrations much larger than currently observed. Methane, a much more potent greenhouse gas than CO_2, could explain the anomalously high temperatures. Extensive volcanism could also have helped to raise the levels of greenhouse gases in the young Earth's atmosphere.

In the very long term, astrophysicists believe that the Sun's output increases by about 10% per billion years. Some scientists speculate that, in about one billion years, the additional 10% will be enough to cause a runaway greenhouse effect on Earth. The rising temperature produced by the increase in solar radiation will cause more water vapor. This in turn will accelerate greenhouse heating to the point where it will evaporate the oceans and destroy life on Earth. This will take place long before the Sun swells into a red giant near the end of its life.

Sunspots and Solar Climate

The slow increase in the Sun's output is not thought to be the cause of the short-term warming over the past several decades. But variation in solar output over shorter periods has been a subject of intense study. In particular, the observed link between sunspot activity and Earth's climate.

Perhaps, the most interesting feature of sunspots is that their number increases and decreases in a regular rhythm over about a decade. This regular cycle was first noticed by the German astronomer Samuel Heinrich Schwabe in 1843. This has become known as the solar magnetic activity cycle, or sunspot cycle. The number of sunspots in each cycle is not

---

† Carl Edward Sagan (1934-1996) American astronomer, astrobiologist and popularizer of the natural sciences.

constant; there have been periods where many sunspots were observed, and others when sunspots seem to disappear altogether. Sightings from China, Korea and Japan between 28 BC and 1743 AD averaged only six sunspots per year. None were observed between 1639 and 1700.[269]

The most famous of these periods is known as the Maunder Minimum, a period of low sunspot activity that lasted from 1645 to 1715. This coincides with the coldest portion of the famous Little Ice Age period discussed in Chapter 3. The Maunder Minimum is named after the English astronomer Edward W. Maunder (1851-1928). From studying historical records of sunspot counts, called the *sunspot number*,[†] Maunder discovered that sunspots were virtually absent during this period, and disappeared altogether during the decade starting in 1670. Astronomers observed only about 50 sunspots during the 70 year period from 1645 to 1715. Normal sunspot activity would have produced 40,000 to 50,000 sunspots.

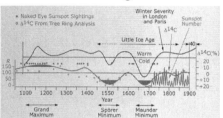

Illustration 88: Sunspot activity and solar output during the Maunder Minimum. Source John Eddy based on Science, no.192, 1189 (1976).

The correlation between sunspot activity and climate during this period is striking (see Illustration 88). Other historical periods of low sunspot activity have been detected either directly, or by the analysis of carbon-14 isotope ratios from ice cores or tree rings. Other periods include the Sporer Minimum (1450-1540), and less dramatic Dalton Minimum (1790-1820). In total, 18 periods of sunspot minima have been found since the beginning of the Holocene. Studies indicate that the Sun spends up to a quarter of its time in these minima.[270]

The behavior of other Sun-like stars have been used to estimate solar irradiance during the Maunder Minimum. Lean, et al., estimated solar irradiance then was 0.15% to 0.35% lower than the present mean value.[271] An independent estimate by Baliunas and Jastrow gave a range of 0.1% to 0.7% based purely on observations of solar-like stars. They concluded that a reduction in irradiance of 0.4% would be enough to explain the cold average temperatures of the Little Ice Age.[272]

---

† The sunspot number, or Wolf Number, measures the number of sunspots and sunspot groups present on the surface of the sun. Named after Swiss scientist Rudolf Wolf.

There are good reasons for believing that changes in irradiance have been a significant factor in the rise in global temperature since the late 17th century. Reid pointed out the similarity between the overall level of solar activity, as expressed by the 11-year sunspot cycle, and globally-averaged sea-surface temperatures over the period from 1860 to the present.[273] However, the Sun's contribution to more recent climate change remains controversial, partially because such links are based on correlations between the global temperature record and proxies for solar irradiance.

The decadal variation in the number of sunspots suggests that they are an integral feature of the solar climate. Short term variations in the Sun's output have been tied to the sunspot cycle by satellite instruments. Comparing the two plots in Illustration 89, it is clear that total solar irradiance is directly proportional to sunspot activity. This relationship was shown empirically by NASA's Solar Maximum Mission/Active Cavity Radiometer Irradiance Monitors (SMM/ACRIM) experiment.

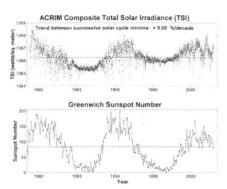

Illustration 89: Solar irradiance and observed sunspot activity. Source NASA.

What troubles scientists about attributing decadal climate change to variation in the Sun's output, is that the actual measured differences in solar irradiance is not sufficient to cause the changes in earthly observed temperatures. Satellite measurements indicate the energy Earth receives from the Sun varies by a small amount. There is a measurable increasing trend, but it amounts to only 0.05% per decade. It has been estimated that increasing solar irradiance accounted for one quarter (0.25°C) of last century's temperature increase.[274]

In conjunction with the sunspot cycle, the magnetic poles of the Sun reverse their polarity. These reversals influence the Sun's interaction with Earth's magnetic field. In addition to the 11-year solar cycle, the sunspot record shows century-scale modulation and periods of low activity. These changes in the Sun have led to speculation that solar variability is a trigger for other mechanisms that regulate Earth's climate. Influence on ozone production, thunderstorms and the influx of cosmic rays, have all been proposed as amplifiers for solar variation.

In contrast, the human-caused global warming scenario, as presented by the IPCC and its adherents, states that the Sun doesn't need to increase the energy it showers on our planet. The IPCC's models indicate that raising the amount of CO_2 in the atmosphere is sufficient to account for global warming. To see if this is possible, we need to examine the greenhouse effect more closely, and determine if basic physics supports the greenhouse gas hypothesis.

Solar Spectrum and Greenhouse Absorption

The light coming from the Sun is spread over a wide spectrum of wavelengths. Wavelength determines the color of a photon, or packet, of light energy. Shorter wavelengths have higher frequencies and more energy in each photon. Light toward the higher frequency, blue and ultra-violet end of the spectrum, is more energetic than light from the longer wavelength, red and infrared end of the spectrum.

The hotter an object, the higher the average frequency of the light it emits. This is why large hot stars appear blue in color while the cooler red giants appear orange or red in color. By examining the color of light a hot object emits, it is possible to measure its temperature. This is how the temperature of red-hot molten lava in volcanoes, and steel in steel mills is measured. But, hot objects do not just produce a single frequency of light, they emit a range of colors fitting a characteristic curve called the *black-body radiation curve.*

Discovering how to calculate this curve was a major challenge to theoretical physicists during the late 19th century. The problem was finally solved in 1901, by German physicist Max Planck,[†] and is formulated in Planck's law of black-body radiation.[275] This is why the black-body radiation curve is also called the Plank curve.

Classical physics failed to explain the observed behavior of glowing hot objects at high frequencies. Classical theory predicted an impossible rise of the energy density towards infinity, dubbed the *ultraviolet catastrophe.* It was Plank's new approach, treating light energy as small, individual packets, that solved the problem. In doing so, he created a new *quantum theory* of physics. Plank's work was extended by Einstein, in 1905, when he published his paper on the *photoelectric effect,* for which Einstein received the Nobel Prize.

---

† Max Karl Ernst Ludwig Planck (1858-1947) German physicist considered to be the founder of quantum theory.

Several Plank curves for different temperatures are shown in Illustration 90, along with the ultraviolet catastrophe curve predicted by classical theory. Notice that the temperatures here are given in degrees Kelvin. A single degree in temperature Kelvin is the same as a single degree Celsius, but the Kelvin scale is an absolute temperature scale. Unlike the Celsius scale, which places zero at the freezing point of water, the

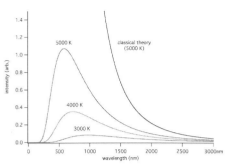

Illustration 90: Plank's black-body radiation curve for several temperatures.

Kelvin scale takes zero (0°K) to be the coldest possible temperature, equivalent to -273.15°C. On the Fahrenheit scale, 0°K is equivalent to -459.67 degrees. All objects that are warmer than absolute zero emit radiation in accordance with the Plank curve for their temperature.

Visible light is a small portion of the electromagnetic spectrum, and only a portion of the light emitted by the Sun. The curve of light frequencies given off by the Sun is shown in Illustration 91. It is not surprising that the light frequencies we see with our eyes are the frequencies of peak output by the Sun.

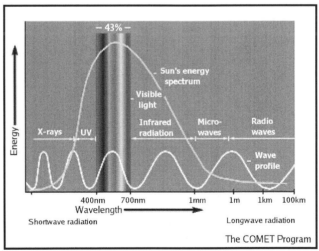

Illustration 91: The Sun's energy spectrum, Source: the COMET Program.

In order to properly understand the greenhouse effect, we must take into account the nonlinearity of the effect of increasing the concentration of atmospheric greenhouse gases. This is caused by different greenhouse gases having different spectra for the absorption of thermal radiation.

Each greenhouse gas has a range, or spectrum, of radiation frequencies it will absorb and re-radiate. For individual atoms, these frequencies are related to the energy levels of electrons orbiting each atom, which only come in discrete steps. These energy levels correspond to the energy of different photons and hence, to specific frequencies of electromagnetic radiation. If molecules only absorbed radiation at precisely those frequencies, very little interaction of molecules and radiation would take place. This is due to the low probability of radiation of exactly the right frequencies striking the gas molecules. However, there are factors which result in the absorption of radiation at frequencies near those in its spectrum. One such factor is the Doppler effect, resulting from the motion of the molecules.

Gas molecules, made up of multiple atoms, can also absorb radiation due to natural vibration modes of their chemical bonds and rotation of the molecule as a whole. Absorption of a photon results in a change in electronic energy accompanied by changes in the vibrational and rotational energies. The molecule's electrons get excited, its bonds vibrate like springs, and the whole molecule spins like a top.

This causes each greenhouse gas to have a characteristic profile, called its absorption spectrum, which describes its potential for absorbing light of different frequencies. The combined vibrational-rotational spectra of a gas can contain tens of thousands to millions of absorption lines. Illustration 92 shows the absorption spectra for the main greenhouse gases, plotted against the spectrum of incoming light from the Sun. Also shown is the outgoing infrared energy radiating from Earth back into space.[276]

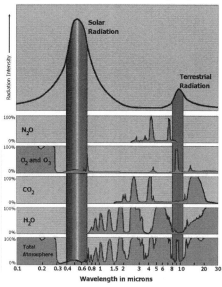

The two peaks in the upper part of Illustration 92 represent two "windows" of almost clear radiation transmission. These windows are a consequence of the total atmospheric absorption spectrum shown at the bottom of the diagram, which sums up the affects of all

Illustration 92: Absorption spectra of the major greenhouse gases.

atmospheric gases on light transmission. The one labeled "solar radiation" represents the transmission of visible sunlight, while the peak labeled "terrestrial radiation" represents the major band of long wavelength, infrared energy radiated back into space.

As mentioned in Chapter 7, only 51% of incoming solar radiation is absorbed by Earth's surface. This light energy is absorbed by the ocean and land, as well as living plants. Plants convert a small portion of this energy into chemical energy by photosynthesis, but most is translated into heat. When all the various factors are considered, only an annual average of ~235 W/m² of light energy is absorbed at a typical earthly location, out of the 1366 W/m² available at the top of the atmosphere. Earth's atmosphere recycles heat coming from the surface and delivers an additional 324 W/m², resulting in the habitable temperatures at the surface.[277] This all forms part of what scientists call Earth's *Energy Budget*, shown in Illustration 93.[278] The portion of Earth's re-radiated infrared energy, which gets trapped by greenhouse gases in the atmosphere, depends on the frequencies of the outgoing radiation and the *radiative efficiency* of the gases involved.

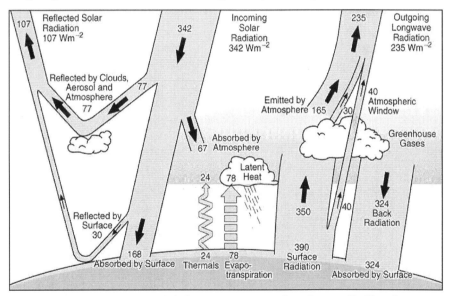

Illustration 93: Earth's Global Mean Energy Budget. Source Kiehl and Trenberth, 1997.

The radiative efficiency of a gas is not based solely on the narrow frequencies of light in its absorption spectrum. It does depend upon the line spectrum of the substance, but that spectrum can be broadened by both atmospheric pressure and temperature. Increased absorption, due to

171

broadening, can cause a positive feedback loop. On Mars the atmosphere is carbon dioxide, but at a low pressure and temperature. Therefore, the absorption spectrum of carbon dioxide on Mars is not significantly broadened and the greenhouse effect is even less than accounted for by the low density.

On Venus, on the other hand, the high pressure and temperature broadens the absorption spectrum of carbon dioxide so it is a more effective greenhouse gas than on Earth. But remember, the atmosphere of Venus is 90 times thicker than Earth's and is 96% carbon dioxide, making the atmospheric carbon dioxide concentration on Venus 300,000 times higher than on Earth. Even so, the high temperatures on Venus are only partially caused by carbon dioxide; a major contributor is the thick bank of clouds containing sulfuric acid.[279]

The conclusion drawn from physics is that there is a limited amount of outgoing heat energy available for greenhouse gases to absorb. Also, the amount of radiation absorbed depends upon the spectrum of the radiation impinging upon the gas. A small increase in a greenhouse gas, under conditions of low concentration, can have more of an impact than a much larger increase under conditions of high concentration. This results in a nonlinear response to increasing atmospheric concentrations of CO_2.

The source of the nonlinearity may be thought of in terms of a saturation of the absorption capacity of the atmosphere in particular frequency bands. The concentration of greenhouse gases can make the atmosphere essentially opaque in a particular band. If the atmosphere absorbs 100% of the radiation in a frequency band, no amount of additional greenhouse gas will increase heat absorption. Under these conditions, the atmosphere is said to be saturated in that particular frequency band.

At lesser concentrations, the absorption of radiation is described by *Beer's Law*.[†] Beer's Law relates the amount of radiation absorbed to a gas's *absorption coefficient*, a frequency-dependent molecular property, the length of the path radiation must travel through the atmosphere, and the concentration of the gas. This relationship produces what is known mathematically as a logarithmic curve, which results in a decreasing warming effect for increasing gas concentration.

As seen in Illustration 94, the increase from A to B produces a much bigger impact on the proportion of radiation energy absorbed than the increase from C to D even though the magnitude of the increase from C to D is larger than

---

[†]   Also known as the Beer-Lambert-Bouguer law;
$-log_{10}(\Phi_i/\Phi_0) = -log_{10}\tau_i = \varepsilon cb = A$.

the increase from A to B. In fact, from point C, no increase in concentration, no matter how large, will produce as much of an impact as the increase from A to B.

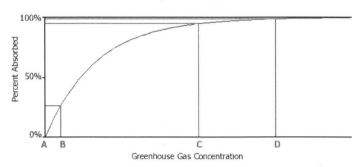

Illustration 94: Nonlinear increase in radiative absorption with increasing greenhouse gas concentration.

Complicating factors include the partial overlap of absorption bands between CO_2 and H_2O, which limits to how much warming an increase in carbon dioxide can cause. The 13,000 Gt of water in the atmosphere (~0.33% by weight) are responsible for about 70% of all atmospheric absorption of radiation.[280] If the CO_2 and H_2O absorption spectra completely overlapped, there would be no significant role for atmospheric carbon dioxide to play in greenhouse warming.

According to data from the Global Historical Climatology Network of land temperatures (GHCN) and the Extended Reconstructed Sea Surface Temperature (ERSST) data sets, the period from 1910 through mid-1940s had a global warming trend of 0.23°F (0.13°C) per decade for a net warming of 0.8°F (0.45°C)—leaving only 0.63°F ± 0.36°F (0.35°C ± 0.2°C) net warming potential for CO_2 emissions from fossil fuel use during the post-WWII period.[281] Ignoring the possible contributions of all other natural drivers of planetary temperature change, a 30% increment in atmospheric carbon dioxide during the later part of the 20th century caused, at most, about 1°F in temperature increase. Based on these data, most empirical calculations yield a projected temperature increase of less than 1.8°F (1°C) for a doubling in CO_2.[282]

Why does carbon dioxide figure so prominently in the IPCC's predictions? The warming levels quoted by the IPCC are generated by GCM computer models, models that have been written based on the assumption that CO_2 is the primary driver of Earth's climate. Climate model results vary widely when the underlying assumptions are changed. In a study of 108 different models, based on doubling CO_2 levels, the temperature predictions ranged from a low of 0.29°F to a high of 15.6°F (0.16- 8.7°C).[283] In order to get their computer models to come close to matching the past century of climate data, a number of assumptions about linkage among climate mechanisms and positive feedback loops have been made by the modelers. These assumptions are not justified by the physics of the greenhouse effect or empirical data from past climate variation.

The scientific evidence is clear: there is a significant effect from increasing CO_2 at low concentrations, with decreasing impact as concentrations rise. But, for CO_2 to play a dominant role at high concentrations, the level of atmospheric CO_2 must rise to levels not seen since the PETM (page 63) or possibly the end of the Precambrian Snowball Earth period, more than 550 million years ago. The net effect of all these factors is that doubling atmospheric carbon dioxide levels today would not double the amount of global warming. The IPCC's models wildly overestimate the impact of CO_2.

If the variation in irradiance is too small to cause the recent observed changes in temperature directly, and greenhouse gases cannot account for the warming, then what does? Other causes of warming must be investigated. Because the statistical fit of historical data, linking the sunspot cycle and climate variation, is so good, other mechanisms coupled to variation in the Sun's activity level have been proposed.

Other Possible Links to the Sun

Since ancient times, the Nile River has been the life blood of Egypt. Beginning as two separate rivers, deep in the heart of Africa, the mighty river flows for some 4,000 miles before it empties into the Mediterranean Sea. The Blue Nile originates in the Abyssinian Mountains of Equatorial Africa, while the White Nile emerges from Lake Victoria. The two merge into one at what is now Khartoum, Sudan, and flow north through the ancient lands of Nubia and Egypt.

The Nile was so central to the lives of Egyptians that they simply called it *Iteru*, meaning River. Without the River and its annual inundation, Ancient Egypt would never have come into being. Its fertile valley was renewed every year by the annual flood, which deposited rich silt along the riverbanks. The River filled all areas of life with religious symbolism: the creator sun-god *Ra* was believed to be ferried across the sky daily in a boat, while hymns to *Hapi*, the river god personifying the Nile, praised his bounty. Egyptian creation myths revolve around a primordial mound rising from the River's flood waters.[284] Amazingly, the linkage between the Sun and River gods in Egyptian mythology are echoed in modern day scientific findings.

Knowledge of the water level of the Nile River was critically important for agriculture in Egypt throughout its history. Measurements of these levels have been carried out since the times of the pharaohs.[285] Shakespeare mentioned a tower used in the measurement of water levels in *Antony and Cleopatra* (Act II Scene VII). These special towers, called *Nilometers*, are used to gauge the rise and fall of the Nile's waters to this day.

Scientists, investigating ways that solar variability influences Earth's climate, have compared the Sun's activity level with the level of water in the Nile. Annual records of the Nile's water level are uninterrupted for the years 622-1470 AD.[286] Using a technique called *Empirical Mode Decomposition* (EMD), which is designed to deal with non-stationary, nonlinear time series, researchers have identified two time scales in the water level data that can be linked to solar variability: an 88 year period and a longer period of about 200 years. According to Alexander Ruzmaikin, Joan Feynman and Yuk Yung, "This suggests a physical link between solar variability and the low-frequency variations of the Nile water level. This involves

Illustration 95: Statue of Hapi.

the influence of solar variability on the North Annual Mode of atmospheric variability and the North Atlantic and Indian Oceans patterns that affect rainfall over Equatorial Africa, where the Nile originates."[287] It seems that the deeper scientists look into the relationship between Earth's climate and the Sun, the more connections they find.

Because of the strong statistical correlation between solar activity and weather, scientists have been looking for other mechanisms that would enable small changes in the Sun to effect measurable changes in Earth's climate. Any mechanism linking solar variability to changing weather and climate must involve influencing the distribution of the energy within Earth's weather system. One possible mechanism is Earth's electric field.

Around the globe, thunderstorms maintain the lowest reaches of the ionosphere at an electric potential of 250,000 volts (250 kV) with respect to the ground. This results in a very weak current flowing from the atmosphere to the ground in the fair-weather regions of the globe. Near the ground, this charge maintains a vertical electric field of around 100 volts per meter. Cosmic ray ionization, which is controlled by solar activity via the solar wind, modulates the resistance of this global electric circuit. In this system thunderstorms are the generators. By controlling the ease by which thunderstorms can dissipate current, solar activity may modulate the intensity of thunderstorm development, thus controlling the distribution of energy within the meteorological system.[288]

Researchers have found other links between climate and solar output. In Portugal, they report the influence of solar variability is strongest in low clouds, pointing to a mechanism involving aerosol formation enhanced by ionization due to cosmic rays.[289] The relation between thunderstorm activity and solar variability was analyzed by using monthly data on regular electromagnetic very low frequency (VLF) radio noises detected at Yakutsk, Siberia. These noises are an indication of thunderstorm activity. In a study covering 1979-1993, it was found that local thunderstorm activity in eastern Siberia and in Africa are in *antiphase* with the solar activity. The more active the Sun, the lower the thunderstorm activity. This result can be interpreted as linking higher levels of cosmic rays to increased thunderstorm activity.[290]

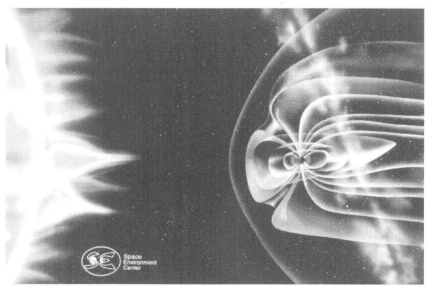

Illustration 96: The space environment around Earth caused by the solar wind. not to scale. Source NOAA.

On longer time scales, going back in time before recorded human history, scientists rely on proxy data. A fairly complete record of solar activity is contained within carbon-14 (^{14}C) and beryllium-10 (^{10}Be) isotope deposits, found in tree rings and ice cores. After correction for other climatic effects, a clear correlation with the 11-year sunspot cycle has been found. On longer time scales, there are characteristic periods of 80, 150, 200 and 500 years and longer. There are even suggestions of cycles lasting millions of years. Clearly, there are influences affecting Earth's climate that do not originate on this planet.

When talking about the wonders of the universe, Carl Sagan was fond of saying that we are all made of "star stuff." Science has shown that the varied elements that make up people and our entire planet were created deep within stars and by supernovae explosions. But the influence of the stars doesn't end there. Astrophysicists have come to suspect there is a direct, ongoing connection between supernovae and Earth's climate. Because exploding stars don't just create heavy elements—they create *cosmic rays*.

Chapter 11 Cosmic Rays

"There are more things in heaven and earth, Horatio, than are dreamt of in your philosophy"

— *W. Shakespeare, Hamlet.*

Scientists have only recently come to suspect that cosmic rays have an important influence on Earth's climate. Cosmic rays are highly energetic charged particles that originate from various sources in outer space. They travel at speeds approaching the velocity of light and strike Earth from all directions. Unlike X-rays and gamma-rays, which are highly energetic forms of light, cosmic rays are actually particles of normal matter. Most cosmic rays are the nuclei of atoms, ranging from the lightest to the heaviest elements in the periodic table. But, cosmic rays also include high energy electrons, positrons, and other subatomic particles.

The term "cosmic rays" usually refers to galactic cosmic rays (GCR), which originate outside the solar system and are distributed throughout the Milky Way galaxy. However, the name cosmic ray is often used to refer to other classes of energetic particles from space. Among these are atomic nuclei and electrons accelerated by energetic events on the Sun—called *solar energetic particle*s. Also included as cosmic rays are particles accelerated by magnetic fields in interplanetary space.

Cosmic rays were discovered, in 1912, by Austrian scientist Victor Hess. Like many scientific discoveries, Hess' discovery built on the work of many earlier scientists. In particular, the recent discovery of X-rays and natural radioactivity.

Strange Particles from Outer Space

On December 22, 1895, Wilhelm Roentgen[†] created a photographic image of his wife's hand, but this was no ordinary photograph. The ghostly image revealed the unmistakable image of his wife's skeleton fingers, complete with wedding ring. Roentgen's discovery of these "mysterious" X-rays, capable of peering through living flesh and producing an image on a photographic plate, greatly excited scientists of his day. His was the first in a series of discoveries that would lead scientists to realize that the Universe was a much more complicated place than they had imagined.

---

† Wilhelm Conrad Roentgen (1845-1923) German physicist and discover of X-Rays. He received the first Nobel Prize in physics in 1901.

Allen Simmons & Doug L. Hoffman

Illustration 97: Wilhelm Roentgen ca. 1895. Inset photo: Photograph of Frau Roentgen's hand.

The discovery of X-rays led to a flurry of activity involving some of the most famous names in science: Marie Sklodowska Curie, Ernest Rutherford and Antoine Henri Becquerel. Antoine Becquerel,[‡] who was studying the related phenomena of fluorescence and phosphorescence, was among those who noticed Roentgen's discovery. After learning how Roentgen discovered X-rays from the fluorescence they produced, Becquerel decided to pursue his own investigations of these mysterious rays. Quite by accident, in the course of his investigations, Becquerel made a remarkable discovery.

While fluorescence and phosphorescence had many similarities to each other and to X-rays, they also had important differences. Fluorescence and X-rays stopped when the initiating energy source was cut off, but phosphorescence continued to emit rays for a time after the energy source was removed. In all three cases, the energy was derived initially from an outside source—or so it was thought.

Becquerel was working with a substance called potassium uranyl sulfate, $K_2UO_2(SO_4)_2$. He exposed the uranium crystals to sunlight and placed them on photographic plates wrapped in black paper. When developed, the plates revealed images of the crystals. At the time, it was believed that absorbing the Sun's energy caused uranium to emit X-rays. In March 1896, during a time of overcast weather in Paris, Becquerel was unable to use the Sun as an energy source for his experiments. He wrapped up his photographic plates and stored them in a dark drawer, awaiting the Sun's return. In one of those fortunate accidents that have so often advanced the development of science, Becquerel also placed some of his crystals in the drawer on top of the wrapped plates.

Much to Becquerel's surprise, when the plates were removed and developed, he found they were strongly exposed. During storage, invisible emanations from the uranium—emanations that did not require the presence of an initiating energy source—had left clear images on the plates. Later,

---

[‡] Antoine Henri Becquerel (1852-1908) French physicist awarded the 1903 Nobel Prize for physics for the discovery of radioactivity.

Becquerel demonstrated that the radiation emitted by uranium, unlike X-rays, could be deflected by a magnetic field and, therefore, must consist of charged particles. This discovery that some natural substances spontaneously emit radiation—that they were *radioactive*—changed science dramatically. For his discovery of radioactivity, Becquerel was awarded the 1903 Nobel Prize for physics.

After the discovery of radioactivity, scientists began building devices specifically to detect radiation. As their devices became more sensitive, several researchers notice that radiation was detected even when the devices were not in the presence of a known radiation source. This led to speculation that radiation was being created in the upper atmosphere or entered the atmosphere from space above, but no one was sure. Whatever the source, the radiation was partially blocked by the thicker atmosphere at lower altitudes. An Austrian physicist, Victor Hess, decided to solve this mystery by going up in a balloon.

Victor Franz Hess was born on the 24th of June, 1883, in Waldstein Castle, near Peggau in Steiermark, Austria. His father, Vinzens Hess, was chief forester in the service of Prince Öttingen-Wallerstein. His mother was Serafine Edle von Grossbauer-Waldstätt, a member of the local aristocracy. He received his entire education in Graz. He attended Gymnasium (1893-1901), and then attended Graz University (1901-1905), where he took his doctorate degree in 1910.[291]

He worked, for a short time, at the Physical Institute in Vienna, where Professor von Schweidler acquainted him with the recent discoveries in the field of radioactivity. From 1910 through 1920, he worked under Stephan Meyer at the Institute of Radium Research of the Viennese Academy of Sciences.

Illustration 98: Victor F. Hess. Source: Nobelprize.org.

While at the Academy, Hess and two assistants ascended in a balloon to 17,500 feet (5,400 m). During the ascent, they measured the amount of radiation increase. Hess used a simple detector called an *electroscope*, that consisted of a doubled-over strip of gold leaf. When given an electrical charge, the two flaps of gold leaf would repel each other, spreading apart in an inverted V shape. Radiation striking the strip caused the charge to be lost, allowing the flaps to come back together. So, the intensity of radiation could be measured by charging the strip and timing how long it took to discharge. Hess found the higher the balloon rose, the faster the electroscope discharged. In Hess' own words:

When, in 1912, I was able to demonstrate by means of a series of balloon ascents, that the ionization in a hermetically sealed vessel was reduced with increasing height from the earth (reduction in the effect of radioactive substances in the earth), but that it noticeably increased from 1,000 m onwards, and at 5 km height reached several times the observed value at earth level, I concluded that this ionization might be attributed to the penetration of the earth's atmosphere from outer space by hitherto unknown radiation of exceptionally high penetrating capacity, which was still able to ionize the air at the earth's surface noticeably. Already at that time I sought to clarify the origin of this radiation, for which purpose I undertook a balloon ascent at the time of a nearly complete solar eclipse on the 12th April 1912, and took measurements at heights of two to three kilometers. As I was able to observe no reduction in ionization during the eclipse I decided that, essentially, the sun could not be the source of cosmic rays, at least as far as undeflected rays were concerned.[292]

Hess received the Lieben Prize, in 1919, for his discovery of the "ultra-radiation." The following year he became Extraordinary Professor of Experimental Physics at the Graz University. Despite Hess' work, there was still a dispute as to whether the radiation was coming from above or from below.

In 1925, Robert Millikan,[†] of Caltech, introduced the term "cosmic rays" after concluding that the particles came from above, not below the cloud chamber he used to study them. For some time, it was believed that the radiation was electromagnetic in nature, resulting in the name cosmic "rays." Millikan became embroiled in a debate with Arthur Compton[‡] over whether cosmic rays were composed of high-energy photons or charged particles.

Millikan put forth the theory that cosmic rays were photons—the "birth cries" of atoms—and pursued the study of cosmic rays for many years, trying to prove his theory.[293] However, by the 1930s, it was proven that cosmic rays were electrically charged. This was discovered by observing the affect of Earth's magnetic field on the incoming radiation.

---

† Robert Andrews Millikan (1868-1953) American physicist and Nobel laureate.

‡ Arthur Holly Compton (1892-1962) American physicist, inventor of X-ray crystallography and Nobel laureate for the Compton Effect.

Compton, an expert in X-rays and their behavior, believed that cosmic rays were charged particles and not a form of electromagnetic radiation. He was eventually proven correct. In 1929, a Russian scientist, D. V. Skobelzyn, discovered ghostly tracks made by cosmic rays in a cloud chamber. Also in 1929, Bothe and Kolhorster verified that the cloud chamber tracks were curved.[294] Since such curved paths were only made by charged particles passing through a magnetic field, cosmic rays had to be particles, not electromagnetic radiation. Still, some older textbooks incorrectly included cosmic rays as part of the electromagnetic spectrum —science evolves.

Illustration 99: Hess and his balloon in 1912. Source CERN.

Compton published his, "positive evidence that the primary cosmic rays consist of electrical particles," in 1936.[295] That same year, Hess was awarded the Nobel prize for his discovery.[296] After a period as head of the Institute for Radiation Research at the University of Innsbruck, he returned to the University of Graz as Professor of Physics and director of the Physics Institute in 1937.

Two months after the *Anschluss*[†] in March 1938, Hess was dismissed from his post because he had a Jewish wife. A sympathetic Gestapo officer warned Hess that he and his wife would be taken to a concentration camp if they remained in Austria. They made their escape to Switzerland four weeks before the order came for their arrest. Shortly thereafter, Hess and his wife settled in the United States. Hess taught at Fordham University in New York City until 1956. He became an American citizen in 1944, and lived in New York until his death in 1964.[297]

From the 1930s to the 1950s, cosmic rays served as a source of particles for high energy physics investigations, and led to the discovery of subatomic particles, including the positron, π meson, μ meson and K meson. Since that time, man-made particle accelerators have reached very high energies, supplanting the use of cosmic rays in basic particle physics. Nowadays, the

---

† The Anschluss (union) was the forced annexation of Austria by Hitler's Germany on 13th March 1938. National Socialist (Nazi) rule was established in Austria.

main focus of cosmic ray research is in astrophysics: investigations of where cosmic rays originate, how they are accelerated to such high velocities, what role they play in the dynamics of the Galaxy, and what their composition tells us about matter from outside the solar system. To measure cosmic rays directly, research is carried out by instruments carried on spacecraft and high altitude balloons. This allows observations to be made before the cosmic rays slow down and break up in the atmosphere.

Cosmic rays include the nuclei of all the naturally occurring elements found in the periodic table. Hydrogen nuclei, which consist of a single proton, account for 89%, helium 10%, and heavier elements account for the remaining 1%. The common heavier elements, carbon, oxygen, magnesium, silicon, and iron, are present in about the same relative abundances as in the Universe. There are, however, important differences in the distribution of rare elements and isotopes. For example, there is a significant overabundance of the rare elements lithium (Li), beryllium (Be), and boron (B) produced when heavier cosmic rays fragment into lighter nuclei during collisions with the interstellar gas.

The existence of isotopes provide information about the origin and history of galactic cosmic rays. An overabundance of an isotope of neon (^{22}Ne) indicates that the nucleosynthesis of cosmic rays and solar system material are different. Other particles are also present: electrons constitute about 1% of galactic cosmic rays and positrons a tenth of that.[298]

The energy of sub-atomic particles is usually measured in electron volts (eV). One eV is the energy gained when a single electron is accelerated through an electrical potential difference of 1 volt. This is equivalent to 4.45 × 10^{-26} kilowatt-hours, a very tiny value indeed. The energy of cosmic rays is usually measured in units of millions of electron volts (MeV), or billions of electron volts (GeV, for giga-electron volts).

Just as cosmic rays are deflected by the magnetic fields in interstellar space, they are also affected by the interplanetary magnetic field embedded in the solar wind. The wind from the Sun is a plasma of ions and electrons blowing from the solar corona at speeds of 250 miles/second (400 km/sec). The Sun regularly emits particles with energies in the 100 MeV range, but solar flares, on occasion, can produce particles with up to 20 GeV.

Another source of energetic particles inside the solar system is the planet Jupiter. Jupiter has a powerful, complex magnetic field of its own and, for reasons not fully understood, its magnetosphere emits floods of high-speed electrons. These electrons have energies ranging from 1 MeV to 20 MeV, and the intensity of particles varies in a 10 hour cycle, Jupiter's period of rotation.[299]

Much like its brood of planets, the Sun rotates on its axis, with one rotation taking around four weeks. Because of the rotation, its magnetic field is twisted around forming a spiral pattern, though there are many irregularities caused by flares and eruptions. When Earth and Jupiter are aligned with one of the spirals in the Sun's magnetic field, the intensity of high-speed electrons reaches a maximum. Due to the orbits of both planets, this happens about every 13 months.

But electrons from Jupiter and other particles from the Sun do not reach the ultra-high energies observed in some cosmic rays. Most galactic cosmic rays have energies between 100 MeV, corresponding to a velocity for protons of 43% of the speed of light, and 10 GeV, corresponding to 99.6% of the speed of light. Single particles with energies as high as 100,000,000,000 GeV (10^{20} eV) have been observed, higher than any other form of natural radiation and far beyond the capabilities of human particle accelerators. A 10^{20} eV cosmic ray, a single atomic nucleus, has as much kinetic energy as a baseball traveling at 100 mph.[300] Considering the incredibly tiny size of an atomic nucleus, the energy contained in these ersatz-baseballs is mind-boggling.

Supernovae and Cosmic Rays

Since most of the cosmic rays, that originate in the Sun or from other sources in the solar system, lack the energy to deeply penetrate Earth's atmosphere, there must be other sources. What sources produce these highly energetic (>100MeV) particles?

The Sun's magnetic field extends outwards as far as 100 *astronomical units* (AU),[†] shielding the inner solar system from galactic cosmic rays. Cosmic rays from interstellar space beyond the solar system must penetrate the Sun's magnetic field and the deluge of particles in the solar wind. As a result, galactic cosmic rays with less than ~100 MeV cannot penetrate as far as Earth's orbit. The Sun, through this process of solar modulation, controls the flux of cosmic rays arriving at Earth.

Spacecraft, venturing out toward the boundary of the solar system, have found that the intensity of galactic cosmic rays increases with distance from the Sun. As solar activity varies over the 11 year solar cycle, the intensity of cosmic rays arriving at Earth also varies, in inverse proportion to the sunspot

---

† An astronomical unit, or AU, is the average distance from Earth to the Sun, about 93 million miles or 150 million kilometers.

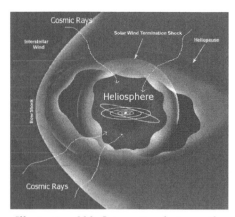

Illustration 100: Interaction between the Heliosphere and Cosmic Rays. Source H. Svensmark.

number (page 166). This caused scientists, particularly astrophysicists, to speculate about the galactic origins of very-high-energy cosmic rays. The logical place to look is at supernovae, the tremendous death explosions of massive stars.

For thousands of years, people have gazed in wonder at the sudden appearance of new, bright stars in the evening skies. The ancient Romans called such stars *novae*, Latin for "new." Over the past 2,000 years, seven stars, visible to the naked eye, have appeared in the heavens: the earliest in 185 AD, the latest in 1604 AD. These are now recognized to have been what scientists call supernovae, and they are among the most energetic events in the Universe. When a supernova erupts, it explodes with an energy equivalent to 10^{28} megatons of TNT, the force of a few octillion nuclear warheads.[301] For comparison, all of mankind's nuclear weapons combined would only amount to 10^5 megatons, not even noticeable on this scale.

On a night in February, 1987, Canadian astronomer Ian Shelton was completing another night's study of the southern sky from the Las Campanas Observatory, in Chile. A routine observation of the Large Magellanic Cloud, a small companion galaxy to our own Milky Way, yielded an unexpected surprise. Where previously there had been only faint stars, a new bright star had appeared. Realizing that he was witnessing a supernova explosion, Shelton quickly spread the word to other observatories around the world.[302]

This supernova, discovered by Shelton, known as SN 1987A, proved to be the brightest supernova event since 1604. The 1604 supernova, known as Kepler's nova, occurred five years before Galileo began using a telescope to make astronomical observations. SN 1987A was the first supernovae scientists had been able to observe with instruments, and is still being studied today. From their observations, scientists theorized about the cause of the explosion.

The star that exploded was identified from star catalogs as Sanduleak (Sk)-69 202, a bright type B red super-giant with a mass of 19 Suns. From Earth, this star was not visible to the naked eye because the Large

Magellanic Cloud is ~160,000 light-years away. A star of this size would spend about 10 million years on the main-sequence.

Once its hydrogen was exhausted, the star that was to become SN 1987A spent about one million years as a red super-giant. When the fusion fires had consumed its helium, and a carbon core had formed, the star had about 1,000 years left to live. When the core became iron, the star's remaining life was measured in hours. Internal temperatures rose rapidly, until the very light the star generated was powerful enough to disrupt the atomic nuclei in the iron core—a phenomenon called *photodisintegration*. Suddenly, the iron core, containing about 1.5 times the mass of the Sun, collapsed from a radius of 600 miles down to less than 6 miles. This released a vast torrent of neutrinos that dissipated most of the energy generated by the collapse. As quickly as it had collapsed, the core rebounded, blowing most of the star's matter into space. The collapse and rebound only took a few milliseconds—the star went from being a placid red giant to a supernova in literally the blink of an eye. For a brief time, SN 1987A shined with the light of 10,000,000,000 Suns.[303]

The collapse of the star's core has been verified by neutrino detectors here on Earth, and subsequent photon emission studies have verified that most of its mass was ejected carrying a kinetic energy of around 10^{51} ergs. What remains today of SN 1987A is a neutron star of about 1.6 solar masses. From photographs taken prior to the explosion, it was known that SN 1987A's predecessor star had two faint companion stars. Both seem to have survived the eruption.

On the twentieth anniversary of the SN 1987A supernova, NASA's Hubble Space Telescope took a picture of the stars remnants. This image shows the entire region around the supernova (see Illustration 101). The most prominent feature in the image is a ring, about a light-year across, with dozens of bright spots. The shock wave of material unleashed by the stellar blast slamming into regions along the ring's inner regions causes them to glow. The two bright objects that look like car headlights are a pair of stars in the Large Magellanic Cloud. The object in the center of the ring is debris from the supernova blast. The debris will continue to glow for many decades.

As a check on the theory that cosmic rays are created by supernovae, we are going to compare the amount of energy found in cosmic rays with that produced by observed supernovae. To quote Douglas Adams, in his famous book *The Hitchhiker's Guide to the Galaxy*, "Space is big. Really big. You just won't believe how vastly, hugely, mind-bogglingly big it is." Because we are dealing with calculations involving the entire Milky Way Galaxy, we are going to use scientific notation, where very large numbers are

*Illustration 101: SN 1987A on its 20th anniversary, Source NASA
Hubble Space Telescope.*

represented as powers of ten.[†] We will also use metric units, since it is only the final result that we are really interested in, not the actual values.

From radioisotope studies of the material in meteorites, it has been determined that cosmic ray intensity has stayed fairly constant for the past several million years. Calculations indicate that the average cosmic ray remains in the galaxy for about 10,000,000 (10^7) years. The energy density of cosmic rays is about 1 eV/cm^3, about the same as from starlight. The disk of the galaxy is ~10^{67} cm^3 in volume.[304] Assuming that the energy density is the same throughout our galaxy, 10^{67} eV must be created every 10^7 years to maintain the energy density level. This translates into a production rate of 10^{60} eV/yr, or 10^{48} ergs/yr. We realize numbers of this magnitude are hard to grasp, but they are included here to demonstrate the unbelievable amount of energy involved in supernovae events. This energy is equivalent to 10^{35} kilowatt-hours per year—think of that the next time you receive your electric bill.

---

† The number 10^6, read "ten to the sixth," is 10 times itself six times or 1,000,000. Another way to interpret powers of ten is that 10^n is a one with n zeros after it. So, 10^{50} would be a one followed by fifty zeros.

The estimated average energy release from a supernova is $\sim10^{51}$ ergs. With an observed rate of two supernovae per century this gives a yearly energy of 10^{49} ergs. Astrophysics estimate 10% of the energy released by supernovae goes into creating cosmic rays, about 10^{48} ergs/yr, the same estimated energy production required to maintain the observed cosmic ray density. Theory and measurement agree, the source of galactic cosmic rays are exploding supernovae.

Earth Showers and Muons

In 1938, French physicist Pierre Auger[‡] noticed that two detectors located several meters apart detected particles at the same time. This led to the discovery of *air showers,* cascades of secondary nuclei produced by the interaction of a cosmic ray particle with air molecules. The term *cascade* means that the incident *primary* particle, which could be an atomic nucleus, a proton, an electron, or occasionally a positron, strikes an atom in the atmosphere producing many high-energy ions. These *secondary* cosmic rays in turn create more, and so on.

When the energy of the primary is high enough, an air shower can produce a widespread flash of light due to excitation of air molecules and the *Cerenkov effect.* When a charged particle passes through an insulator at a speed greater than the speed of light in that medium, electromagnetic radiation is emitted. The characteristic "blue glow" of nuclear reactors is due to this *Cerenkov radiation.* It is named after Russian scientist Pavel Alekseyevich Cerenkov, the 1958 Nobel Prize winner who discovered it. The Cerenkov detector, which can detect the presence of high-energy particle radiation using the Cerenkov effect, has become a standard piece of equipment in atomic research. Such a device was installed on the Sputnik III satellite.

Cosmic rays that strike Earth hardly ever hit the ground. As they enter Earth's atmosphere, they collide with a nucleus of the air, usually several tens of kilometers high. In such collisions, many new particles are usually created and the colliding nuclei shatter. When a primary cosmic ray produces many secondary particles, it is called an air shower. When many thousands, millions or even billions of particles arrive at ground level, it is called an *extensive air shower* (EAS). Most of these particles will arrive within a hundred yards of the axis of motion of the original particle, now the

---

‡   Pierre Victor Auger (1899-1993) French physicist who worked in the fields of atomic, nuclear and cosmic ray physics. One of the founders of the European Space Agency.

shower axis. But some particles can be found miles away. A large EAS can rain down particles over several acres.

Illustration 102: Simulation of a cosmic ray air shower. Source COSMUS and Sergio Sciutto for AIRES.

We mentioned in the previous chapter that scientists had found a link between cosmic ray levels and thunderstorms. There is also a positive correlation between cosmic ray flux (CRF) and low-altitude cloud formation. However, correlation does not always imply causation. It is also known that the Sun is slightly brighter when it is more active, which also may affect cloud formation on Earth. But, the correlation is strong so cosmic rays could be involved.

When scientists observe a close correlation between phenomena, they start looking for a reason, some mechanism that causes the linkage. There is a possible mechanism linking cosmic rays and low-level cloud formation: elevated levels of ionization seem to facilitate the coagulation of such molecules as sulfuric acid (H_2SO_4) in the atmosphere into tiny droplets, which then form condensation nuclei for water vapor. The condensed droplets of water then form clouds.

The main problem with linking low-level cloud formation to cosmic rays was identifying particles that could penetrate to ground level. When *muons* were discovered 70 years ago, by Carl Anderson and Seth Neddermeyer at Caltech, just such a particle had been found. It had exactly the mass predicted for Yukawa's meson,[‡] but it did not undergo strong nuclear interactions at all. Since all mesons are affected by the strong nuclear force, the new particle could not be a meson. The discovery of the muon was a compete surprise to particle physicists at the time, because their theories did not predict or allow for such a particle. When the muon was spotted in 1937, Isadore Rabi[†] reportedly remarked, "Who ordered that?"

As it turns out, muons are charged particles that are identical to electrons, with the exception that they weigh 200 times as much as an electron. Muons live for about 2.2 microseconds, and often survive to

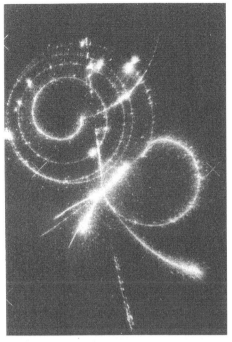

Illustration 103: The pion, muon, electron decay cycle captured by a steam chamber. Source CERN.

ground level, before changing into electrons and neutrinos. In 1947, ten years after the muon discovery, Cecil Powell's group at Bristol University discovered that the muons are produced by other particles, called pions.

Pions are even heavier versions of the electron, which live for only a few hundredths of a microsecond. In Illustration 103, pions fly out from a collision in the steam chamber. One of the pions makes the looping track to the right, before it decays into a muon, which then curls anticlockwise four times, and eventually changes into an electron which moves off towards the upper right. From similar collisions, an EAS produces large quantities of muons that penetrate the atmosphere to Earth's surface and even below.

---

‡ Hideki Yukawa (1907-1981) Japanese physicist who predicted a meson 200 times as heavy as an electron in 1937. The π meson, discovered in 1947, was the particle he predicted.

† Isidor Isaac Rabi (1898-1988) Nobel Prize-winning, Austrian-born, American physicist known for work on the strong nuclear force that holds atomic nuclei together.

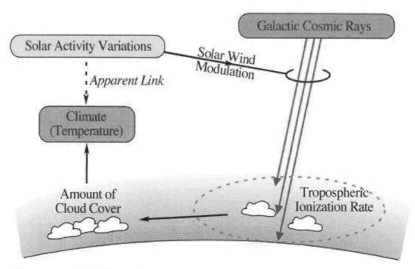

Illustration 104 Effect of Cosmic Rays on Cloud formation. Source N. Shaviv.

Cosmic Rays and Cloud Formation

At the end of the 20th century, a new theory about the Sun, the stars, and our solar system's path around the Milky Way galaxy, affects cloud cover on Earth had been postulated by a number of astrophysicists. But claims that solar variation is not sufficient to cause climate change have prevailed because a solid, scientific explanation of how such a link would work was lacking. The hypotheses has only been supported by historical records and statistical associations. The mechanism causing the effect had not been empirically demonstrated.

The primary proponent of cosmic ray induced low-level cloud formation is Danish physicist Henrik Svensmark, of the Danish Space Research Institute. Svensmark and Eigil Friis-Christensen reported their discovery in a cogent paper in 1997: "Variation of Cosmic Ray Flux and Global Cloud Coverage — a Missing Link in Solar-Climate Relationships."[305] In it, they describe how ions created in the troposphere by cosmic rays could provide a mechanism for cloud formation. And, since the level of cosmic rays is controlled by the solar cycle, they suggested that the Sun is controlling Earth's climate variation by changing low-level cloud cover.

They were not the first to propose such a mechanism. In 1959, Ney pointed out that cosmic ray induced ions could be an important variable in climate regulation.[306] Aerosol particles are widely held to be the primary source for nucleation in cloud formation. Others reported how charged water droplets combine with aerosol particles 10 times, or even 100 times more efficiently than uncharged ones.[307]

A strong link between long-term variations in solar activity and Earth's climate had been reported by Friis-Christensen and Lassen in 1991. They showed that an empirically constructed measure of solar activity called the *filtered solar cycle length*, matched very closely variations in northern hemispheric temperature during the last 400 years.[308] All this, claimed Svensmark and Friis-Christensen, pointed to a major role for cosmic rays in cloud formation, and hence, climate regulation. How this control mechanism works is shown in Illustration 104.

In 2000, Nigel Marsh and Svensmark followed up on the earlier work by Svensmark and Friis-Christensen with a paper showing that the influence of galactic cosmic ray modulation was strongest on low-level clouds.[309] If this theory is true, then the Sun controls Earth's climate. When the Sun is active, its magnetic field is stronger and as a result fewer global cosmic rays (GCR) arrive in the vicinity of Earth. When solar activity is low, the magnetic field is weaker, and more GCR arrive. The modulation of GCR is dependent on their energy. The higher energy the GCR particles, the less they are modulated by the solar cycle.

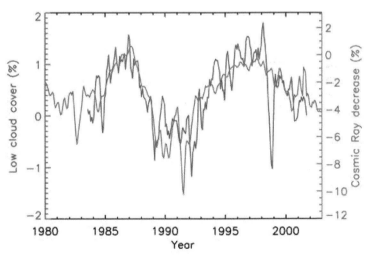

Illustration 105: The correlation between cosmic ray flux and low altitude cloud cover , Marsh & Svensmark, 2003.

This new theory, proposed by astrophysicists, was not without its critics in the climatological community. To answer criticisms that their earlier work didn't use global cloud coverage data, Marsh and Svensmark further refined their work using data from more extensive infrared satellite observations. In 2003, they updated their calculations based on more extensive cloud data.[310] The correlation between cosmic ray flux, as measured in neutron count monitors in low magnetic latitudes, and the low altitude cloud cover using International Satellite Cloud Climatology Project (ISCCP) satellite data is shown in Illustration 105.

This additional data helped strengthen the theory, but the most troublesome criticism was that the actual physical process of GCR induced cloud formation had not been demonstrated experimentally. This situation changed when Svensmark and the team at the Danish National Space Center experimentally demonstrated the very mechanism they proposed a decade ago.

In a basement at the Danish National Space Center an experiment was set up to verify that cosmic rays could cause low level clouds to form under controlled conditions. The SKY experiment used a *cloud chamber* to mimic conditions in the atmosphere. This included varying levels of background ionization and aerosol levels, and sulfuric acid (H_2SO_4), in particular. The SKY experiment demonstrated that more ionization implies more particle nucleation, as shown in Illustration 106.

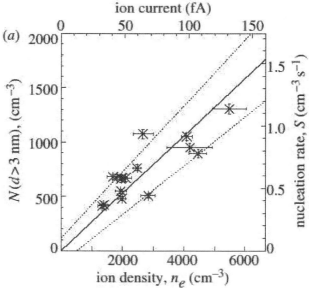

Illustration 106: Condensation nuclei density as a function of ion density.

In 2006, Svensmark and colleagues reported the results of their laboratory experiments and published them in the *Proceedings of the Royal Society*. This work showed that highly ionizing radiation can create ultra-small aerosol particles. Critics, however, continue to attack Svensmark's results. Dr. Svensmark does not claim that human activity isn't a factor in climate change. He said in a recent interview:

> "Humans are having an effect on climate change, but by not including the cosmic ray effect in models it means the results are inaccurate. The size of man's impact may be much smaller and so the man-made change is happening slower than predicted."[311]

At this point, the primary focus was on short-term effects caused by the various solar variability cycles (page 176). Svensmark's theory had attracted the attention of an Israeli physicist, Nir Shaviv. Shaviv decided to look farther back into the past, to see if cosmic rays could have played a part in the much longer term fluctuation in Earth's climate. As we will see, his intuition was rewarded.

Subsequently, a debate over this new and controversial theory has raged in climatological circles. During the decade since its introduction, many criticisms have been leveled at the theory's assumptions and calculations. This is normal and part of the way science works. Unfortunately, given the high profile of global warming, the back and forth scientific arguments have often filtered into the popular press.[312,313,314,315,316]

To summarize, the link between cosmic rays and climate regulation, we know the following: The Sun and Earth both have magnetic fields that deflect some incoming cosmic rays. This divides cosmic rays into three categories; low-energy rays that are deflected by Earth's magnetic field, medium-energy rays that are deflected by the solar wind and magnetic field, and high-energy rays that are not deflected by either magnetic field.

As energy levels rise, the number of GCR drop dramatically. But, very energetic GCR create a majority of extensive air showers, and hence, a majority of the muons that penetrate to Earth's surface. About 60% of GCR created muons are created by high-energy ray category. These are the GCR that cannot be blocked by the Sun's protective magnetic field. About 40% of GCR muons fall into the middle or low categories, caused by primary particles that can be blocked by the Sun's field.

These are the muons that vary with changes in the Sun's activity, causing decadal and centuries long cycles of temperature variation. Earth's magnetic field plays almost no role in regulating GCR generated muons. It has been estimated that the total disappearance of Earth's magnetic field would only result in a 3% increase in ground level radiation.[317]

This new explanation of Earth's climate history is presented for the non-scientist in Henrik Svensmark and Nigel Calder's book, *The Chilling Stars*. This new work is a severe blow to proponents of the enhanced greenhouse hypothesis and advocates of anthropogenic global warming who have worked so hard to deny solar influence on global climate.

Our Vagabond Sun

The astrophysical explanation for the major ice age periods in Earth's past has been elaborated on by Nir J. Shaviv, Associate Professor of Physics at the Racah Institute of Physics in Israel. Shaviv decided to take a look at data from the distant past, back through time, to the beginning of complex life in the Cambrian Period. On this research he collaborated with Jan Veizer, a researcher at the Institut für Geologie, Mineralogie und Geophysik, Ruhr Universität, in Germany.

Jan Veizer had originally set out to reconstruct the tropical temperature over the past 550 million years because he wanted to find the CO_2 fingerprint. It was a disappointment to him that he couldn't quantify the effect of CO_2.[318] Working together, he and Shaviv found something much more exciting. The results of their study were revealed in a paper published in 2003, entitled "Celestial driver of Phanerozoic climate?" In it they clearly point out the close correlation of CRF with temperature variation.

Their reconstructed record of climate variations during the Phanerozoic, based on climate sensitive sedimentary indicators, shows intervals of tens of millions of years duration characterized by cold and warm episodes, called icehouses and greenhouses, respectively. Superimposed on these are higher-order climate oscillations, such as the waning and waxing of ice sheets during the past million years. These variations can be clearly seen in Illustration 107.

The dark bars across the top of the figure represent ice-house periods, including the great prehistoric ice ages. The lighter shading for the Jurassic-Cretaceous icehouse reflects the fact that true polar ice caps have not been documented for that period. Notice the extremely high prehistoric CO_2 levels, with a noticeable dip during the Carboniferous, 300 mya, the time of great coal swamps and giant insects discussed in Chapter 4 (page 57). The rise and fall of temperature, represented here by oxygen isotope ratios ($\delta^{18}O$) from calcite shell deposits, are noticeably out of phase with the variation in CO_2 levels (the upper shaded area).

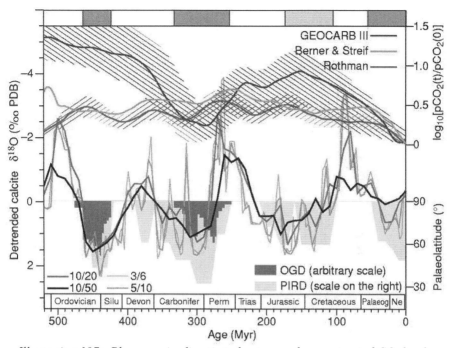

Illustration 107: Phanerozoic climatic indicators and reconstructed CO_2 levels. Shaviv & Veizer 2003.

Other information presented include the amount of drifting sea ice and the extent of glaciation. The paleolatitudinal distribution of ice-rafted debris (PIRD), on the right-hand vertical axis, is an indicator of sea ice. Glaciation is indicted by the presence of other glacial deposits (OGD), such as tillite and glacial marine strata. Both sets of histograms are for relative comparison and represent no actual physical values.

The main result of this research is that the variations of the flux, as predicted from the galactic model and as observed from the Iron meteorites, is in sync with the occurrence of ice age epochs on Earth. The agreement is both in period and in phase: The observed period of the occurrence of ice-age epochs on Earth is 145 ± 7 million years, compared with 143 ± 10 million years for the cosmic ray flux variations. The mid-point of the ice age epochs is predicted to lag by 31 ± 8 million years and is observed to lag by 33 ± 20 million years.

In short, when the historical temperature pattern was matched against a reconstruction of the cosmic ray intensity, a much better fit was found. A graph revealing the relationship is shown in Illustration 108. The inverse relationship between temperature and CRF is clear; when CRF rises,

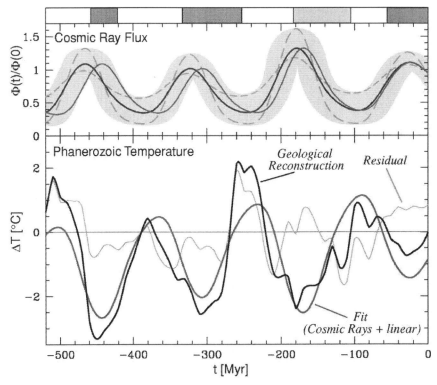

Illustration 108 Cosmic Ray Flux and Temperature. Source Shaviv & Veizer 2003.

temperature falls, when CRF drops off, temperature climbs. Their conclusions are concisely stated in this excerpt from the paper:

> "One interpretation of the above result could be that the global climate possesses a stabilizing negative feedback. A likely candidate for such a feedback is cloud cover. If so, it would imply that the water cycle is the thermostat of climate dynamics, acting both as a positive (water vapor) and negative (clouds) feedback, with the carbon cycle "piggybacking" on, and being modified by, the water cycle."

The evidence of correlations between paleoclimate records and solar and cosmic ray activity indicators, suggests that extraterrestrial phenomena are responsible for climatic variability on time scales ranging from days to millennia. Dr. Shaviv's theory is that the movement of the solar system in and out of the spiral arms of the Milky Way galaxy is responsible for changes in the amount of cosmic rays impacting Earth's atmosphere. Quoting again from Shaviv and Veizer, "We find that at least 66% of the variance in the paleotemperature trend could be attributed to CRF variations likely due to solar system passages through the spiral arms of the galaxy."[319]

A Grand Tour of the Galaxy

The Milky Way galaxy is a barred spiral galaxy with a diameter of about 100,000 light years containing more than 200 billion stars. Besides stars, the galaxy is composed of gaseous interstellar medium that sometimes concentrates into dense gas clouds. These clouds, made up of atoms, molecules, and dust, come from the explosions of older stars and are the source of new ones. All of the matter rotates around the galaxy's central axis. Our Sun resides about two-thirds of the distance from the center of the galaxy to the edge of its disk. Seen from above, we are located in one of the Milky Way's outer spiral arms, known as the Orion Arm.[320]

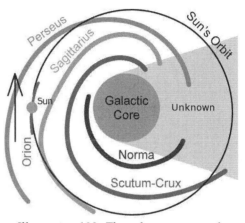

Illustration 109: The solar systems path through the spiral arms of the Milky Way.

Just as Earth orbits the Sun, the Sun and its brood of planets orbit the Milky Way Galaxy's center of mass. The direction of the Sun's path through interstellar space is in the direction of the bright star Vega, near the constellation of Hercules. The solar system's average orbital speed is about 132 miles per second (212 km/sec). The Sun lies 28,000 light years from the Galactic Center, so it completes one revolution every 226 million years.[321] Note the relationship of the Sun's orbit to the arms in Illustration 109.

It is thought that the Solar System's location in the galaxy was a factor in the emergence of life on Earth. The Solar System lies well outside the star-crowded environs of the galactic center. If it were closer to the center, the combination of gravitational forces and intense radiation levels might have prevented life from developing on our planet. The solar system's current location in the Orion Arm is not safe from supernovae. Some scientists have hypothesized that recent supernovae may adversely affect life by pummeling Earth with radioactive dust grains and larger bodies.[322]

The galaxy's spiral arms are home to large concentrations of massive, young blue stars—the type of stars that result in supernovae. The arms themselves are actually density waves rippling through the dust and gas of interstellar space. At the same galactic orbital radius as the Sun, the arms travel at a rotational velocity of 82 miles per second (130 km/sec), about half the velocity of the solar system.[323] This difference in velocity, between the solar system and the density waves that form the arms, has given Earth long periods of interstellar stability for life to evolve.

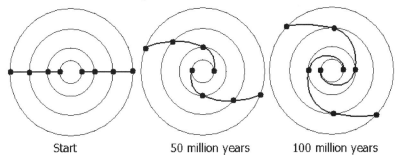

| Start | 50 million years | 100 million years |

Illustration 110: Differential Rotation caused by differing orbital velocities.

To understand why scientists think the spiral arms are not simply formed by orbiting stars, consider what would happen to the arms if they were. Just as satellites orbiting Earth take different amounts of time to circle the planet, stars or anything else that orbit the galaxy, do so at different rates of speed. This is called differential rotation, and it is the reason that the spiral arms cannot be formed by orbiting stars, dust and gas. The difference in angular speeds of different parts of the galactic disk cause stars closer to the galactic center to complete a greater fraction of their orbit in a given time (Illustration 110).

Over billions of years, differential rotation leads to the *winding dilemma,* first described by Bertil Lindblad in 1925. Though differential rotation provided a ready explanation for the spiral pattern of stars, he realized that a permanent spiral arrangement of stars was untenable. This is because differential rotation is too efficient at making the spiral arms. After only 500 million years, the arms should be so wound up that the spiral structure would disappear (see Illustration 111). Anything left of the spiral pattern would occupy only a small part of the disk. Observations of other galaxies contradicts this: the spiral arms in galaxies

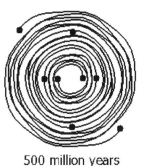

500 million years

Illustration 111: The winding dilemma.

rarely have more than two turns. Since galaxies are billions of years old, and the spiral pattern must be a long-lasting feature, some other mechanism must be at work.

One theory says that the spiral structure is a wave that moves through the disk causing the stars and gas to clump up along the wave—a density wave. The spiral arms are where the stars pile up as they orbit the galactic center. The greater density of stars in a spiral region causes greater gravity, which concentrates the stars and gas. The spiral regions rotate about half as fast as the stars move. Stars behind the region are pulled into the region by gravity, at the same time speeding up—almost as though they are hurrying to pass through the dangerous congestion of the arm. Stars leaving the region of greater gravity are pulled backward and slow down (Illustration 112).

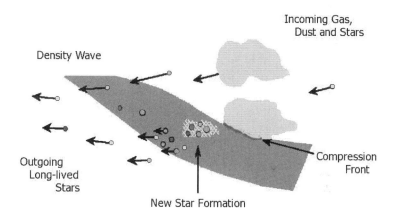

Illustration 112: Density wave causing compression and new star formation. After Strobel.

Gas entering a spiral region density wave is compressed, leading to the formation of new stars (page 160). On the downstream side of wave, there are many star formation regions. The star formation region nearest to Earth is an open cluster in the constellation of Taurus called the *Pleiades,* also known as the Seven Sisters. The cluster is about 12 light years away and is dominated by hot blue stars, thought to have formed within the last 100 million years. The observed bright reflection nebula and dark nebular filaments in the Pleiades indicates they are interacting with a dense molecular cloud.[324]

There are many stellar nurseries in the solar system's neighborhood. One of the most spectacular star-birth regions is the Orion Nebula, which lies 1,500 light-years away, in the direction of the constellation Orion the Hunter. It is one of the nearest regions of recent star formation, with stars believed to have formed only 300,000 years ago. The nebula is a giant gas cloud illuminated by bright, young stars as shown in Illustration 113. The great plume of gas in the lower left in this picture is the result of the ejection of material from a recently formed star.

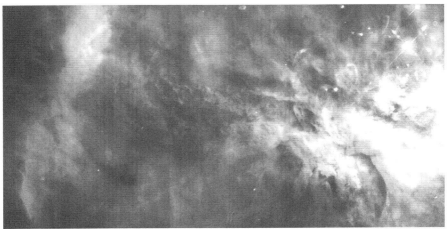

Illustration 113: A view of the Orion Nebula, an active stellar nursery. Source NASA/HST

The other popular theory of spiral arm formation depends on shock waves from supernova explosions to shape the spiral pattern. When a supernova shock wave reaches a gas cloud, it compresses the cloud to stimulate the formation of stars. Some of them will be massive enough to produce their own supernova explosions to keep the cycle going. Coupled with the differential rotation of the disk, the shock waves will keep the spiral arms visible. Whatever the mechanism, the spiral arms are areas of elevated cosmic ray radiation, due to the births and subsequent deaths of hot massive stars.

Ice Ages and Spiral Arms

Given the estimated locations of the Milky Way's spiral arms and the difference in rotational velocity between them and the solar system, the Sun should pass through an arm every 134 ± 25 million years on average. This fits rather well with both the fluctuation in cosmic rays and the occurrence of ice ages (page 197).

A record of the long-term variations of the galactic cosmic ray flux can be extracted from Iron meteorites. It was found that the cosmic ray flux varied periodically, with flux variations greater than a factor of 2.5. These variations had an average period of 143 ± 10 million years. This is consistent with the expected spiral arm crossing period and with the theory that the cosmic ray flux should vary.[325]

A second area of agreement is in long-term activity. There were no ice age epochs on Earth between 1 and 2 billion years ago. During this same period, it appears that the star formation rate in the Milky way was about half of its average during the past billion years or prior to 2 billion years ago. The correlation is shown in Illustration 114.

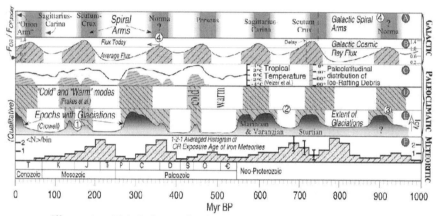

Illustration 114: Relationship between meteorite records, cosmic ray flux, ice ages, and spiral arm passage. Source N. Shaviv 2003.

Marsh and Svensmark and Shaviv and Veizer are not alone in their conclusions. In 2005, R. G. Harrison and D. B. Stephenson explored the robustness of this chain of events by examining relationships between diffuse solar radiation and cloud amount measured at ten United Kingdom solar radiation-recording sites.[326] In the words of Harrison and Stephenson, "our data analysis confirms the existence of a small, yet statistically robust, cosmic ray effect on clouds, that will emerge on long time scales with less variability than the considerable variability of daily cloudiness."

Shaviv has concluded that Cosmic Ray Flux variations explain more than two-thirds of the variance in the reconstructed temperature, making CRF variability the dominant climate driver over geologic time scales. But we are more interested in shorter periods of change. In particular, what was the effect of cosmic rays on Earth's climate over the past century.

In a 2005 paper on the calculated effects of cosmic ray flux on climate, Shaviv reported, "CRF over the previous century should have contributed a warming of $0.47 \pm 0.19°K$, while the rest should be mainly attributed to anthropogenic causes. Without any effect of cosmic rays, the increase in solar luminosity would correspond to an increased temperature of $0.16 \pm 0.04°K$."[327] This is in agreement with empiricist calculations, as discussed in the previous chapter, that indicate CO_2 levels account for less than half of the recent temperature rise. In fact, Shaviv's estimate for the contribution of irradiance is less that the commonly attributed $0.25°C$.[328] If these predictions are correct, CO_2 may have contributed only $0.25°C$ to last century's warming trend.

If correct, this theory implies that we are at the end of a 10 million year long ice age, during which glacial episodes come and go. Gradually, over the next few millions of years, the severity of the glacials should diminish, until they disappear altogether. But, to quote Dr. Shaviv, "I wouldn't buy real estate in Northern Canada just yet."

A final wrinkle in the story of cosmic rays affecting events on Earth has been reported by researchers from the University of Kansas.[329] It had been known for some time that a 62 ± 3 million-year cycle in fossil diversity has persisted over the past 542 million years. There have been efforts to link this cycle to global mass extinctions, but no satisfactory mechanism has been found to explain the phenomenon. Recently, Mikhail V. Medvedev and Adrian L. Melottpropose have proposed that the cycle is caused by modulation of CRF due to the solar system's vertical oscillation in the galaxy, which has a period of around 64 million years.

Much like the Sun and Earth generate shock waves in space, due to their magnetic fields, the Milky Way also generates a shock wave in intergalactic space. The production of cosmic rays that takes place in the galactic halo varies from north to south due to our galaxy's motion toward the Virgo cluster. This new research has shown that CRF can vary by a factor 4.6 and reaches a maximum at the northern-most displacement of the Sun.[330]

The addition of this 64 million year cycle to the arm transit cycle could increase the complexity of CRF variability. This would be similar to the way eccentricity, obliquity and precession combine to create the Croll-Milankovitch cycles.

When asked by the authors about the impact of Medvedev and Melott's idea on climate, Shaviv said: "Although I haven't checked it, my suspicion with the idea of Medvedev and Melott is that it will not be consistent with the Be isotope ratio age of the cosmic rays... Another point is that there is no

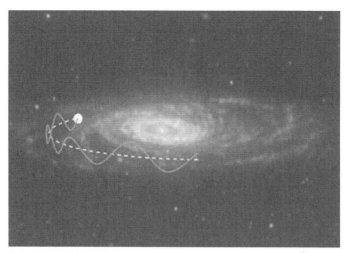

Illustration 115: The solar system's wobbly trajectory. Source M. Medvedev, University of Kansas.

notable climate variations on the 60 Million year time scales." He added, "Anyway, Medvedev is my friend, we overlapped as post-docs at CITA. Since he is sharp and original, I wouldn't discard his ideas without thoroughly checking them."[331] Clearly, this will be an area of ongoing scientific investigation for some time.

Cosmo-Climatology

This concludes the case for cosmic rays influencing Earth's climate. As we have seen, the same underlying mechanism—regulation of low-level, tropospheric clouds by ionizing radiation caused by cosmic ray generated muons—explains two puzzling climatological questions. These questions are: what causes short term, decadal temperature change, and what causes long-term climate change over millions of years? Regulation by the Sun of 40% of the GCR induced muon showers explains decadal variation where varying solar irradiance cannot. And the solar system's transiting of the spiral arms of the Milky Way galaxy explains why Earth has experienced long warm periods, punctuated by ice ages tens of millions of years long.

If we combine the effects of all the forcings that derive from extraterrestrial sources, we find a compelling set of explanations for climate change during Earth's long history. The forcings and associated time scales are as follows:

- Decadal — cosmic ray muons regulated by the solar cycle. This accounts for temperature variability in sync with the 11 year sunspot cycle.

- Hundreds to thousands of years — Solar regulation of cosmic rays plus changes in solar irradiance. This variability includes historical climate change as witnessed in the Little Ice Age and Medieval Warm Period.

- Tens to hundreds of thousands of years — The Croll-Milankovitch cycles that combine Earth's attitudinal and orbital variations. This variability drives the glacial-interglacial cycles during ice ages.

- Millions to hundreds of millions of years — The solar system's transit of the galactic spiral arms, causing variation in overall cosmic ray intensity. This variability regulates the cycles of ice ages and hot-house periods.

Add to these cyclic climate factors the complex interactions within Earth's atmosphere—including the effect of greenhouse warming, regulation of the carbon cycle by Earth's biota and active volcanism—the complexity of the heat engine that is Earth's climate is placed into perspective. Further complexity is added to the chore of deciphering the mechanisms of climate regulation by the slow movement of the continents. The constant rearrangement of Earth's land and oceans is reflected in changes in atmospheric circulation. It is as though Earth's climate system is constantly being re-designed, making understanding its structure even more difficult. No wonder climatologists cling to the relatively simple explanation that CO_2 is responsible for recent climate change.

This chapter on cosmic rays completes our survey of the science underlying climate change. As we have seen there are may different factors involved and several competing explanations that do not get mentioned in the publicly presented case for human-caused global warming. We need to return to our examination of the IPCC's case for human-caused global warming. But, before we do that, we need to take a look at the way science itself works. To understand how so many people can be lured into supporting the IPCC's simplistic explanation of global warming requires an understanding of the scientific method and scientists themselves.

Chapter 12 How Science Works

"The most exciting phrase to hear in science, the one that heralds new discoveries, is not 'Eureka!' but 'That's funny...'"

— *Isaac Asimov*

As great as the impact of science has been on the lives of human beings, it still remains a mystery to the average person. Scientists are partly to blame for this sad state of affairs, failing to communicate the results of their work in ways that non-scientists can understand. The public's perception of scientists has never been accurate and it certainly hasn't been helped by the popular media. Scientists are portrayed as intelligent bumblers, like the "Doc" in *Back To The Future;* cold, emotionless automatons like "Mr. Spock" on *Star Trek;* or crazed, power-mad villains like "Doctor Octopus" of *Spider Man* fame.

The simple truth is that scientists are just normal people, like every one else on the planet. They have the same range of personalities, the same strengths and weaknesses as other human beings. They can be stubborn, willful and prideful. Because they possess a great deal of intimate knowledge about a particular area of scientific inquiry, they often think they know more than they do in other areas as well. This is usually not the case.

Though trained to be objective, careful observers, scientists can be as easy to fool as non-scientists. James "The Amazing" Randi, a well-known magician and debunker of paranormal claims, has said that scientists are easy to fool because they are trained to observe nature, and nature doesn't lie.[332] Similarly, scientists can be as vulnerable to falsehoods hyped by the media, and accepted as "common knowledge," as non-scientists. These normal human tendencies are not expected in scientists, but they are present just the same. Ask scientists who may not be familiar with climatological study at all, what they think about "global warming" and they are apt to repeat the same stories presented in the media.

This is not to say that scientists are untrustworthy, foolish or deluded—just that they are human. When scientists are asked a question about a field that they have studied, they will usually provide as honest and accurate an answer as they can. But scientists are cautious, and it can be notoriously hard to get a simple answer from them. Sometimes, you need a translator to understand that the answer is really, "I don't know."

If scientists have the same weaknesses that normal people have, what does it mean to be a scientist? How can we interpret what science has to say about hard, complicated subjects like climate change and global warming? Where did science come from in the first place? To answer these questions, we will again start with a bit of history.

The Invention of Science

Science is usually traced back to the ancient Greeks, in particular, the early Greek mathematicians and philosophers of nature. This is because their writings have been preserved and passed down, forming the basis of the western scientific tradition. But science and invention was not restricted to the Greeks or Europeans. Printing, the magnetic compass and gunpowder weapons were all Chinese in origin. We have already mentioned the history of detailed astronomical observation in pre-European Meso-America. India, Persia, and Mesopotamia all have rich traditions of early scientific inquiry. But it was science in the western tradition that fueled the Scientific and Industrial Revolutions, so we will start our story where western science started, Ancient Greece.

Archimedes of Syracuse was the greatest mathematician of his age. His contributions to geometry revolutionized the subject and his methods anticipated integral calculus 2,000 years before Newton and Leibniz. He was also a thoroughly practical man who invented a wide variety of machines including pulleys, cranes and other ingenious devices.

Archimedes was born in 287 BC in the Greek city state of Syracuse, on the island of Sicily. His father was Phidias, an astronomer. We know nothing else about Phidias other than this one fact stated in Archimedes' book, *The Sandreckoner.*

When he was a young man, it is thought that Archimedes visited Egypt and studied with the successors of Euclid in Alexandria. Later in life, he often sent his results to the mathematicians working in Alexandria with personal messages attached. While he was in Egypt, he invented a device known as Archimedes' screw. This is a device with a revolving screw-shaped blade inside a cylinder for pumping water. It is still used in many parts of the world today. It was this practical side of Archimedes that distinguished him from many Greek mathematicians and philosophers of his day.

Unlike philosophers, such as Socrates, Plato and Aristotle, Archimedes didn't just think up theories to explain the natural world, he actually conducted experiments to test his ideas. It was his willingness to experiment that made Archimedes the forerunner of modern day scientists. His talent for solving practical problems also led to the most repeated story told about him.

It involved a crown of gold made for his friend and relative, Hieron, the ruler of Syracuse.

According to the Roman historian, Vesuvius, a new crown had been made for King Hieron in the shape of a laurel wreath. Archimedes was asked to determine whether it was of solid gold, or whether other metals had been added by a dishonest goldsmith. Archimedes knew the density of the crown would be lower if cheaper and less dense metals had been added. The simplest solution would have been to melt it down and measure its density, but he had to solve the problem without damaging the

Illustration 116: Archimedes of Syracuse (287 BC-212 BC).

crown. While taking a bath, he noticed that the level of the water rose as he got in. He realized that this effect could be used to determine the volume of the crown, and therefore its density after weighing it. So excited by his discovery that he forgot to dress, Archimedes took to the streets naked, crying "Eureka!" —"I found it!"

The story about the golden crown does not appear in the known works of Archimedes, but in his treatise, *On Floating Bodies,* he described the law of buoyancy, known in hydrostatics as *Archimedes' Principle.* This states that a body immersed in a fluid experiences a buoyant force equal to the weight of the displaced fluid.

Archimedes died in 212 BC during the Second Punic War, when Roman forces under General Marcus Claudius Marcellus captured Syracuse after a siege that lasted more than two years. Plutarch, in his *Parallel Lives,* gives the popular account of Archimedes' death. Supposedly, Archimedes was contemplating a mathematical diagram when the city was captured. A Roman soldier commanded him to come and meet General Marcellus, but he declined, saying that he had to finish working on the problem. Despite orders from Marcellus that Archimedes not be harmed, the enraged soldier killed the elderly mathematician with his sword.

Ancient Roman historians wrote several biographies about Archimedes' life and works. Unfortunately, only a few copies of Archimedes' own writings survived through the Middle Ages. Those that did were an influential source of ideas for scientists during the Renaissance and Enlightenment. But before the Enlightenment swept over Europe in the seventeenth and eighteenth centuries, bringing with it the Scientific Revolution, there were many dark times. In the backward lands of Europe following the fall of the Roman Empire, one shining light was a Franciscan Friar named Roger Bacon.

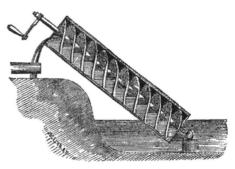

Illustration 117: Archimedes' screw. Source Chambers's Encyclopedia, 1875.

Roger Bacon, also known as *Doctor Mirabilis* (Latin for "wonderful teacher"), was the most famous cleric of his time. An English philosopher, who placed considerable emphasis on empiricism, he was one of the earliest European advocates of the modern scientific method. He is often considered a modern experimental scientist who emerged 500 years before the Scientific Revolution burst upon the European mind.

Bacon is thought to have been born near Ilchester in Somerset, though he has also been claimed by Bisley in Gloucestershire. His date of birth is uncertain, the only source of information being his statement in the *Opus Tertium,* written in 1267. In it, he wrote that forty years had passed since he first learned the alphabet. It is generally assumed that this meant 40 years had passed since he matriculated at Oxford at the age of 13, placing his birth around 1214. His family was wealthy, but his parents sided with Henry III against the rebellious English barons and lost nearly all their property. Several members of the family were driven into exile.

Roger pursued his studies at Oxford and Paris, and later became a professor at Oxford. There is no evidence he was ever awarded a doctorate—the title Doctor Mirabilis was bestowed by scholars after his death. His thinking was greatly influenced by his Oxonian masters and friends. Their influence created in him a predilection for languages, physics and the natural sciences. Bacon became an early advocate of experimental science, in an age generally thought to be hostile toward scientific ideas. Later, in Paris, he met the Franciscan Petrus Peregrinus de Maricourt, whose influence led to Bacon entering the Franciscan Order.

In 1256, he became a Franciscan Friar hoping to be assigned to a teaching post, but this was not to be. Instead, his superiors imposed other duties on him. A restless spirit, he was often in trouble with church authorities for his theological writings. In 1260, the Franciscan Order forbade him to publish any work outside of the order without special permission from higher authorities "under pain of losing the book and of fasting several days with only bread and water."[333] Despite these restraints, Bacon managed to leave behind a remarkable legacy of independent thought and inquiry.

Bacon possessed one of the most commanding intellects of his age and made many discoveries. He performed many varied experiments, which were among the earliest instances of true experimental science. His *Opus Majus* contains treatments of mathematics, optics, alchemy, the manufacture of gunpowder, and the positions and sizes of the celestial bodies. In it, he anticipated such modern inventions as microscopes, telescopes, spectacles, flying machines, hydraulics and steam ships.

But Bacon was also a man of his time, not totally immune to the superstition and mysticism of the day. He studied astrology and believed that celestial bodies had an influence on the fate and personalities of human beings. Even so, he made discoveries that anticipated those later made by the giants of the Scientific Revolution. It was Bacon who first reported the visible spectrum of light created by a glass of water, four centuries before Isaac Newton discovered that a prism could split white light into a rainbow of colors. Reportedly, he planned to publish a comprehensive encyclopedia, but only fragments ever appeared.

Bacon is thought to have died around 1294, but the date of his death is as uncertain as the date of his birth. Two years before his death, he composed his *Compendium Studii Theologiæ*, where he set forth a last scientific confession of faith. In it, he described the ideas and principles which had driven him during his long life. According to historian Theophilus Witzel: "he had nothing to revoke, nothing to change."[334]

When asked about the origins of modern science, many people recall the leaders of the Scientific Revolution: Galileo, Francis Bacon, and Isaac Newton or, going back at bit further, Nicolaus Copernicus or even Leonardo da Vinci. But it was from early experimentalist like Archimedes and Bacon, that modern science evolved. As Isaac Newton put it, "If I have seen further it is by standing on ye shoulders of Giants."[335] Generations of natural philosophers gradually came to reject supernatural influences and magical explanations of the natural world. Eventually, astrology and alchemy would become astronomy and chemistry, casting off their mystical past and embracing what has come to be known as the scientific method.

The Scientific Method

According to the Miriam-Webster dictionary, the scientific method consists of "principles and procedures for the systematic pursuit of knowledge involving the recognition and formulation of a problem, the collection of data through observation and experiment, and the formulation and testing of hypotheses." Or, as more succinctly put by Meg Urry, "Scientists observe nature, then develop theories that describe their observations."

The scientific method is a body of techniques for investigating natural phenomena and acquiring new knowledge. It also provides mechanisms for correcting and integrating previous knowledge. The scientific method is based on gathering *empirical* evidence. This is accomplished by collecting data through observation, experimentation, and the testing of hypotheses. Empirical means simply what belongs to or is the product of experience or observation. The *Science Fair Handbook* puts it this way: "The scientific method involves the following steps: doing research, identifying the problem, stating a hypothesis, conducting project experimentation, and reaching a conclusion."[336]

The advantage of the scientific method is that, if followed faithfully, it is unprejudiced. An hypothesis can be tested through experiment and its validity determined. The conclusions must hold regardless of the state of mind, or bias of the investigator. In fact, the cornerstone of modern science is the testability of theories. This means that a theory must make predictions about the way the physical universe behaves, so that it may be tested by investigators other than the theory's author.

The dual requirements of testability and empirical evidence disqualify mystical or religious arguments from scientific consideration. Such arguments are based on forces outside of nature, and science is only concerned with the natural world. You cannot test or measure God, so attributing some phenomenon to an act of God is not a scientific theory.

Religious truth is *revealed* to individuals, and must be taken on *faith* by others. Scientific knowledge is *discovered* through observation, and can be *tested* through experiments repeatable by anyone. Some religions are based on secret or hidden knowledge[†] that must be accepted without proof, science is based on shared knowledge open to question. Religion requires acceptance of that which is unseen (see *Hebrews 11* for an example), science is based only on that which can be observed. But, this does not mean that religion and science need to be in opposition.

Religion answers questions that science cannot, science answers questions that religion should not. Just as religious teachings cannot be viewed as a

---

† From the Greek *gnosis* (γνῶσις), described as direct knowledge of the supernatural or divine. An *agnostic* is someone who denies the existence of such knowledge, believing all things are knowable without divine revelation.

valid source of scientific knowledge, science has no authority in spiritual or ethical realms. Science is the study of nature and nature is neither moral nor immoral. Nature, at best, can be viewed as amoral, and even that is dangerously close to viewing nature as a sentient being. It is not.

Nature is a collection of physical processes, possessing no intelligence, no conscience, and no moral compass. Nature does not mourn the passing of a single creature or the extinction of entire species. There is nothing in nature that provides a scientific foundation for morality, though some have sought one. Galileo is credited with saying that religion "tells us how to go to heaven, not how the heavens go." The opposite also holds true, science does not provide moral guidance or satisfy the human longing for an underlying meaning to existence. Religion is religion, science is science and the two should never be confused.

Hypotheses, Theses and Laws

Scientific explanations are known by a number of different names. An initial scientific idea is called a *working hypothesis*, which consists of a brief statement of the explanation. After testing by experiment, an hypothesis that proves to be accurate becomes a *theory*. Sometimes theories pull together a number of hypotheses into a single, larger explanation. There is a great deal of misunderstanding about what scientists mean when something is called a theory.

Often, people will dismiss a scientific idea by saying "it's just a theory," as though a theory is just someone's opinion or something made up on a whim. This could not be more wrong. To be accepted as a scientific theory means that the ideas expressed have been examined and tested by many scientists, not just the one who first proposed it. Theories that have endured the test of time come as close to "fact" or "truth" as anything known to science. Scientists tend to shy away from absolute terms like fact and truth, because they would give the impression that a particular theory is absolute and never subject to change. In science, nothing is above challenge or immune to modification. When a theory has survived for several hundred years, and its author has departed this life, it may be elevated to being a *law*—but in science, even a law is subject to change.

This is not to say that old, well-established theories are often discarded. As new information becomes available, old theories often remain valid, but the regions over which they are valid become more narrowly defined. For example, Newton's laws of motion were not "overthrown" by Einstein's Theory of Relativity. Instead, it was recognized that Newton's laws were limited to objects traveling at velocities much less than the velocity of light.

In this way, the accumulated body of knowledge that is science continues to grow. New, better theories replace or supplement older ones. But always new theories must be in agreement with others that are accepted as valid. A new idea cannot contradict a large volume of accepted theory. Not because the weight of the old theories makes them inviolate, but because the new theory would have to offer satisfactory explanations of all the things the old theories had explained. As Marcello Truzzi† put it, "extraordinary claims require extraordinary proof."

Regardless of the field of inquiry, the process of investigation must be objective in order to reduce biased interpretations of results. Another basic expectation is that scientists document their work and share all data and methodology with other scientists. This is to allow other researchers to verify experimental results by attempting to reproduce them. This practice, called "full disclosure," also allows statistical measures of the reliability of these data to be established. It was lack of full disclosure by Mann, et al, that initially led to the flap over the "hockey stick" temperature graph promoted by the IPCC.

The Aha! Moment

Scientists' challenge is to create new theories, based on observation, that build on and are in general agreement with the existing body of scientific knowledge. In pursuit of this goal, scientists spend many years studying and learning to do research. In the American academic system, the highest degree granted is Doctor of Philosophy, or PhD, from the Latin *Philosophiæ Doctor,* meaning "teacher of philosophy." Though requirements vary, a PhD candidate must submit a thesis or *dissertation* consisting of a suitable body of original academic research. This work must be deemed worthy of publication after peer-review, and the candidate must defend the work before a panel of expert examiners appointed by the university.

Often there is coursework associated with getting the degree, but not always. If all a student desires is mastery of a particular field there is a lesser degree, the Masters degree, that can be earned with significantly lower investment of time and effort. What the Doctorate degree signifies is not mastery of a particular area of study, that is assumed, but a demonstration that the candidate can do scientific research. To earn a Doctorate an aspiring new scientist must be able to frame an hypothesis, construct experiments to demonstrate its soundness, and finally, defend his work before scientific

---

† Marcello Truzzi (1935-2003) Danish-American Sociologist and co-founder of the Committee for the Scientific Investigation of Claims of the Paranormal (CSICOP).

peers. A PhD in a scientific discipline signifies that one has learned, and is a practitioner of, the scientific method.

After the years of preparation, hard work and testing, that a scientist in training is subjected to, it might be assumed that life becomes easier after earning the title Doctor. This is not the case. A working scientist will continue to formulate new hypotheses, do research, and publish papers that are reviewed by peers. Given the training and work environment, the volume of knowledge to be mastered and the rigors of the scientific method, it is easy to assume that scientists are logical, methodical individuals. This implies that the moment of discovery in science is arrived at in a slow, methodical way. More often than not, the exact opposite is true.

We began this chapter with a quote from Isaac Asimov[†] that made reference to Archimedes exclamation, "Eureka!" What Asimov meant was that, despite careful planning and experiment, it is often the unexpected, unanticipated result that brings scientific insight. Accidental discoveries have always played an important role in science.

Sir Horace Walpole[‡] coined the term *Serendipity* for such accidental discoveries. In 1754, he wrote a letter to a friend, Sir Horace Mann, describing a story that had made a profound impression on his life. The story was a "silly fairy tale, called The Three Princes of Serendip; ... as their highnesses traveled, they were always making discoveries, by accidents and sagacity, of things which they were not in quest of." [337] Serendip was an old name for Ceylon, nowadays called Sri Lanka, but then a mysterious island in the East. The tale described the fate of three princes who left their home to travel the world seeking great treasures. They rarely found the treasures they were looking for, but discovered others of equal or greater value, which they were not seeking.[338]

More than a century ago, Louis Pasteur said, "Chance favors only the prepared mind." By this, he meant that sudden flashes of insight don't just happen, but are the product of careful preparation. Pasteur was a master of experimental research. Though he wasn't greatly interested in theory, he made many important discoveries through careful observation. Pasteur didn't always know what he was looking for, but he was capable of recognizing something important when it occurred.

[†] Isaac Asimov (1920-1992) Russian born American author and biochemist, best known for his works of science fiction and for his popular science books.

[‡] Horace Walpole (1717-1797), fourth Earl of Orford, son of Prime Minister Robert Walpole, connoisseur, antiquarian and author.

The lesson in all of these observations is that scientists should be as prepared as possible when investigating nature, but above all, they must keep an open mind. Deep insights can occur at the most unexpected times. Archimedes had such a moment when he discovered the Archimedes Principle in his bath. Herschel must have thought "that's funny" when his thermometer unexpectedly registered increasing temperature even though placed outside of the visible spectrum of light. Moments of sudden insight are often called "Aha!" moments.

An "Aha!" moment occurs when a key concept, mechanism, or relationship suddenly comes into focus. It is the moment that scientists hope for all their professional careers—clear thoughts crystallize in "Aha!" moments. Often, scientists aren't even sure what they are looking for, as in the case of Pasteur, or as Wernher von Braun said, "Research is what I'm doing when I don't know what I'm doing."

Luigi Galvani discovered the true nature of the nervous system when an accident in his laboratory made a dead frog's legs twitch. Based on the serendipitous observations by Galvani, Alessandro Volta designed the first modern electric battery in 1800. Becquerel's discovery of radioactivity (page 180), Tombaugh's discovery of Pluto on the basis of Lowell's flawed calculations, and Fleming's discovery of penicillin when it contaminated a bacterial culture are only a few of the unexpected discoveries that have changed science.

To be able to benefit from an unexpected discovery, from serendipity, a scientist must be prepared to accept change. Experiments do not always yield the expected results, and theories must sometimes be modified or discarded when the universe reveals its true nature. Just as Luis and Walter Alvarez recognized the importance of the unexpectedly high concentration of iridium at the KT boundary, new discoveries constantly arise to challenge scientific dogma. Unfortunately, even correct new theories are seldom easily accepted.

Accepting New Ideas

In many chapters of this book, we have presented the stories of scientists and their discoveries. Though we hope the stories of Cuvier, Agassiz, Wegener, Milankovitch, the Alvarezs and all the rest have been entertaining, we had a deeper purpose for describing their labors. There is a common theme to the stories of discovery we chose to present, and that theme is how difficult it can be to overcome weak, erroneous, but commonly accepted theories.

Cuvier fought religiously inspired dogma by claiming species went extinct. It took Agassiz's theory of ice ages thirty years to overcome resistance and be accepted by the geological establishment. Wegener did not live to see his theory of continental drift proven. Both Croll and Milankovitch witnessed the initial, tentative acceptance of orbital cycles' influence on climate, only to see their theories fall out of favor. Neither lived to see the ultimate acceptance of the Croll-Milankovitch Cycles as the main driver of glacial-interglacial variation. It is hard to unseat accepted theory, compelling proof must be provided—that is the scientific way. In all of the cases presented, final acceptance depended on experimental proof of the physical mechanisms underlying the theory. This is how it should be, even if the proof takes more than a lifetime to arrive.

What is necessary for a new theory to supplant an existing one is for the new explanation to be more compelling than the old. More accurate, more demonstrable, more straightforward than the old theory. In short, the old theory must be unable to provide as good an explanation as the new one. Sometimes, as mentioned earlier, the new theory adds to the old—extending, refining or augmenting its predictions. Other times, the new theory is simpler, more elegant than the old.

In science, there is a principle known as *Ockham's Razor,* named after William of Ockham, a fourteenth century English monk and philosopher. Also called the *law of economy,* or *law of parsimony,* the way Ockham stated it was "Pluralitas non est ponenda sine neccesitate," Latin for "entities should not be multiplied unnecessarily." Today, some translate this into "keep it simple, stupid," or the KISS principle. But that is an over-simplification of Ockham's rather subtle rule for judging ideas.

Illustration 118:
William of Ockham
(1288-1348).

Suppose there are two competing theories which describe the same phenomenon? If these theories produce different predictions, it is a relatively simple matter to find which one is better. Experiments are performed to determine which theory gives the most accurate predictions. For example, Copernicus' theory said the planets move in circles around the Sun, in Kepler's theory they move in ellipses. By carefully measuring the path of the planets it was determined that they move in ellipses—Copernicus' theory was then replaced by Kepler's.

Sometimes things are not so clear cut. The adoption of Kepler's theory of elliptical orbits was really just an improvement on Copernicus' theory. The initial adoption of Copernicus' theory was a much more radical change. It

217

was not the shape of orbits that formed the central idea of that theory —it was what was at the center.

Before Copernicus, the widely held view of the cosmos was that described by Ptolemy[†] more than 1,400 years earlier. The Ptolemaic universe placed Earth in the center with the Sun, the planets and the stars traveling around it. His geocentric planetary system represented the universe as a set of nested spheres, with heavenly bodies embedded in the spheres. But the observed motion of some of the objects could not be represented by simple circular orbits around Earth.

Illustration 119: Ptolemy's heliocentric model of the Universe.

For instance, the path of Mars across the sky doesn't progress in a single smooth ark. Its motion can be seen to stop, reverse course for a time, and then resume its forward progress. This *retrograde* motion cannot be reconciled with simple circular orbits in an Earth centric system. A more complicated model is needed and Ptolemy's solution was ingenious. It was possible to approximate the planets' observed motion by having them travel on smaller circular paths, called *epicycles,* imposed on top of the main orbit. Apply enough additional circles and you can come arbitrarily close to the observed path.

Copernicus took a more radical approach. He placed the Sun at the center, with all the other heavenly bodies, including Earth, orbiting it. This accounted for the observed apparent motion of the planets using simple circular orbits. No complicated epicycles were needed. Under these conditions, Ockham's Razor allows us to decide which theory is best. The circles within circles theory is much more complicated than the *heliocentric* theory, so Copernicus wins. All things being equal, the simplest solution tends to be the best one.

Kepler's modification improved on Copernicus' original idea, making the orbits more accurate with only a modest increase in complexity. As Einstein said in his version of Ockham's Razor, "So einfach wie möglich und so

---

† Claudius Ptolemaeus (ca. 90-168 AD) a Greek-Egyptian mathematician, geographer, astronomer, and astrologer in Roman Egypt.

kompliziert wie nötig," or "As simple as possible and as complicated as necessary." This criteria for judging competing theories is widely accepted in modern science.

Karl Popper[†] argued that preferring simple theories does not need to be justified on practical or aesthetic grounds. Simpler theories are preferred to more complex ones "because their empirical content is greater; and because they are better testable."[339] Popper called this the *falsifiability criterion:* A simple theory applies to more cases than a more complex one, and is thus more easily refuted.

In light of these principles, when the CO_2 theory of climate change is examined, it is found wanting. We have shown that known mechanisms involving greenhouse warming do not account for the amount of temperature increase the IPCC's climate scientists attribute to it. The total amount of warming is not in dispute, but the magnitude of carbon dioxide's contribution is questionable. The IPCC's case fails to correctly account for all the physical observations.

The second failing of the IPCC theory is that, in order to account for the observed warming, a number of assumptions about feedback mechanisms must be made. These assumptions are not supported by experimental data or observations, yet they are included in the GCM, the computer models on which the IPCC's case rests. Because of this added baggage, the IPCC theory of human-caused global warming fails Ockham's Razor. They have multiplied their entities unnecessarily—in this case, assumptions about climatic feedback—instead of searching for more fundamental explanations for the observed temperature rise.

A huge question remains—why the climatological establishment has not looked further afield for better explanations for global warming? As we have shown in the previous two chapters, theories have been proposed by astrophysics that offer explanations for much of the warming. When taken into account, these theories reduce the amount of warming attributable to CO_2 to empirically supportable levels. Yes, humans are causing some global warming by way of greenhouse gas emissions, but the levels are much lower than those proclaimed by the IPCC reports.

Even though their existing theory is weak, climatologists cling to the belief that their computer models can accurately predict the future. Their resistance to change is similar to that faced by Agassiz, Wegener, and Alvarez. This resistance is strengthened by the fact that the proponents of change are from

---

† Sir Karl Raimund Popper (1902-1994) an Austrian and British philosopher of science.

outside of the insular climatological community. Just as Agassiz, known for working on fossil fish, was an outsider to geologists when he proposed his ice age theory, and Wegener, a meteorologist, was an outsider when he proposed continental drift. The Alvarezs—one a physicist and the other a geologist—were outsiders to paleontology when they declared that an asteroid killed the dinosaurs. Today, climatological dogma is being challenged by outsiders.

In the words of American philosopher Charles S. Peirce, "Doubt is an uneasy and dissatisfied state from which we struggle to free ourselves and pass into the state of belief; while the latter is a calm and satisfactory state which we do not wish to avoid, or to change to a belief in anything else. On the contrary, we cling tenaciously, not merely to believing, but to believing just what we do believe."[340] It is human nature to resist change. To accept change and embrace a new theory requires what Thomas Kuhn[†] called a *paradigm shift*.

The term paradigm shift was first used by Kuhn in his 1962 book, *The Structure of Scientific Revolutions,* to describe a change in basic assumptions within the ruling theories of science. It is more than simply changing your mind. It is more like a revolution, a sudden transformation, a sort of metamorphosis. Paradigm shifts do not just happen, but rather must be driven by agents of change. Agents in the form of determined, stubborn scientists who believe they have found better solutions than the current ones. This situation is absolutely normal for science. What is abnormal, and deeply harmful, is that the course of scientific debate has been warped by being catapulted onto the world political stage by the IPCC.

Why Consensus is Meaningless

The most commonly heard argument during public debate of the global warming question is "there is scientific consensus" agreeing with the IPCC's conclusions. This is an example of what logicians call *argumentum ad verecundiam*, Latin for "argument from authority." Logically, this is a fallacy because the validity of a claim does not follow from the credibility of the source. Nonetheless, this type of weak argument is used all the time in advertising (e.g. "Four out of five doctors recommend..."). In simple terms, it means "all these smart people say it's true, so it is."

In science, theories are strengthened when other parties can repeat an experiment providing evidence that the theory's predictions are accurate. This helps validate the theory's correctness. But, outside of providing

---

† Thomas Samuel Kuhn (1922-1996) American historian and philosopher of science.

specific empirical data or experimental verification, asking scientists what they think is just an opinion poll.

It may be hard to understand why asking a large group of experts their opinion is an unreliable way of deciding a question. Recall the examples of scientific discovery cited in the preceding chapters. In almost all of those episodes, scientists had to fight existing opinion, the consensus of the time, to get their new theories accepted. Here are some past examples of scientific consensus:

- Consensus said the Sun, the planets and the stars orbited Earth, which was the unmoving center of the universe.

- Consensus said Earth was no more than 10,000 years old.

- Consensus said no animal species had ever gone extinct.

- Consensus said that masses of glacial ice could not have covered Europe or other temperate parts of the Northern Hemisphere.

- Consensus said that the continents were fixed in rock and could not move.

- Consensus said changes in Earth's orbit could not affect climate.

- Consensus said that the planet and its creatures were only changed by slow gradual processes, an asteroid could not have killed off the dinosaurs.

And now, in the absence of solid, definitive proof, *consensus* says that people are causing dangerous global warming. More specifically, that rising CO_2 levels are going to cause an unprecedented temperature rise that will threaten all Earth's creatures. As we have shown, the theory is uncertain, incomplete and far from definitive. Every day, new criticisms of existing climate orthodoxy arise yet claims of consensus are widespread. Many scientists are not in agreement with the IPCC's public pronouncements but, for some reason, the proponents of human-caused global warming find it important to claim unanimity exists within the scientific community.

Scientists argue—that's the way science works. Attempting to shut down debate by claiming that consensus has been reached is a sure sign that something other than science is at work. Michael Crichton, best known for his novels but also a graduate of Harvard Medical School and a former postdoctoral fellow at the Salk Institute for Biological Studies, warned of the dangers of "consensus science" in a 2003 speech:

"Historically, the claim of consensus has been the first refuge of scoundrels; it is a way to avoid debate by claiming that the matter is already settled. Whenever you hear the consensus of scientists agrees on something or other, reach for your wallet, because you're being had. Let's be clear: the work of science has nothing whatever to do with consensus. Consensus is the business of politics. Science, on the contrary, requires only one investigator who happens to be right, which means that he or she has results that are verifiable by reference to the real world. In science consensus is irrelevant. What is relevant is reproducible results. The greatest scientists in history are great precisely because they broke with the consensus."

As Anatole France said, "If fifty million people say a foolish thing, it is still a foolish thing." If any number of scientists believe an erroneous theory to be correct, it is still an erroneous theory. Scientific consensus is what people fall back on when there is no clear-cut evidence or compelling theoretical explanation. To borrow a phrase from that great Texan, John "Cactus Jack" Garner,[†] consensus is "not worth a bucket of warm spit."

The First Pillar of Climate Science

As we stated in Chapter 1, scholars often refer to the *three pillars of science:* theory, experimentation, and computation.[341] Now that we have completed our survey of the science behind Earth's climate and the natural causes of climate change we can return to analyzing the IPCC's theory of human-caused global warming. Having spent the last six chapters discussing the theories that try to explain climate change, we will begin here with the first pillar—theory.

It should be obvious from the number of times we have quoted scientists, declaring the causes of one aspect of climate change or another as "unknown" or "poorly understood," that the theoretical understanding of Earth's climate is suspect. The detailed and tortuously defined levels of uncertainty presented in the IPCC reports themselves is an admission of fact: the theoretical understanding of Earth's climate is incomplete in fundamental ways.

In Chapter 7, we discussed the "missing sink" of carbon that has been under study for thirty years without being found. We have cited the recent realization by the European Parliament that animal emissions are more potent than human CO_2 and that, for large portions of Asia, it is particulate

† John Nance Garner (1868-1967) Speaker of the U.S. House of Representatives and Vice President of the United States.

pollution in the "brown clouds" causing most of the atmospheric warming. We discussed statistical links between climate and the sunspot cycle that are not explained by conventional climate theory.

In Chapter 11, we discussed the astrophysical based theories, linking climate change to solar cycles and even the solar system's path through the Milky Way. These theories, yet to find wide acceptance among climatologists, have been strengthened by new findings regarding *ion initiated nucleation* (IIN) in the troposphere and lower stratosphere.[342] Recent research has also found a link between the ozone layer and global cooling. A report in the Proceedings of the National Academy of Sciences (PNAS) states that current global warming would be substantially worse if not for the cooling effect of stratospheric ozone.[343] Every day, science uncovers new relationships, new factors influencing Earth's climate. Theoretical understanding of how Earth's climate functions can only be called incomplete. Theory—the first pillar of climate science—is weak at best.

In the context of climate science, experimentation involves taking measurements of the oceans and atmosphere and collecting historical climate data in the form of proxies. In the next chapter, we will examine the second pillar, experimentation.

Chapter 13 Experimental Data and Error

"Doubt is not a pleasant condition, but certainty is absurd."

— Voltaire

Papers discussing global warming often include historical temperature records going back hundreds or thousands of years. At first glance, this seems a bit puzzling, since the thermometer is a relatively recent invention. The first thermometers were called *thermoscopes* and several scientists invented versions around the same time. Galileo invented a water-filled thermometer in 1593. Italian inventor Santorio Santorio was the first to put a numerical scale on the instrument, allowing precise measurements. Gabriel Fahrenheit[†] invented the alcohol thermometer in 1709, and the first modern mercury thermometer in 1714.

Fahrenheit also devised a temperature scale to use with his thermometers. He set zero (0°F) to the coldest temperature he could attain under laboratory conditions. This was created using a mixture of water, ice and ammonium chloride. Fahrenheit adjusted his scale so the high end was the boiling point of water. Fahrenheit's final temperature scale had 180 degrees between the freezing and boiling points of water, which he put at 32°F and 212°F, respectively. But Fahrenheit's scale was not the only one devised.

The Celsius scale was invented by Swedish Astronomer Anders Celsius[‡] in 1742. Originally called the centigrade scale because it had 100 degrees between the freezing point (0°C) and boiling point (100°C) of water, the term "Celsius" was officially adopted in 1948 by an international conference on weights and measures. For scientific measurements, Fahrenheit's scale has been replaced by the Celsius scale.

Illustration 120: Anders Celsius (1701-1744).

---

† Daniel Gabriel Fahrenheit (1686-1736) German physicist who invented the Fahrenheit temperature scale.

‡ Anders Celsius (1701-1744) Swedish Astronomer and polar explorer.

As scientists and explorers spread across the world they began collecting temperature readings from other lands, from the tropics to the frozen poles. Even so, we only have reliable readings from Europe and eastern North America going back a little over 200 years. Worldwide records have only been available for the past half century. These records are mostly surface temperature readings. Nowadays, satellites and weather balloons report temperatures at various levels in the atmosphere.

In addition to temperature, many other parameters are of interest to climatologists: insolation levels, cosmic ray flux, atmospheric CO_2 levels, the amount of dust and particulates in the atmosphere, to name a few. Since modern instruments have only been available for the last quarter century or so, this poses a problem for scientists wanting to examine Earth's climate in the past. Paleoclimatologists have to rely on data gathering techniques that involve stand-ins for actual direct instrument readings—so called proxy data.

In order to understand the second pillar of the climate science—experimentation—we need to examine data collection using modern instruments, as well as historical data collection using proxies. We will start with modern data collection techniques first.

Satellites and Radiosondes

How reliable are modern climate data? For temperature measurements there are three main sources; remote sensing data from orbiting satellites, direct temperature readings from radiosondes attached to balloons, and surface temperature readings taken at local weather stations and ships at sea. With the launch of TIROS I (Television and InfraRed Observation Satellite) on April 1, 1960, climate science entered the space age. From that day on a large number of spacecraft have observed Earth's weather conditions on a regular basis. Today, most of the world is monitored from the vantage point of outer space.

In 1963, following the experimental TIROS series, NASA formed an in-house satellite design team at the Goddard Space Flight Center (GSFC). This team was led by William Stroud, Rudolf Stampfl, John Licht, Rudolf Hanel, William Bandeen, and William Nordberg. Together, these men developed the satellite series called NIMBUS (Latin for rain cloud). These spacecraft subsequently led to the swarm of satellites that currently observe Earth's climate and weather conditions.

Today, most of the world is monitored by orbiting spacecraft and it was the Nimbus pioneers who helped make this possible. The 1960s were a golden age at NASA—the race with the Soviets to put a man on the moon, and competition in Earth-orbiting satellites spurred development. There were a total of seven Nimbus Satellites launched between 1964 and 1978. It was during this period that Simmons worked with Dr. Rudy Hanel, principal scientist for Nimbus 3 and 4. Hanel developed a modified Michelson infrared interferometer spectrometer, dubbed IRIS. IRIS was designed to produce vertical profiles of temperature, water vapor, ozone, chemical species, and interferograms/spectral measurements. The prototype device was assembled in a small, closed room where the instrument looked at a "black body"—a three foot circle of plywood painted black—which Simmons called the "black gong."

One day, while Hanel and Simmons worked on the prototype in the closet, something happened. After thirty minutes of adjusting wires and calibrating the beam-splitter with a laser beam, Simmons said, "There's something wrong. The CO_2 level is climbing." Excitedly, Dr. Hanel replied, "Yes! Yes! The instrument sees our CO_2 breaths." It was an "Aha! Moment" when IRIS first saw its creators' breath. Simmons realized that the device had measured the CO_2 from his body and would soon measure the CO_2 in Earth's atmosphere. After Hanel's first IRIS successfully flew on Nimbus, another IRIS traveled to Mars on Mariner 9. After the Mars mission, an IRIS flew on Voyager 2 and took measurements as the spacecraft navigated the rings of Saturn. The valuable lessons learned exploring other planets were soon reapplied to monitoring Earth's climate.

In 1978, the first NOAA polar-orbiting satellite, TIROS-N, was launched. It was followed by the spacecraft of the NOAA series, the latest of which is NOAA-N, the 15th to be launched. NOAA uses two satellites, a morning and afternoon satellite, to ensure every part of the Earth is observed at least twice every 12 hours. Each satellite has a lifetime of around six years, so constant replacement is required to maintain coverage.

These satellites monitor severe weather, which is reported to the National Weather Service, and even assist international search and rescue efforts. Among the instruments carried by these

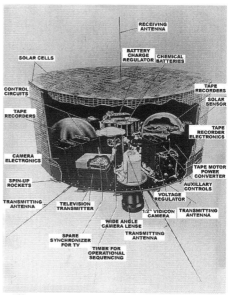

Illustration 121: The TIROS 1 weather satellite from 1960. Source NASA/NOAA.

orbiting weather stations is a passive microwave radiometer known as the Microwave Sounding Unit (MSU). The MSU monitors microwave emissions from atmospheric oxygen at several frequencies. By watching different frequency bands, or channels, the temperatures from three overlapping zones of the atmosphere are constantly monitored. These atmospheric zones are known as the lower troposphere (LT), from the surface to 5 miles (8 km), the mid-troposphere (MT), from the surface to 11 miles (18 km), and the lower stratosphere (LS), from 9 miles to 14 miles (15 km to 23 km). With the launch of the NOAA-15 spacecraft, in 1998, the MSU was replaced by the Advanced MSU (AMSU), which provides expanded monitoring capability.

There are a number of factors that impact the accuracy of satellite temperature readings including differences between MSU instruments, the varying temperature of the spacecraft themselves, and slow changes in the spacecraft's orbit. One scientist who has been intimately involved with gathering and correcting satellite temperature data is John Christy, Professor of Atmospheric Science and director of the Earth Systems Science Center at the University of Alabama in Huntsville. Over the years, Christy has published numerous papers regarding the accuracy of satellite data and is responsible for generating unified satellite temperature histories at UAH. When the temperature trends were analyzed, he found decadal error rates of ±0.05°C, ±0.05°C and ±0.10°C for the LT, MT and LS respectively.[344]

These may seem like low error rates but, as Christy has pointed out, "below the stratosphere the anticipated rate of human-induced warming is on the order of 0.1°C to 0.3°C decade^{-1} so that errors or interannual impacts of 0.01°C decade^{-1} approach the magnitude of the signal being sought."[345] Meaning that even modern satellite data is too uncertain to base decade long temperature predictions on. Another interesting point made by Christy is that during the 30 years of "good" satellite data collected by the NOAA spacecraft there have been two major volcanic eruptions (Mount St. Helens and Mount Pinatubo) and two exceptionally strong El Niño events (1982-83 and 1997-98). The impact of these events on global temperatures skewed the data collected over the past three decades in ways that cannot be fully accounted for. In short, the only good recent data we have is not typical, so any long-term projections based on that data will be biased.

Much of Earth's surface is covered by water and there are few weather stations scattered about the ocean's surface. To provide more complete coverage, spacecraft are used to monitor the temperatures of surface water around the globe. Satellite data are calibrated using ship observations of surface temperature from the same time and place. In a study of sea surface temperatures (SST), Reynolds et al. summed up the situation saying, "The globally averaged guess error was 0.3°C; the globally averaged data error was 1.3°C for ship data, 0.5°C for buoy and daytime satellite data, and 0.3°C for nighttime satellite data and SST data generated from sea ice concentrations. Clearly SST analyses have improved over the last 20 years and differences among analyses have tended to become smaller. However, as we have shown, differences remain."[346] If satellite data alone is not good enough to predict the future, what about other data collection methods? Next, we examine the other major method of gathering temperature data from different levels in the atmosphere—radiosondes.

A radiosonde, from the French word *sonde* meaning "probe," is an expendable package of instruments sent aloft attached to a weather balloon. As the balloon ascends, the radiosonde measures various atmospheric parameters and transmits them to receivers on the ground. The first radiosonde was launched   by Soviet meteorologist Pavel Molchanov on January 30, 1930.[347]

Modern radiosondes measure a number of parameters: altitude, location, wind speed, atmospheric pressure, relative humidity and temperature. Worldwide there are more than 800 radiosonde launch sites and most countries share their data. In the United States, the National Weather Service is tasked with providing upper-air observations for use in weather forecasting, severe weather watches

Illustration 122: Launching a radiosonde c. 1936. Photo NOAA.

and warnings, and atmospheric research. There are 92 launch sites in North America and the Pacific Islands and 10 more in the Caribbean. Each site launches two radiosondes daily. Radiosonde data is freely available from NOAA on the web.

Measuring temperature using radiosondes involves a number of complications that must be compensated for. As the balloon rises, the instrument that measures temperature, usually a form of thermistor, experiences a time lag. This lag makes accurately matching temperatures with correct altitudes difficult. Another complication is the Sun heating the instrument package as it rises, which biases the temperature readings. Worse still, different instrument packages from different manufacturers respond in different ways to these factors. Changing instrument brands can cause a shift in recorded temperatures by as much as 5.5°F (3°C).[348]

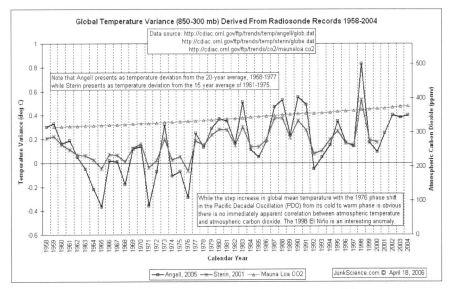

Illustration 123: Radiosonde temperature data. Source JunkScience.com.

Over time, changes in instrumentation and lack of documented comparison data among sites has made trying to construct global temperature histories very difficult. Regardless, radiosonde data is quite useful, particularly when readings are averaged over time. Variation caused by the El Niño-Southern Oscillation (ENSO) can be seen in Illustration 123. When scientists are trying to measure decadal variations on the order of 0.05°C, these data inaccuracies totally overwhelm any detectable trend. Given the uncertainties in both methods of data collection, it is not surprising that there have been a number of controversies regarding disagreement between satellite and radiosonde data.

If both satellite data and radiosonde measurements have had problems in the past, what about temperature readings from ground sites? It would be natural to assume that thermometer readings from weather stations would be reliable considering that thermometers have been around for more than 200 years. Quickly checking the temperature to see if an extra jacket is needed is one thing—reliably recording temperature readings, accurate enough for climate study use, turns out to be much more difficult.

Recently, NASA became aware of a glitch in their historical temperature data. It seems that a volunteer team, investigating problems with US temperature data used for climate modeling, noticed a suspicious anomaly in

NASA's historical temperature graphs. A strange discontinuity, or jump, in temperature readings from many locations occurred around January, 2000.

The original graphs and data available on the NASA/GISS website were created by Reto Ruedy and James Hansen. Hansen has been a vocal supporter of the IPCC claims, and gained notoriety by accusing the Bush administration of trying to censor his views on climate change. When contacted, Hansen refused to provide the algorithms used to generate the graph data, a position reminiscence of Mann's refusal to release the data and algorithms behind the hockey stick graph. Faced with Hansen's refusal, one of the volunteers reverse-engineered the data algorithm. After analyzing the results, what appeared to be a Y2K bug in the handling of the raw data was found.

For those too young to remember, Y2K refers to the "year 2000 computer crisis." At the end of the 20th century, computer programmers realized that some software might have problems when the year rolled over from 1999 to 2000. This was anticipated, and steps were taken to correct the problems before they occurred. Even so, the news media was filled with stories of impending disaster; people trapped in elevators, airplanes falling from the skies, the power grid failing and bank accounts emptying overnight. None of these things happened, of course, but there was a run on the banks as the panicked public withdrew money in case the financial system collapsed.

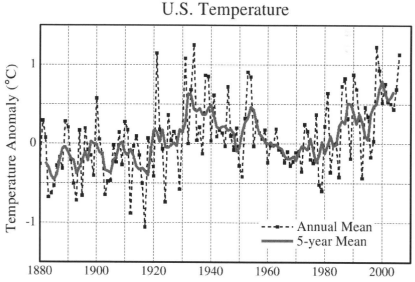

Illustration 124: Revised US Temperatures since 1880. Source NASA.

When Steve McIntyre, the volunteer who deciphered the data, notified NASA of the bug, Ruedy replied. He acknowledged the problem as an "oversight" that would be corrected in the next data release. Without fanfare, NASA quietly released corrected figures and graphs on the GISS website.[349] When the corrected data, shown in Illustration 124, are compared with the previous data from Illustration 2 (page 16), the changes are truly astounding.

The warmest year on record is now 1934. The year 1998, trumpeted by human-caused global warming proponents and the media as record-breaking, now moves to second place. The year 1921 takes third. In fact, 5 of the 10 warmest years on record now occur before World War II. The overall effect of the correction on global temperatures is minor, only 1-2% less warming than originally thought, but this serves as an example of why temperature data must be cross-verified. It is also a warning to those who insist on claiming global warming is demonstrated by a single, or even a cluster of warm yearly temperatures.

Though the problem reported by McIntyre seems to have been a programmatic one, there are other problems with collecting temperature measurements. Over time, the environment surrounding a weather station can change dramatically. Sites with long histories often show temperature trending upwards, but this may have nothing to do with global warming. A half century ago, a station may have been situated in a grassy field—today the same station could be located in a paved parking lot in the middle of a dense urban area. The station has not moved, but present day temperature readings will be higher due to a phenomenon called *heat islands*. A heat island is an area of increased temperature caused by buildings and pavement found in cities and urban areas. Often, weather station sites are not well-situated and their data are biased. This is clearly shown by the station in Illustration 125.

How important are adjustments for heat islands and sensor placement? In a 2003 study, Kalnay and Cai found effects from land-use changes and urbanization were three times greater than previously thought. When they applied their corrections to what was previously thought to be an accurate temperature history for the continental United States, they found no trace of a significant warming trend for the previous century.[350] As temperature increases across the continental US have generally been much lower than in other parts of the world, this observation does not imply that global warming is not a reality. It does suggest that the rate of warming may not be as rapid as suggested by the IPCC.

Illustration 125: Detroit Lakes, MN. Photo by Don Kostuch.

Proxies and Paleoclimate

When reliable, directly measured climate data are not available, scientists must turn to proxy data. Proxies come in many forms: silt deposits in lakes and oceans, plant pollen trapped in sediment, rings in trees and coral, and the skeletons of dead organisms. Some proxy data only provide the most general indicators of climate. Other proxies provide more precise measurements.

Among the indicators used by paleoclimatologists are plant macrofossils, charcoal, diatoms, chrysophytes, phytoliths, biogenic silica and pollen analysis.[351] Surprisingly, pollen turns out to be very tough stuff, able to survive where other plant matter decays and disappears. By analyzing pollen that collects in lake bed sediment layers, called *varve,* it is possible to gain insight into past climatic conditions. The word varve is derived from the Swedish word *varv* meaning "in layers." From the pollen of trees, grasses and flowering plants, scientists can tell if past temperatures were warm or cold, the weather wet or arid.

It was the presence of pollen in northern lake bed silt deposits that first identified the onset and end of the Younger Dryas period (page 139). *Dryas* is the genus name of a small flowering plant that likes cold weather. When the last ice age waned, dryas pollen vanished from European lake sediments. But then, it suddenly reappeared in the pollen record, marking the return of glacial conditions during the Younger Dryas.

Climatic variations we can observe directly today, such as El Niño and the North Atlantic Oscillation, are visible in climate proxies such as tropical corals and European tree rings. Volcanic eruptions and inferred solar variability, which can have significant cooling effects, are reflected in historical records and ancient writings.

Many proxies involve measuring the ratios between isotopes, different versions of atoms belonging to the same element. Elements are identified by the number of protons present in their atomic nuclei. The number of protons is matched by the number of electrons surrounding each nucleus. Since atoms form chemical bonds by sharing electrons, it is the number of protons/electrons, the *atomic number,* that determines an element's chemical properties. But there is another type of subatomic particle present in the nuclei of atoms, neutrons.

Neutrons weigh slightly more than protons, but they carry no electric charge. Atoms can have different numbers of neutrons in their nuclei without changing the number of protons and electrons. For example, all oxygen atoms have eight protons and eight electrons. The isotopes ^{16}O, ^{17}O and ^{18}O are each forms of oxygen with 8, 9 and 10 neutrons in their nuclei, respectively. They are the same chemically, but have slightly different weights. Of the three isotopes, ^{16}O is the most abundant comprising over 99.7% of the oxygen on Earth. By measuring slight variations in the ratios of isotopes, scientists can uncover important clues about Earth's climate in the past.

Henry C. Fricke, of the University of Michigan, tested teeth from dead Vikings for oxygen isotopes. His study analyzed the tooth enamel from 29 human teeth excavated at three archaeological sites in Greenland and one in Denmark. Fricke explained, "the ratio of heavy (O-18) to light (O-16) isotopes in the calcium phosphate that comprises tooth enamel is directly related to this isotopic ratio in rain or snow falling on a local area, because the oxygen in this precipitation is incorporated into the tooth enamel of growing children who drink from local groundwater supplies like springs, lakes and rivers."[352] Comparing teeth from skeletons buried in 1100, with those buried in 1400, he documented a 1.5°C drop in temperatures, confirming the onset of the Little Ice Age in northern Europe.

Measuring Time

Some isotopes are radioactive, meaning they are unstable and decay into other elements over time. This fact can be used to establish the dates of some substances. Plants absorb carbon from the atmosphere when they are alive. Carbon comes in two stable, nonradioactive forms: ^{12}C and ^{13}C. But there is another form of carbon present in Earth's atmosphere, the unstable isotope ^{14}C. Carbon-14 has a half-life of 5730 years and would have long ago vanished from Earth were it not for the shower of cosmic rays that bombards our planet. Cosmic rays striking nitrogen atoms in Earth's atmosphere continually replenish the amount of terrestrial ^{14}C. Thus, the ratio of ^{14}C to the stable isotopes of carbon remains fairly constant.[353]

As plants grow, they absorb all forms of carbon equally, so the level of ^{14}C present in their living tissues remains constant. But when a plant dies, this exchange stops, and the amount of ^{14}C gradually decreases through radioactive *beta decay*. Radioactive isotopes are subject to *exponential decay*, the rate of which is expressed in half-life time. The half-life of ^{14}C is 5730 years, meaning that in 5730 years, half of the ^{14}C atoms present in dead plant matter will decay into atoms of nitrogen. Atoms of ^{14}C become ^{14}N atoms as neutrons in the carbon nuclei become protons, and electrons are ejected. Eventually, the ^{14}C will disappear altogether. So, by measuring the ratio of ^{14}C to the stable forms of carbon present in dead plants, it is possible to assign a date to when they died.

This technique of radiocarbon dating was discovered by Willard Libby[†] and colleagues in 1949. Because of its short half-life, Carbon-14 dating is only viable for organic matter younger than ~50,000 years before the present (BP), and usually much younger than that. Despite this restriction, carbon-14 dating revolutionized the field of archeology, and for its discovery Libby was awarded the Nobel prize in Chemistry. This technique, while very useful, has built-in uncertainties that grow larger as the material being dated becomes older. Radiocarbon labs generally report an uncertainty of around 3000 ± 30 BP, or about 1% for things 3000 years old. All proxy measurements that involve isotope decay have similar built-in uncertainties.

For longer time scales, isotopes of potassium (K) and uranium (U) can be used to extend our view of the past. Atoms of ^{40}K decay into ^{40}Ar through beta decay much like ^{14}C is transformed into ^{14}N. Uranium, however, follows a different path, undergoing radioactive decay through *nuclear fission*. There are two naturally occurring isotopes of uranium, ^{235}U and ^{238}U, which decay through a series of fission transformation to become the lead isotopes ^{207}Pb and ^{206}Pb respectively.[354] As with all of these radioisotope dating techniques, the key is knowing the material's half-life and starting with known proportions of the isotopes in question.

For uranium dating, the half-lives of ^{235}U and ^{238}U have been measured with great accuracy and are found to be 703,800,000 and 4,460,000,000 years,

---

† Willard Frank Libby (1908-1980) American physical chemist and Nobel laureate.

respectively. Because of these long half-lives, uranium allows dating of rock from much further in the past than carbon-14 dating of organic material—the problem is finding a good sample to perform the analysis.

Fortunately, nature has provided tiny time capsules in the form of the mineral zircon. Zircon sand consists of nearly microscopic needles of silica glass. When these tiny grains form, they contain small amounts of uranium, but the physical process of formation excludes lead atoms. This results in a material that initially contains uranium and no lead to confuse the dating. Over time, the trapped uranium decays into lead and the proportion of lead to uranium reveals the elapsed time since the zircon formed. It has been through the careful analysis of uranium contained in zircon grains that the timing of the great Karoo Ice Age and the subsequent Permian-Triassic Extinction have been accurately established. Uncertainties of less than 1% can be attained, but 1% of 100,000,000 years is still 1,000,000 years.

Aside from dating rock and organic material with isotopes, other, more accurate methods exist, at least for the past million years or so. These dating methods apply to ice, mud and tree rings and involve a simple concept—counting. As trees grow, they form wood in plainly discernible bands that mark the alternating season of summer and winter. Sediments form yearly seasonal layers on both the ocean floor and the bottoms of lakes. The ice found in glaciers is formed from annual snowfalls. All these natural processes provide enterprising and patient scientists with accurate natural time lines stretching back into the past. When layer counting is combined with other proxy measurements, climate history can be revealed.

Indicators of Climate

Some natural processes create isotopes at constant or measurable rates. Radiation from the Sun and cosmic rays create beryllium (^{10}Be) and chlorine (^{36}Cl) isotopes in Earth's atmosphere, which then decay at known rates. Consequently, these proxy indicators can be used to determine past solar activity. Other processes discriminate against different isotopes of the same element, providing different insights. One such process, involving tiny ocean plankton called *foraminifera,* provides the basis for one of the main methods of tracking temperature over ages past.

Foraminifera, or forams for short, are single-celled marine organisms with hard shells. Their shells are commonly divided into chambers which are added during growth, though the simplest forms are open tubes or hollow spheres. Depending on the species, the shell may be made of organic compounds, particles cemented together, or crystalline calcite. When forams die, their shells drift to the bottom of the ocean, becoming part of the ocean floor sediment. In large enough concentrations, these tiny shells can form large beds of chalk and limestone.

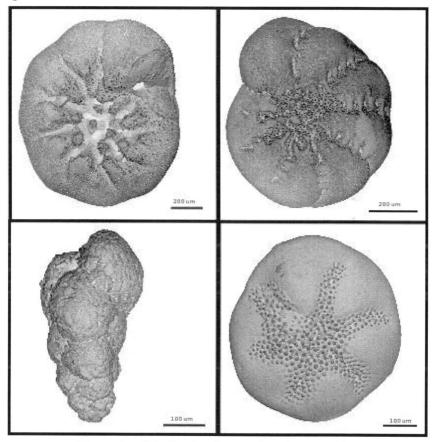

Illustration 126: Benthic foraminifera. Source USGS.

Forams have been found in the fossil record as far back as the Cambrian. Early forams were much larger than their modern relatives, but their small, almost microscopic size, makes modern forams much more useful than the larger fossils. In the 1920s, it was discovered that many species of foraminifera were geologically short-lived, and others are only found in specific environments. This allowed paleontologists to determine when rock formed, and the environmental conditions at the time, by examining the specimens in a rock sample. As a result, the oil industry became a major employer of paleontologists who specialized in these microscopic fossils.

Foraminifera are divided into two primary groups based on their mode of life; *planktonic* marine floaters, and *benthic* sea floor dwellers. Benthic

237

foraminifera are found at all latitudes and occupy a wide-range of marine environments, from brackish estuaries to the deep ocean basins. Much like plant pollen on land, they are very useful as environmental indicators. Particular species and assemblages can be used to identify ancient environmental conditions. Forams found in deep-sea core data have been utilized to identify changes in intermediate and deep water circulation, and to determine past sea level changes in coastal regions.[355]

Forams' role in paleoclimatology became even more prominent when biologists made the observation that the portion of [18]C forams absorbed varied with water temperature. In 1947, American nuclear chemist Harold Urey[†] combined that observation with newly-developed techniques that allowed precise measurements of very small quantities of atomic isotopes to create a way of determining temperatures in the ancient ocean. Cesare Emiliani, an Italian geologist who was a graduate student in Urey's lab, became interested in the problem as it applied to samples of foram shells taken from deep sea sediment cores. He is the same Dr. Emiliani who was interviewed by Time magazine about the impending new ice age in 1972 (page 42).

To obtain a sediment core, a long tube is sunk into the sea floor and then extracted with the layers of sediment intact. The date of each layer is established by counting down from the top layer or other methods. Test material is recovered from a single layer of sediment by carefully picking out a few hundred pinhead sized shells. The shells are cleaned, ground to powder and then heated to release CO_2 gas. The ratio between carbon isotopes is then measured using a mass spectrometer. From this ratio, known as $\delta^{18}O$,[‡] the temperature of the sea water that the forams lived in is estimated.

By 1955, Emiliani had applied this technique to many sea bed cores, compiling a record of ocean temperature change going back nearly 300,000 years. It was from these data that Milankovitch's theory of astronomically driven climate change rose, phoenix-like, from the ashes. At that time, the geological mainstream had rejected the Croll-Milankovitch cycles and settled back into their traditional beliefs regarding glacial cycles. Consensus among geologists said that there had been only four long glacial episodes in the Pleistocene Ice Age, separated by long warm periods. When Emiliani

---

† Harold Clayton Urey (1893-1981) American physical chemist and Nobel prize winner.

‡ The Greek character δ is a lower case delta (Δ). Scientists use delta as shorthand for difference or ratio. $\delta^{18}O$ is short for "delta oxygen eighteen."

looked at his data he found a dozen cycles, all in close agreement with Milankovitch's predictions.

Of course, this new case for Milankovitch's old theory was not easily accepted. Debate raged for more that 10 years, with traditional geologists defending their turf vigorously. Emiliani's data was rejected when it was shown that water containing ^{18}O and ^{16}O evaporated at different rates depending on temperature. The lower evaporation rate for $H_2(^{18}O)$ would tend to bias the $\delta^{18}O$ from the foram shells. Additional evidence provided by the mixture of foram species present in the core samples helped validate Emiliani's original result. To finally end the debate, corroborating data from ancient coral beds collected by Wally Broecker settled the argument.[356]

Interestingly, the argument that nearly derailed Emiliani's analysis, the different temperature-dependent evaporation rates of water containing different oxygen isotopes, led to the development of the other major source of paleoclimate temperature data—glacial ice cores.

Ice Cores, Gas Bubbles and Isotopes

The other major source of paleoclimate data combines several different proxies, including measurements of multiple isotopes and layer counting. That source is ice core data retrieved from mountain glaciers and the long-term ice sheets of Greenland and Antarctica. Samples are often used to establish temperature and CO_2 levels in times past. To do this, three values must be accurately identified; date, atmospheric CO_2 level, and temperature. Each of these data points are determined as follows.

There are four distinct methods for determining the ages of ice cores. Three are direct experimental tests and the fourth rests on somewhat uncertain theories. The methods for dating ice core layers are; counting annual layers, using predetermined ages as markers, radioactive gas dating, and calculating ice flow rates. Each of these methods is complex and presents unique problems. We will concentrate on the most popular method, annual layer counting.

The annual layer counting method looks for items in the ice that vary with the seasons in a consistent manner. There are a number of approaches to this as well. Among these are looking for indicators that depend on the temperature (colder in the winter and warmer in the summer) and solar irradiance (less sunlight in winter and more in summer). Regardless of the marker used, going back tens of thousands of years requires identifying a proportional number of yearly transitions in the ice core sample—this can be very time consuming.

Both the temperature and irradiance methods are based on measuring isotopes of certain elements; hydrogen or oxygen for temperature, and beryllium or chlorine for irradiance. Since we are interested in how temperature is determined anyway, we will concentrate on the temperature method using oxygen isotopes.

Of the temperature dependent markers, the ratio of ^{18}O to ^{16}O is the most important, though measurements based on isotopes of hydrogen are also used (δ^2H or $\delta D^\dagger$). Water molecules composed of $H_2(^{18}O)$ evaporate less rapidly and condense more readily than water molecules composed of $H_2(^{16}O)$. Thus, water evaporating from the ocean starts off $H_2(^{18}O)$ poor. As the water vapor travels towards the poles, it becomes increasingly poorer in $H_2(^{18}O)$ since the heavier molecules tend to precipitate out first. This depletion is a temperature-dependent process so in winter the precipitation is more enriched in $H_2(^{16}O)$ than is the case in the summer. Therefore, each annual layer starts ^{18}O rich, becomes ^{18}O poor, and ends up ^{18}O rich. Once the H_2O is captured in the crystalline ice of snowflakes, the $\delta^{18}O$ ratio is fixed.

This ratio can be measured very accurately using a mass spectrometer. Though accuracies of better than 1% can be achieved with modern instruments,[357] the quantities involved are often close to the limits of measurability to start with. Great care must also be taken to avoid contaminating the samples as contact with the modern atmosphere or the investigator's breath can bias the results.

The $\delta^{18}O$ ratio is expressed in parts per 1,000 compared with a reference sample called Standard Mean Ocean Water (SMOW). The more negative the value of $\delta^{18}O$, the lower the implied temperature when the snow fell. There are numerous other factors that can affect $\delta^{18}O$, such as altitude,[358] atmospheric circulation patterns,[359] and the timing of storms.[360] These complications add more uncertainty to determining temperature. Generally quoted uncertainty figures from studies going back to the start of the Holocene have temperature uncertainties in the range of ±3.0°C.[361] Even in more recent time frames, data quoted by the IPCC show temperature uncertainties that exceed the measured temperature increases for the last century. In fact, the IPCC's projected increase falls within the uncertainty range of the data they based their predictions on.

To determine historic atmospheric CO_2 levels, readings are taken from gas bubbles trapped in glacial ice. As snow layers build up year by year, they compact into progressively denser layers of *firn*. Firn is old snow found on

---

† The hydrogen isotope ^2H is also called heavy hydrogen or deuterium, hence δD.

top of glaciers, granular and compact but not yet converted into solid ice. Also called névé, it is a transitional stage between snow and glacial ice. Until the firn has become solid ice, the gaps between the ice crystals are still in contact with the atmosphere. The gas ratios do not become fixed until the gaps become bubbles in ice—a process that can take centuries.

The trapped gas bubbles are not the same age as the ice that encases them. The ice was formed as crystals when the snow fell, while the gas bubbles were formed later, when the snow eventually turned to ice. The time lag involved could be a few hundred years or more than a thousand. Just how long the bubble trapping process took cannot be accurately known, but the time lag used can change the relationship between CO_2 levels and temperature levels. A short lag shows CO_2 leading, a longer lag has temperature leading. Scientists have come to the conclusion that the later case is true. From studies of Antarctic ice cores going back half a million years, the average CO_2 to temperature lag is 1,300 years ±1,000 years.[362] Samples taken from around the end of the last glacial period indicate that the CO_2 levels did not begin to rise until after the warming began.

Illustration 127: Snow compacting into firn, and finally glacial ice.

There is uncertainty in the date, in the temperature readings, and the CO_2 levels, as well as the temporal relationship among all three. How a particular study chooses to resolve the uncertainties affects the interpretation of the results. The location of the ice may also result in a distorted view of global average temperature. We know that the circumpolar current keeps Antarctica thermally isolated from the rest of the world. Historical records show that temperature differences between the tropics, temperate latitudes and the poles have varied over time. This is not to say that ice core data is bad, or even wrong in a scientific sense. It simply means that there are open questions as to what such data means. As always, with proxy data, the levels of uncertainty must be taken into account when interpreting results.

Ice cores have been taken from Antarctica, Greenland, and mountain glaciers around the world. Since Antarctica has been covered by glacial ice longer than any other place on Earth, it is quite popular with scientists looking for samples (see Illustration 128). The oldest continuous ice core ever taken was retrieved from Dome Concordia in Antarctica. Initial dating confirmed that the 104,000 ft (3200 meter) core dated back at least 750,000 years.[363] For comparison, the oldest Greenland ice cores date back only around 125,000 years.[364]

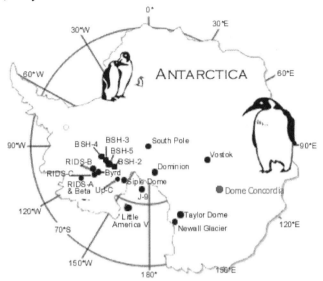

Illustration 128: Antarctic ice core locations, source USGS.

The annual layers in the Dome Concordia core are too thin to count visually, so the researchers are forced to use markers such as dust, gas and electrical conductivity to match different layers to known events. By matching layers to events that have already been dated, such as volcanic eruptions or ice ages, the entire ice core can be aligned with the past. The core was made by the European Project for Ice Coring in Antarctica (EPICA) and is being analyzed at the Alfred Wegener Institute in Bremerhaven, Germany.

In results published in *Science,* a high-resolution deuterium profile is now available for the entire EPICA Dome C ice core. This profile allowed the construction of a climate record that extends back to 800,000 years before the present. The ice core has provided temperature data covering 11 glacial, and corresponding interglacial, periods. The authors used an atmospheric GCM to calculate an improved temperature record for the entire interval, finding temperatures during warm intervals as much as 8°F (4.5°C) warmer, and, during cold intervals, as much as 18°F (10°C) lower, than pre-anthropogenic Holocene values.[365]

Living with Error and Uncertainty

Errors can be divided into two general categories: systematic and random. Systematic errors are errors which tend to shift all measurements in a systematic way, shifting data values in a consistent way. This may be due to incorrect calibration of equipment, consistently improper use of equipment or failure to properly account for some external effect. The time lag in temperature data from an ascending radiosonde and the offset between CO_2 and proxy temperature in ice core samples are examples of systematic error. Large systematic errors can often be identified and measured, allowing corrections to be applied to the experimental data, but small systematic errors will always be present. For instance, no instrument can ever be calibrated perfectly.

Random errors are errors which vary from one measurement to the next. They cause data measurements to fluctuate about some average value. Random errors can occur for a variety of reasons.

- Lack of sensitivity—an instrument may not be able to respond to or to indicate a sufficiently small change. Most instruments come with resolution limitations noted by their manufacturers.

- Noise—extraneous disturbances which cannot be quantified or accounted for. Short duration local climate variation can distort long-term climate trends.

- Random fluctuation due to statistical processes—such as the average rate of radioactive decay. Such randomness is an inherent property of the phenomenon being measured.

- Imprecise definition—lack of understanding of the property being measured.

Random errors displace measurements in arbitrary ways, whereas systematic errors displace measurements in a single direction. Some systematic error can be taken into account and substantially eliminated. Random errors are unavoidable and must be lived with.

As previously stated, proxy data is secondhand data, not actual measurements of the property being evaluated. Proxies are used because direct measurements are not available. All proxy measurements contain errors from at least two sources: uncertainty inherent in the measurements themselves and error in interpreting the data. Errors in measurement involve not only the difficulties in taking the proxy readings but uncertainty in dating the samples used to take the readings. The farther back in time we go, the more uncertainty enters into the process—uncertainty increases with the age of the sample. For the Holocene, ice core data has an uncertainty of about 300 years—for the early Pleistocene, the uncertainty can be much as 9,000 years.

Errors in ice core temperature proxies also become larger with age. They are affected by the hydrological conditions at the time the sample precipitated, which become harder to ascertain with age. Ice core samples from the Holocene are quoted with uncertainties ranging from ±0.8°C to as much as ±4.0°C.[366,367] Even data collected from satellites and direct measurement can contain significant errors. Accuracy of the combined ship and satellite data set, the Reynolds Optimum Interpolation Sea-Surface Temperature maps is about ±0.3°C. Experimental uncertainty is simply a fact of life.

Another type of error comes from how one type of measurement is translated into another. Scientists will tell you that this translation requires judgment. In other words, guesswork. In order to remove errors—and to provide more continuous historical coverage—combining multiple proxies into a single record has become standard procedure.

At the end of the 20th century, three teams of researchers carried out the time consuming and painstaking task of combining multiple proxy records into uninterrupted climate records for the recent past. In 1998, the teams that performed the statistical analyses reported their results in separate journal articles: Briffa et al., in *Nature*, vol. 393, p. 350; Jones et al., in *Holocene,* vol. 8, p. 445; and Mann et al., in *Nature,* vol. 392, p. 779. This leads the discussion back to questions of methodology and interpretation surrounding the Mann/IPCC hockey stick temperature history.

The Hockey Stick Revisited

Michael Mann's findings were arguably the single most influential factor in convincing the public that human-caused global warming is real. To construct the hockey stick plot, Mann, Raymond S. Bradley of the University of Massachusetts Amherst and Malcolm K. Hughes of the University of Arizona analyzed paleoclimate data sets from tree rings, ice cores and coral. Joining proxy data with thermometer readings from the recent past, they created a "reconstruction" of Northern Hemisphere temperatures going back 600 years. A year later, in 1999, they had extended their analysis to cover the last 1,000 years. In 2001, Mann's revised climate history became the official view of the IPCC, superseding previously accepted historical climate records. The IPCC placed the hockey-stick chart in the Summary for Policymakers section of the panel's Third Assessment Report, thrusting it into the public eye. But Mann's work had also attracted critics, particularly two Canadians, Ross McKitrick and Steve McIntyre. The same Steve McIntyre who recently discovered the Y2K discontinuity in NASA's GISS temperature records.

McIntyre, a businessman involved in financing speculative mineral exploration, became interested in Mann's hockey stick because of the public debate about global warming. Not a scientist, but very familiar with statistical methods and promotional material, he was struck by how similar Mann's graph looked to a typical "sales pitch" chart. McIntyre attempted to reproduce the hockey stick graph from the data and methodology presented in the 1998 paper by Mann, Bradley and Hughes (MBH98), but could not. So began a statistical odyssey that would last for several years and end up in one of the most rancorous public debates in modern science.

Eventually, McIntyre made contact with Ross McKitrick, an Associate Professor of Economics at the University of Guelph. Together, they began a correspondence with Mann, et al, that eventually led to a running debate in a number of scientific journals. The tortuous history of this public squabble has been well documented[368] and will not be recounted here. What will be presented are the arguments, voiced by McKitrick and McIntyre, as stated in the abstract from their 2005 paper:

> The "hockey stick" shaped temperature reconstruction of Mann et al. [1998, 1999] has been widely applied. However it has not been previously noted in print that, prior to their principal components (PCs) analysis on tree ring networks, they carried out an unusual data transformation which strongly affects the resulting PCs. Their method, when tested on persistent red noise, nearly always produces a hockey stick shaped first principal component (PC1) and overstates the first eigenvalue. In the controversial 15th century period, the MBH98 method effectively selects only one species (bristlecone pine) into the critical North American PC1, making it implausible to describe it as the "dominant pattern of variance." Through Monte Carlo analysis, we show that MBH98 benchmarks for significance of the Reduction of Error (RE) statistic are substantially under-stated and, using a range of cross-validation statistics, we show that the MBH98 15th century reconstruction lacks statistical significance.[369]

Eventually, the argument came to a full boil, bringing both the scientific authorities and the US Congress into the fray. First to weigh in was the American National Academy of Sciences (NAS), a society of distinguished scholars engaged in scientific and engineering research sanctioned by the American government. The NAS was signed into being by President Abraham Lincoln in 1863, with a mandate to "investigate, examine, experiment, and report upon any subject of science or art" whenever called upon to do so by any department of the government.[370]

Under its legislatively established mandate, the NAS was asked to investigate the veracity of Mann's work. A panel of experts was convened and, after due deliberation, a report was issued. Mann's supporters have claimed that the report fully exonerated his conclusions. Others found that the report, written in typically careful scientific style, was more like "damning him with faint praise." Here are the actual conclusions from the NAS report:

> Based on the analyses presented in the original papers by Mann et al. and this newer supporting evidence, the committee finds it plausible that the Northern Hemisphere was warmer during the last few decades of the 20th century than during any comparable period over the preceding millennium. The substantial uncertainties currently present in the quantitative assessment of large-scale surface temperature changes prior to about A.D. 1600 lower our confidence in this conclusion compared to the high level of confidence we place in the Little Ice Age cooling and 20th century warming. Even less confidence can be placed in the original conclusions by Mann et al. (1999) that "the 1990s are likely the warmest decade, and 1998 the warmest year, in at least a millennium" because the uncertainties inherent in temperature reconstructions for individual years and decades are larger than those for longer time periods, and because not all of the available proxies record temperature information on such short timescales.[371]

In short, it is seemingly a valid statement (i.e. "plausible"[†]) to say that things are warmer now than in the Medieval Warm Period. A stronger statement would have said they concur or agree with Mann's statement, not that it was plausible. The panel expressed a "high level of confidence" for cooling during the Little Ice Age and warming in the 20th century, conclusions that were not in dispute. But the claims made by Mann, et al., that 1998 was the hottest year "in at least a millennium" are questionable due to lack of supporting data and flaws in their work. As we have seen, 1998 was not even the hottest year in the last century. An even blunter analysis of the Mann work is offered by the Wegman Report.

This 2006 report was the result of an ad hoc committee of independent statisticians who were asked by a congressional committee to assess the statistical information presented in the Mann papers. Dr. Edward Wegman, a prominent statistics professor at George Mason University and chairman of

---

†   plausible (adj) 1: apparently reasonable and valid [ant: implausible] 2: likely but not certain to be or become true or real. Usage: Plausible denotes that which seems reasonable, yet leaves distrust in the judgment. Source http://www.dict.org.

the National Academy of Sciences' (NAS) Committee on Applied and Theoretical Statistics, headed the panel of experts. Few statisticians in the world have qualifications to rival his. Also included on the committee were Dr. David Scott of Rice University, Dr. Yasmin Said of The Johns Hopkins University, Denise Reeves of MITRE Corp. and John T. Rigsby of the Naval Surface Warfare Center.

Wegman found that Mann et al, made basic errors that "may be easily overlooked by someone not trained in statistical methodology." Further, "We note that there is no evidence that Dr. Mann or any of the other authors in paleoclimate studies have had significant interactions with mainstream statisticians." Instead, this small group of climate scientists were working on their own, largely in isolation, and without the academic scrutiny needed to ferret out false assumptions. When Wegman corrected Mann's statistical mistakes, the hockey stick disappeared. Wegman's committee found Mann's conclusions unsupportable, adding that "The paucity of data in the more remote past makes the hottest-in-a-millennium claims essentially unverifiable." In the words of the report:

> It is important to note the isolation of the paleoclimate community; even though they rely heavily on statistical methods they do not seem to be interacting with the statistical community. Additionally, we judge that the sharing of research materials, data and results was haphazardly and grudgingly done. In this case we judge that there was too much reliance on peer review, which was not necessarily independent. Moreover, the work has been sufficiently politicized that this community can hardly reassess their public positions without losing credibility. Overall, our committee believes that Dr. Mann's assessments that the decade of the 1990s was the hottest decade of the millennium and that 1998 was the hottest year of the millennium cannot be supported by his analysis.[372]

Mann's supporters rushed to criticize the Wegman report, some even claiming that the report was invalid because it was not peer-reviewed. They missed the point—the Wegman report *was* a peer-review. In fact, Wegman's analysis of the peer-review process within the close-knit climatological community calls the validity of that process into question.

More troubling is how widespread the misapplication of statistical techniques is in the broader climate-change and meteorological community. An indication of the depth of the problem is evident in the American Meteorological Society's Committee on Probability and Statistics. Again quoting Dr. Wegman: "I believe it is amazing for a committee whose focus is on statistics and probability that of the nine members only two are also members of the American Statistical Association, the premier statistical association in the United States, and one of those is a recent PhD with an assistant-professor appointment in a medical school."

While Wegman's advice—to use trained statisticians in studies reliant on statistics—may seem obvious, Mann's supporters and the IPCC faithful find the suggestion objectionable. Mann's hockey stick graph may be wrong, many experts now acknowledge, but they assert that he nevertheless came to the right conclusion.

For those who demand that climate science be held to more rigorous standards, this is the ultimate betrayal of scientific integrity. As Wegman stated in testimony before the energy and commerce committee: "I am baffled by the claim that the incorrect method doesn't matter because the answer is correct anyway. Method Wrong + Answer Correct = Bad Science." For some climate scientists, the end seems to justify the means.

The Second Pillar of Climate Science

Aside from the uncertainties in current and historical data there are many gaps in scientists' current data collection capabilities. Many factors affecting climate change are poorly understood and not captured well by current proxy or data gathering techniques. Here are comments regarding climate data from a number of experts in the field. These quotes are taken from the Geotimes web site.[373]

> "Some major gaps in our understanding of past and future climate are left by existing proxies. For example, cloud properties and atmospheric composition are poorly characterized by proxies, but that may change in the future. Recently, techniques targeted on understanding the role of sulfur in the climate system have begun to make exciting progress on these issues." — **Matt Huber, Danish Center for Earth System Science, Niels Bohr Institute in Copenhagen.**

This reinforces the often voiced criticism that cloud cover is not being properly accounted for in GCM climate predictions. In particular, the mention of new work on understanding sulfur's role in cloud formation is central to the mechanism through which cosmic rays affect climate presented in Chapter 11.

> "Measurements of water vapor and clouds and precipitation and temperature on time scales of hours and a spatial resolution of 5 kilometers in the horizontal and 1 kilometer in the vertical would

be great, so would radiosond measurements of wind throughout the tropics. Unfortunately, we don't know how to get the water vapor data. Satellite orbits rarely meet the resolution requirements. As a result, any currently envisaged global observing system would be inadequate for at least some essential components." — **Dick Lindzen, Massachusetts Institute of Technology.**

Again, the subject of clouds and water vapor is mentioned along with the spatial resolution available for such measurements. Interestingly, the 5 kilometer resolution mentioned here echoes the statement regarding the importance of sub-5 kilometer resolution in accurately modeling hurricanes presented in Chapter 14. The question of resolution is critical in accurately modeling cloud effects in GCM. Also note that water vapor, a greenhouse gas that dwarfs the contributions of CO_2, is also inadequately monitored. And, according to Lindzen, there is no prospect of achieving adequate monitoring anytime in the foreseeable future. Another related factor, also on the IPCC poorly understood phenomenon list, is the affect of aerosols, tiny particles suspended in the atmosphere.

"As for climate forcings, the big uncertainties are with aerosols, both their direct forcing and their indirect effects via clouds. These depend sensitively upon aerosol characteristics, particularly the composition and size distribution. We must have detailed monitoring of aerosol microphysics including composition specific information. It is not enough to measure the optical depth or back-scattering coefficient of the aerosols. I strongly advocate making global satellite measurements that use the full information potential in observable radiance." — **James Hansen, NASA Goddard Institute for Space Studies.**

Here, Hansen reinforces the IPCC's admission of ignorance regarding aerosols and their impact. Furthermore, he points out how incomplete modern satellite coverage actually is. Finally, Jeff Kiehl points out what this paucity of data means for climate modeling.

"There is very little data on oceans, things like a long time series of global ocean temperature. There are limited data sets that people are using, but that's an area that we certainly need a lot more data on. There are some observations on surface energy exchanges between the land surface and the atmosphere, but again that is just at certain points; it is not a global data set, which is what we would need for doing the best job evaluating the models. The models produce a lot more information than we have observations for, and this is not a satisfactory situation. You'd like to have more observations than things you are modeling, but unfortunately, it is just not the case for global modeling." — **Jeff Kiehl, National Center for Atmospheric Research.**

As we have seen, all climate data are inexact, modern measurements and historical proxies all come with margins of error. Historical records, from any period but the recent past, are inherently incomplete, unreliable, and even more recent data are subject to multiple interpretation. This is not to say that such data is bad or erroneous, just that uncertainty must always be taken into account when analyzing the data. The data on which predictions are based must always come under the closest scrutiny.

All experimental data contain some uncertainty, but the uncertainties in climate data are often larger than the predictions published by the IPCC. This is due to the extensive use of historical proxy data to try and predict future climate trends. IPCC experts have testified that "temperatures inferred using such methods have greater uncertainty than direct measurements."[374] If the first pillar of climate science, the theory, is incomplete, then the second pillar, the experimental data, must be called uncertain. Starting from this unsteady foundation, climate modelers have proceeded to construct imposing computational edifices—global climate models. These GCM computer programs are the basis of the third and final pillar, computation.

Chapter 14 The Limits of Climate Science

"There is something fascinating about science. One gets such wholesale returns of conjecture out of such a trifling investment of fact."

— Mark Twain

There are three fundamental problems that limit the effectiveness of climate science. These are lack of understanding of Earth's climate system, inherent uncertainty in baseline data, and reliance on conceptual computer models for prediction of future climate. These three problem areas correspond to the three pillars of climate science: theory, experimentation, and computation. The previous two chapters addressed the incompleteness of climate theory and the inherent uncertainty in climate data. In this chapter, we will address the third and most misunderstood pillar, computation. In the context of climate science, computation is primarily represented by climate models— GCM, complex computer programs that have been under continual development for at least a quarter of a century.

To the layperson, and even many scientists, the pitfalls and problems with modeling of any kind are unknown and unappreciated. To computer scientists who make a study of such things, modeling is fraught with danger for the uninitiated and the unwary. Nevertheless, computer modeling presents a seductive trap for many other wise skeptical scientists. The appeal of running experiments in a clean, mathematically antiseptic world from the comfort of an office can be overwhelming.

Much of the IPCC's case for rapid and accelerating temperature rise in the future is based on the predictions of computer models. Most people unquestionably accept that these results are accurate—after all, they sound very scientific and run on big super computers. What is really not discussed in the public announcements, but is well known by scientists who do computer modeling, is that models are not very accurate. Particularly when asked to make long-term projections based on limited, short-term data. To understand why this is so, we need to look at how computer modeling is done.

Why Models Aren't Reliable

A model is a simplified stand-in for some real system; a computer network, a protein molecule, the atmosphere, or Earth's entire climate. A modeler tries to capture the most important aspects of the system being modeled. For a computer network, this could be the amount of network traffic, the speed of the communication links, the way the various computers are connected to each other. For a protein molecule, it could be the types and number of different atoms, the bonds between them, the kinetic energy of the atoms, etc.

During the 1990s, Hoffman was a research professor of Computer Science at the University of North Carolina at Chapel Hill, working on biosequence and protein structure analysis funded by the Human Genome Project. A number of his colleagues were working on the related problem of molecular dynamics simulation (MD), modeling virtual protein molecules with computers. The goal of MD is to calculate the time-dependent behavior of a molecular system. To run these models, some of the fastest super-computers of the time were used. Even so, running the model programs could take weeks and yield only a few milliseconds of simulated time.

Even relatively small protein molecules consist of several hundred atoms. Molecular dynamics simulation is a type of problem called an N-body problem. This is because every atom in the molecule affects all the other atoms. Computer scientists have a concept called *computational complexity,* a way of formally stating how hard it is to solve a computational problem. More specifically, how the amount of time required to perform the computation changes with increasing problem size. N-body problems have a computational complexity of $O(N^2)$, pronounced "big O, N squared." In practical terms, this means if you double the number of atoms in your molecule, the time needed to run the simulation will be two squared (2^2) or four times the previous value. Four times the number of atoms would require 16 times (4^2) the computer time.

Because the number of atoms in a specific protein cannot be reduced, computer scientists look for other ways to make their models run faster. One parameter that can be adjusted is called the *time step.* All computer models execute in a number of discreet steps. In an MD program, starting with the relative positions of all the atoms, all of the forces acting on each atom are calculated. The effects of these forces acting on the atoms over a short period of time, the time step, are calculated. This results in new positions for each atom in the protein molecule. The process then repeats for the next time step. So, if you double the length of the time step, you can cut the required calculations in half.

Of course, nothing is free and the cost of lengthening the time step is a loss of accuracy. This introduces error into the calculations. What is worse, computer programs also suffer from error propagation. This means that any calculation, where the values used contain errors, will result in answers that contain errors as well. Models that simulate systems over time, like MD and GCM programs, use the output of one time step as input to the next time step's calculations. The result can be ever increasing error that eventually causes the model output to become totally useless.

Hoffman witnessed an example of this at a conference held at the North Carolina Supercomputing Center, in Research Triangle Park. Several of his colleagues presented the results of their efforts to simulate a simple protein molecule surrounded by water at body temperature. Showing the model output as a cartoon movie of the molecule pulsing and vibrating, interacting with the surrounding water molecules, their first example did something spectacular—the protein molecule exploded into several pieces.

This was obviously not correct, since the protein in question was known to be stable at body temperature. The second example was much better, at least the protein didn't tear itself apart. What was the difference between the two simulations? The time step used. As it turned out, if a time step greater than two femtoseconds was used, propagated error built up until the molecule self-destructed. A femtosecond is an extremely short period of time. For a computer with a clock rate of 1GHz, every tick of its clock takes one nanosecond, or one billionth of a second. A femtosecond is one millionth of a nanosecond. If a femtosecond took one second, a second would last about 32 million years.

This story illustrates some of the problems inherent in computer modeling, regardless of the physical system that is being studied. Even very small changes to the model's parameters can cause the output to change from realistic to catastrophically wrong. Identifying the source of the problem in a model is often a matter of trial and error—this was the case with the MD simulation. The MD researchers were lucky, their model was obviously giving the wrong answers because the real molecule didn't act like the simulated one. They had volumes of reliable baseline data to work from—this is often not the case for other models. There is an old truism in computer science: "garbage in, garbage out." The trick is being able to recognize garbage.

Modeling the atmosphere is even more complex, requiring knowledge of incoming solar radiation, the movement of air currents over the land and sea, heat convection, the amount of water vapor, the effects of clouds, and on and on. People have been trying to model Earth's atmosphere for decades, primarily to predict the weather. The weather forecasts you hear on your evening news are all based on computer models. How accurate are these models? In the near term, a few days from today, local weather forecasts are about 60% accurate when predicting high temperatures.[375]

253

Hurricane models suffer from the same problems as GCM programs. Because hurricanes often intensify or lose strength quickly, models have trouble accurately predicting their strength. According to Hugh Willoughby, an atmospheric scientist at Florida International University, if a model's data points are not closer than 5 km apart, the simulated storms end up "larger, weaker cartoons of their counterparts in nature."[376]

Storm track prediction is an example of *quantitative modeling*, where the expected results of a model are hard numbers. In the case of hurricanes, a storm's track and changes in strength over time are the desired results. Climate change modeling is usually an example of *qualitative modeling*. These types of models result in general trends and overall effects of parameter modification. They are used to provide insight into processes where scientists' intuition fails. A qualitative model can tell us that adding more CO_2 to the atmosphere will cause warming. But qualitative models should not be used to make concrete predictions of future conditions, such as the global average temperature for the next 100 years. Or, as Richard W. Hamming[†] put it, "The purpose of computing is insight, not numbers."

Climate modelers will protest that their models are not the same as short-term weather forecasting models or hurricane path models. They are correct, longer term models are more complicated. There are a number of services, both governmental and commercial, that do longer term predictions. Long-term, meaning for the next season or the next year. These claim to be 80-85% accurate, but they usually concentrate on trend predictions, how many days will it be dry or rainy, how many days will have above-average temperature. An 85% accuracy sounds pretty good—but this is only for a year or so into the future. What do the professional climate predictors say about looking farther into the future? According to the *Weather 2000* web site:

> "Trends can be misleading. Examining 30, 40 or even 50 years worth of historical data might only encompass 10 - 20% of the full potential of climate variability. Since quality data only goes back 50 years at best, standard deviations and extreme records based on that data can be gross underestimations, and trends can overestimate the true climate state."[377]

The same source goes on to say, "It is very dangerous to draw conclusions based on the most recent 5 or 10 years worth of historical data." Remember that this is with regard to one year predictions, the IPCC models are trying to predict 100 years or more into the future.

---

† Richard Wesley Hamming (1915-1998) American mathematician and computer scientist, Turing Prize winner most noted for developing error correction codes.

Sources of Modeling Error

There are numerous sources of error that can impact the accuracy of a model. Peter Haff, Professor of Geology and Civil and Environmental Engineering at Duke University, lists seven sources:

1. Model imperfection.

2. Omission of important processes.

3. Lack of knowledge of initial conditions.

4. Sensitivity to initial conditions.

5. Unresolved heterogeneity.

6. Occurrence of external forcing.

7. Inapplicability of the factor of safety concept.

Several of these points seem obvious; if the model is flawed, if important parts have been left out, or the initial conditions are incorrect, no model can provide trustworthy answers. Pilkey states, "it is an axiom of mathematical modeling of natural processes that only a fraction of the various events, large and small, that constitute the process are actually expressed in the equations."[378] The other points may not be so self-evident.

Sensitivity to initial conditions is a result of non-linear responses in the system being modeled. In engineered systems, these types of unexpected and unplanned for responses are called *emergent* behavior. In natural systems, this type of behavior is often called *chaotic.* The discovery of chaotic behavior in 1960, by meteorologist Edward Lorenz while developing a weather model, led to the establishment of a new scientific discipline— chaos theory. Lorenz had constructed a computer model, with a set of twelve equations, to study weather prediction. One day, he attempted to restart a

Illustration 129: Lorenz's experiment: the difference between the starting point of the two curves is 0.000127. Source Ian Stewart.

prediction run from an intermediate point using values from a printout. To his surprise, the second curve deviated wildly from the initial run. The difference was eventually traced to the number of digits in the parameters used to restart the simulation—by dropping the last few digits from the parameter values, Lorenz had inadvertently discovered the non-linearity lurking in his model.[379] Since Lorenz's experiment, chaotic behavior has been found in many natural systems.

Heterogeneity, the quality of being diverse or made up of many different components, tends to increase with the size of the system being modeled. A model that may work well for some geographic regions, may fail simulating others. Desert environments are different from woodlands, open ocean different from coastal areas, mountains different from grassy planes. Heterogeneity tends to make large systems hard to model, and Earth's climate system is very large.

Earth's climate system is definitely affected by external forcing factors; radiant energy from the Sun most obviously. As we have seen, physicists suspect that complex interactions involving cosmic radiation, the solar system's path through the galaxy, and the Sun's magnetic field, have significant impact on low-level cloud formation. This, in turn, affects Earth's albedo, one of the most sensitive regulators of climate. Models that cannot account for cloud cover, cannot begin to address the effects of this external forcing factor. Yet, many climate scientists reject the possibility that these factors are important and refuse to consider them.

The concept of a safety factor comes from engineering, where more robust systems are built by designing in margins of error—"over engineering" them. Bridges are built to support more than their expected load, aircraft wings are designed to safely bend far beyond the range expected in normal flight, scuba tanks are constructed so they won't burst even if they are filled to pressures higher than their rated level. But, environmental systems are not designed, they are created by the interactions of natural forces.

In environmental modeling, the equivalent of a factor of safety is the "worst-case scenario" prediction. Unfortunately, "where relationships between variables are highly nonlinear, and a clear understanding of the controlling variables does not exist, as in many natural and environmental systems, the factor of safety concept is inapplicable."[380] As we have seen, climate models are highly nonlinear and many relationships among the controlling variables remain unknown.

Even addressing these factors may not be sufficient to ensure an accurate model. Dr. Haff has stated, "sources of uncertainty that are unimportant or that can be controlled at small scales and over short times become important in large-scale applications and over long time scales."[381] These factors, combined with the nonlinear nature of natural systems, which leads to emergent (unexpected) behavior, may mean climate systems cannot be modeled at all.

Computational Error

Even if the data used to feed a model was totally accurate, error would still arise. This is because of the nature of computers themselves. Computers represent real numbers by approximations called *floating-point* numbers. In nature, there are no artificial restrictions on the values of quantities but in computers, a value is represented by a limited number of digits. This causes two types of error; representational error and roundoff error.

Representational error can be readily demonstrated with a pocket calculator. The value 1/3 cannot be accurately represented by the finite number of digits available to a calculator. Entering a 1 and dividing by 3 will yield a result of 0.33333... to some number of digits. This is because there is no exact representation for 1/3 using the decimal, or base 10, number system. This problem is not unique to base 10, all number systems have representational problems.

The other type of error, roundoff, is familiar from daily life in the form of sales tax. If a locale has a 7% sales tax and a purchase totals $0.50 the added tax should be $0.035 but, since currency isn't available for a ½ cent, the tax added is only $0.03 dropping the lowest digit. The tax value was truncated or rounded down in order to fit the available number of digits. In computers, multiplication and division often result in more digits than can be maintained by the machine's limited representation. Arithmetic operations can be ordered to minimize the error introduced due to roundoff, but some error will still creep into the calculations.

After the uncertainties present in data are combined with errors introduced by representation and roundoff, a third type of computational error arises— propagated error. What this means is that errors present in values are passed on through calculations, propagated into the resulting values. In this book, values have often been presented as a number followed by another number prefaced with the symbol "±" which is read "plus or minus." This notation is used to express uncertainty or error present in a value: 10±2 represents a range of values from 8 to 12, centered on a value of 10.

When numbers containing error ranges are used in calculations, rules exist that describe how the error they contain is passed along. The simplest rule is for addition: the two error ranges are added to yield the error range for the result. For example, adding 10±2 to 8±3 gives a result of 18±5. There are other, more complicated rules for multiplication and division, but the concept is the same. When dealing with complicated equations and functions, like sines and cosines, how error propagates is determined using partial differential equations. The mathematics of error propagation rapidly becomes very complex and, as seen in the MD example related above, errors can build up until the model is overwhelmed.

```perl
#!/usr/bin/perl
#
for ($n=1; $n<11;$n++) {
    $x[$n] = 1/$n;
    $results[$n][1] = $x[$n];

    for ($j=0; $j<10; $j++) {
        $x[$n] = ($n+1)*$x[$n] - 1;
    }
    # --- 10x result
    $results[$n][2] = $x[$n];

    for ($j=0; $j<20; $j++) {
        $x[$n] = ($n+1)*$x[$n] - 1;
    }
    # --- 30x result
    $results[$n][3] = $x[$n];
}

for ($n=1; $n<11;$n++) {
    printf "%2d  %-18.16g  %-18.16g  %-18.16g\n",
        $n,$results[$n][1],$results[$n][2],
        $results[$n][3];
}
```

Text 1: Perl source code to demonstrate computational and propagated error.

To demonstrate how these sources of computational error can overwhelm the results of even simple calculations, consider the following example. This example cannot be explained without using equations, but understanding them is not essential—only understanding the final result is important. Having said that, let n be a positive integer and let $x(n)=1/n$. Instructing a computer to compute $x=(n+1)*x-1$ should not change the value of x. Source code for this program, written in the Perl language, is given in Text 1.

This program will generate a table of numbers, with the number n varying from 1 to 10. The value the function x is calculated for each value of n. Then the equation $x=(n+1)*x-1$ is computed first ten and then thirty times. Each time the equation is computed, the newly calculated value of x replaces the old value, so that the new value of x from each step is used as the input value of x for the next step. This process is called iteration. If the computer's internal representation of x is exact, the values produced by the second equation should not change.

```
n    x = 1/n               10 iterations         30 iterations
---  -------------------   -------------------   -------------------
1    1                     1                     1
2    0.5                   0.5                   0.5
3    0.3333333333333333    0.3333333333139308    -21
4    0.25                  0.25                  0.25
5    0.2                   0.2000000017901584    6545103.021815777
6    0.1666666666666667    0.1666666606931329    -476641800.7969146
7    0.1428571428571428    0.1428571343421936    -9817068105
8    0.125                 0.125                 0.125
9    0.1111111111111111    0.1111111604543567    4934324553889.695
10   0.1                   0.1000002094278254    140892568471739.2
```

Text 2: Results from running the example program.

The output from running the example program is shown in Text 2. The value of n is given in the first column and the initial value of x is in the second column. The values of x after 10 and 30 iterations are given in columns three and four, respectively. Notice that for some values of n the computed values of x do remain unchanged, but for others the results diverge—slightly after 10 iterations, then wildly after 30.

The reason that the x values for 1, 2, 4, and 8 do not diverge is because computers use binary arithmetic, representing numbers in base 2. The rows of results that did not diverge began with numbers that are integer powers of 2. These numbers are represented inside the computer exactly, and the iterative computation does not change the resulting values of x. The other numbers cannot be represented exactly so the computation blows up. The same thing happens in any computer program that perform iterative computations—like climate simulation models.

If all of these complicating factors—data errors, incomplete and erroneous models, non-linear model response, roundoff and representational error, and error propagation—are not daunting enough, different computer hardware can introduce different amounts of error for different arithmetic operations. This means that running a model on a Sun computer can yield different results than an Intel based computer or an SGI computer.[382] To say the least, computer modeling is not an exact science.

Modeling Earth's Climate

Earth's climate is far too complex for the human intellect to fully encompass —no set of equations can capture it completely. A normal procedure in science is to isolate parts of a more complex system in hopes of understanding the smaller, simpler pieces. Computer scientists call this type of approach "divide and conquer," breaking down a problem into sub-problems until they become simple enough to be solved directly. The fundamental assumption with this problem solving approach is that the parts, when reassembled into a whole, will accurately reflect the original problem. This is often an invalid assumption—it certainly is when dealing with Earth's climate.

Modeling inaccuracy occurs because many of the processes that influence climate also influence each other. We saw examples of positive feedback mechanisms in earlier chapters:

- Increasing CO_2 levels cause increasing temperatures, resulting in the release of more CO_2.

- A colder climate increases the area covered with highly reflective ice and snow, causing Earth to cool, causing more ice and snow.

- Melting sea ice, which is highly reflective, exposes more open ocean, which has a much lower albedo, causing greater warming.

But, we have also seen examples of negative feedback where warming creates more water vapor that increases snowfall in colder regions leading to cooler climate. Sometimes the same response of nature can result in

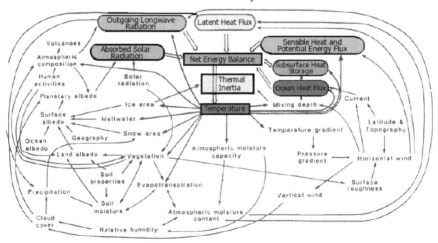

Illustration 130: Flow diagram of the climate system illustrating the massive and complicated physical processes and multiple feedback loops. Source: Robock, 1985.

contradictory effects. For example; increasing forests, which absorb CO_2, cooling the planet, also decreases Earth's albedo, leading to greater warming. So, are forests good or bad with respect to global warming? Illustration 130 shows some of the feedback relationships in Earth's climate system.

A simpler view of the main climate factors is shown as a block diagram in Illustration 131. The actions of feedback linkages are represented by the heavy black arrows. Even this simplified view demonstrates how complex and confusing Earth's climate mechanisms are.

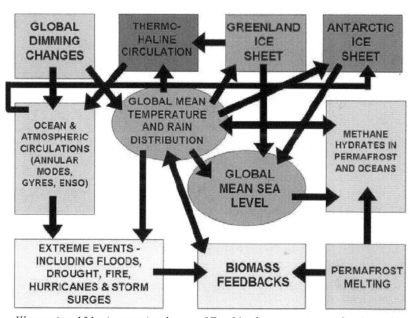

Illustration 131: interacting loops of Earth's climate system. After Pittock.

We have also seen that some mechanisms do not respond gradually, but are instead non-linear. The great ocean conveyor belt, the global circulation of water that redistributes heat around the globe, is known to have been disrupted by glacial melt water during the early years of the Holocene warming. The results of these disruptions were several brief, sharp returns to colder conditions. Scientists fear this might happen again if the climate continues to warm, but they have no way of predicting such an event.

Many climate mechanisms have sensitive dependence on initial conditions, the so-called "butterfly effect." Over time, systems that show this type of response become unpredictable. This concept is often illustrated by a butterfly flapping its wings in one area of the world, causing a storm to occur much later, in another part of the world. Such nonlinear systems are

261

central to chaos theory, exhibiting fantastically complex and unpredictable behavior. According to the IPCC:

"The physical climate system is highly complex, has aspects that are inherently chaotic, and involves non-linear feedbacks operating on a wide variety of time scales. Our empirical knowledge of how these operate ranges from being good on decadal time scales, moderate over time scales of 100-1000 years, to being quite limited at 10000 years and longer."[383]

From the shaky foundation of incomplete theory, with input and defining data from an uncertain and error laden climate history, climate scientists have boldly constructed GCM computer programs, confidently basing their public pronouncements on the models' dubious predictions.

GISS modelE

The GCM used to provide predictions for the IPCC's latest report is called *GISS modelE*, specifically model III, a version of modelE frozen in 2004. ModelE is actually a composite of an atmospheric model and four different ocean circulation models, along with other factors. ModelE is driven by ten measured or estimated climate forcings. A large team of NASA investigators, led by Hansen and Ruedy, produced an in-depth, eye-opening study of modelE's strengths and weaknesses in their 2006 paper titled "Climate Simulations for 1880-2003 with GISS Model E."

As Jeff Kiehl noted, "models produce a lot more information than we have observations for, and this is not a satisfactory situation." This has not stopped the IPCC, and others, from using models to produce a deluge of data with which to inundate the public and the media. Some of the illustrations in the Hansen report are revealing. For instance, Figure 1 from the report shows the relative strengths and growth in the primary forcing used to drive

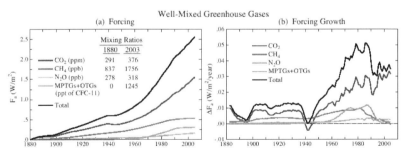

Illustration 132: Growth of GHG, the primary forcing for modelE. Source Hansen et al, 2007.

modelE, well-mixed greenhouse gases (Illustration 132). It should come as no surprise that CO_2 concentration dominates.

When all the driving forcings are viewed together, the result is still clear—CO_2 is the main driving factor of modelE. This is shown in Figure 5 from the report (Illustration 133). Notice that there is a conspicuous lack in forcing variability during the 1940s, a period of time when global temperature hit a peak and then went into decline. "The model's fit with peak warmth near 1940 depends in part on unforced temperature fluctuations," states the report, adding "It may be fruitless to search for an external forcing to produce peak warmth around 1940." This is an admission that modelE is incapable of accurately reproducing the observed temperature fluctuations of the past 120 years. Yet we are asked to accept that the IPCC's projections for the next 100 years, using modelE, are accurate.

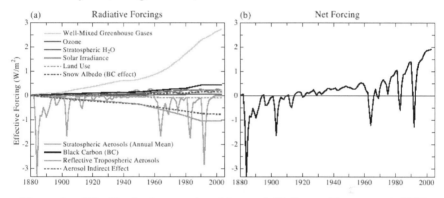

Illustration 133: Net forcings used to drive modelE. Source Hansen et al, 2007.

They also report that the "greatest uncertainties in the forcings are the temporal and spatial variations of anthropogenic aerosols and their indirect effects on clouds." This statement reinforces the concerns of others regarding the inability of current models to accurately represent cloud cover and cloud formation mechanisms due to weak theory and even weaker experimental data. Recall that the IPCC's report listed understanding of aerosols' role in climate regulation at only 10% (Illustration 5, page 24).

Graphs of output from modelE illustrate the model's problem areas. This is shown in Figure 6 from the Hansen report, reproduced in Illustration 134. Atmospheric temperatures, shown in the three graphs on the left, are mostly in agreement with historical measurements but become less accurate in the most recent two decades. It is interesting that these values become more inaccurate just when CO_2 levels start rising. Surface temperature, shown in the upper right hand graph, performs significantly worse. The model does not accurately reproduce the variability of recorded surface temperature and there are decade-spanning periods where it consistently underestimates or over-estimates surface temperatures. But worst of all are the predictions for ocean ice coverage, which overestimates sea ice for one hundred years, and then underestimates it for the past twenty years. Still, the true accuracy of the model is hard to discern from the data plots.

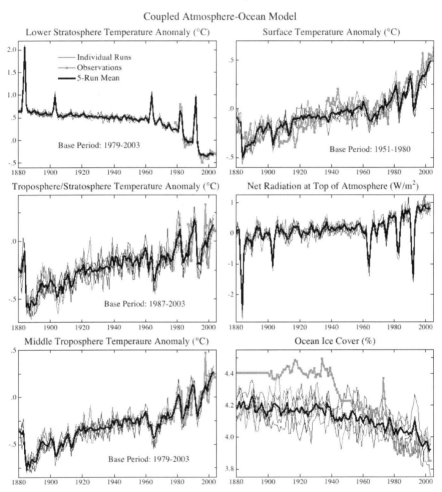

Illustration 134: Results from modelE taken from Hansen et al, 2007, figure 6.

A better idea of the accuracy of modelE's backcast predictions for the past 120 years can be found in the Hansen report's written narrative. This fifty page report contains a frank assessment of the accuracy of the IPCC's favorite GCM. Here are some of the major problems cited in the study:

- The atmospheric model produces polar temperatures as much as 5-10°C too cold in the lower stratosphere during winter and the model produces sudden stratospheric warming at only one quarter of the observed frequency.

- A 25% regional deficiency in summer stratus cloud cover of the west coasts of the continents—resulting in excessive absorption of solar radiation of as much as 50 W/m².

- A net deficiency in solar radiation absorbed over tropical regions of 20 W/m².

- Sea level air pressure readings are too high by 4-8 hPa[†] during winter in the Arctic and 2-4 hPa too low year-round in the tropics.

- A 20% shortfall in rain over the Amazon basin.

- A 25% deficiency in summer cloud cover in the western United States and central Asia, causing summer temperatures to be ~5°C too high in these regions.

- Global sea ice cover is deemed realistic but the distribution of sea ice is not, with too much sea ice in the Northern Hemisphere and too little in the Southern Hemisphere.

- Too much sea ice remains in the Arctic summer affecting climate feedback rates.

- Unrealistically weak tropical El Niño-like variability.

Despite all these shortcomings, modelE "fares about as well as the typical global model in the verisimilitude of its climatology" in IPCC model comparisons.[384] We have already mentioned the study of 108 different models that found predicted temperature increases of 0.29-15.6°F (0.16-8.7°C)[385] so perhaps modelE does fare about as well as other models.

In all, Hansen's report is a blunt admission that, after 30 years of development and more than $50 billion in funded research, current climate models are just not up to the job. In fact, if the predictions of the Third Annual Report are compared with the predictions in Annual Report 4, the IPCC's efforts are moving in reverse—the new predictions in AR4 have a wider range of uncertainty than those in the TAR (1.1-6.4°C vs 1.5-5.8°C).

---

† Hectopascal (hPa), a unit used worldwide for air pressure. One hectopascal corresponds to about 0.1% of atmospheric pressure at sea level.

Modeling Invalidated

In a famous paper, entitled "Ground-water models cannot be validated," Leonard F. Konikow and John D. Bredehoeft, state that some natural systems cannot be accurately modeled. They state, "testing the predictive capability of a model is best characterized as calibration or history matching; it is only a limited demonstration of the reliability of the model."[386] If testing a model cannot prove its ability to provide accurate predictions then there is no way climate modeling, as a technique, can be trusted. They summarize their findings, saying the emphasis in trying to understand natural processes should shift "away from building false confidence into model predictions."

Konikow and Bredehoeft are not members of the scientific fringe. Their paper won the Meinzer prize from the Geological Society of America, the highest honor in the field of hydrogeology in the US.[387] They are not the only modeling experts to warn about the fundamental problems of modeling complex natural processes.

Hendrik Tennekes, Professor of Aeronautical Engineering at Pennsylvania State University and the former director of research at the Royal Dutch Meteorological Institute, was a pioneer in multi-modal weather forecasting. Though a strong proponent of scientific modeling, he has challenged the use of unproven scientific models to predict the future course of global warming. Pointing to the complexity of Earth's climate and the incomplete nature of climate models, Tennekes has said, "if I try to look at climate modeling from this perspective, I'm almost fainting."[388]

An expert in modeling complex systems, Tennekes scoffs at those who think that computer models which implicate CO_2 levels as the primary control for Earth's climate can be trusted, or that trying to regulate the world's temperature by manipulating CO_2 levels is a rational idea. In a newspaper op-ed piece, printed in the *Amsterdam De Volkskrant* on March 28, 2007, Professor Tennekes takes those who believe that CO_2 can be used to control climate to task.

> "I protest vigorously the idea that the climate reacts like a home heating system to a changed setting of the thermostat: just turn the dial, and the desired temperature will soon be reached. We cannot run the climate as we wish. That is fortunate, because a bad season for farmers may be a boon for the tourist industry, deteriorating conditions for French farmers may mean improving conditions for their Polish colleagues, what is good for winter wheat may make things worse for corn, and so on. We are not dealing with a machine, but with Nature herself, and she is not easily mocked."[389]

The limitations of computer models are well recognized by many of the scientists working on climate predictions. This includes scientists associated with the IPCC reports, such as Michael Oppenheimer, Professor of Geosciences and International Affairs at Princeton University. Dr. Oppenheimer was a lead author and contributing author to the IPCC Fourth Assessment Report, and was a lead author or contributing author to various chapters of the Second and Third Assessment Reports of IPCC. Here are some comments taken from an article by Oppenheimer in *Risk Analysis*:

"As with WAIS, the ice mass balance depends on the difference between the ice loss rate and ice accumulation, but the models have not predicted the current loss rate correctly, particularly at the GIS periphery. It is noteworthy that a local acceleration of GIS in response to surface melting that appears to lubricate the ice sheet base has been observed. This process is not incorporated in ice sheet models, and we have no idea how generally it might function over the ice sheet."[390]

The abbreviation WAIS stands for "Western Antarctic Ice Sheet" and GIS for "Greenland Ice Sheet." Note that this comment concerns ice sheet models—only one component of building a comprehensive model for Earth's entire climate system. Even addressing a relatively simple sub-problem, the models used are incomplete, give incorrect answers, and the researchers have no idea how the missing factors would affect their results.

Perhaps the best synopsis of climate modeling comes from a scientific outsider, Freeman Dyson, Professor of Physics at the Institute for Advanced Study, in Princeton:

"My first heresy says that all the fuss about global warming is grossly exaggerated. Here I am opposing the holy brotherhood of climate model experts and the crowd of deluded citizens who believe the numbers predicted by the computer models. Of course, they say, I have no degree in meteorology and I am therefore not qualified to speak. But I have studied the climate models and I know what they can do. The models solve the equations of fluid dynamics, and they do a very good job of describing the fluid motions of the atmosphere and the oceans. They do a very poor job of describing the clouds, the dust, the chemistry and the biology of fields and farms and forests. They do not begin to describe the real world that we live in. The real world is muddy and messy and full of things that we do not yet understand. It is much easier for a scientist to sit in an air-conditioned building and run computer models, than to put on winter clothes and measure what is really happening outside in the swamps and the clouds. That is why the climate model experts end up believing their own models."[391]

The IPCC Report Reexamined

Computation is widely recognized as one of the three pillars of modern science, along with theory and experiment. For climate science, theory translates to understanding how Earth's climate system works, experiment is the collection of data from various sources, and computation is computer modeling. As we have seen, all three pillars of climate science are weak and wobbly.

Here we see the limits of climate science: incomplete, subjective data are used to feed simplistic, unverifiable models in an attempt to make predictions about a complex system that is only partially understood. Climate science is simply too immature to be relied on for definitive predictions about future climate change. Despite widespread recognition of the pitfalls and limitations of modeling, the IPCC would ask us to base worldwide technological, environmental, economic and political policy on model predictions.

In an issue of the Royal Society's journal, *Philosophical Transactions,* an article appeared that discussed the difference in Pascal's approach to statistics and the Bayesian approach. Pascal considered each statistical event, such as a coin toss, as independent of any previous events, while Bayes allowed prior events to influence subsequent ones. Scientists almost always use the statistically independent model, in order to avoid having bad assumptions bias their models. Unfortunately, including prior assumptions into climate models is impossible to avoid. By their nature, climate models contain a multitude of variables that are interrelated in complex ways—effectively, Bayesian assumptions. To quote from the *Economist* regarding this situation:

> "Climate models have hundreds of parameters that might somehow be related in this sort of way. To be sure you are seeing valid results rather than artifacts of the models, you need to take account of all the ways that can happen. That logistical nightmare is only now being addressed, and its practical consequences have yet to be worked out."[392]

The use of computer models has been passed off as science fact when it is actually a technique used when real observations and genuine understanding are not available. As fundamentally weak as the IPCC's methodology is, the press and the public are convinced that the facts have been established and consensus has been reached in the scientific community. What do scientists think about model-based climate science? Again quoting from Dr. Pilkey and Dr. Pilkey-Jarvis:

"We believe that global change modelers fall into two categories. There are the true believers who take no prisoners, believe every word, every model prediction, and feel that criticism is unwarranted or even un-American. A much larger group is uncomfortably aware of the insurmountable nature of the complexities in global change models."[393]

Consensus? Hardly. Still, a continual parade of boffins appears before government agencies and the public urging action. On National Public Radio, on May 31, 2007, NASA Administrator Michael Griffin commented on the American space agency's role in fighting global warming:

"I have no doubt that … a trend of global warming exists. I am not sure that it is fair to say that it is a problem we must wrestle with. To assume that it is a problem is to assume that the state of Earth's climate today is the optimal climate, the best climate that we could have or ever have had and that we need to take steps to make sure that it doesn't change. First of all, I don't think it's within the power of human beings to assure that the climate does not change, as millions of years of history have shown. And second of all, I guess I would ask which human beings — where and when — are to be accorded the privilege of deciding that this particular climate that we have right here today, right now is the best climate for all other human beings. I think that's a rather arrogant position for people to take."[394]

In all, a thoughtful and measured reply to the host's question, "do you have any doubt that this is a problem that mankind has to wrestle with?" Before the day was out, Steven Edwards, writing on the *Wired Magazine* blog, asked, "Does this guy deserve his position as NASA's administrator?"[395] We would guess that Edwards, a non-scientist, falls into the true believer, take no prisoners category.

The subject has become so politicized that rational public debate is no longer possible. Battle lines are drawn, positions cast in concrete, and the charge is led by the IPCC. Bob Carter, a research geologist and Professor of Geology at James Cook University, Townsville, offers his opinion of the IPCC:

"It's received advice from many excellent scientists and they're still involved with the IPCC, but it's not primarily a scientific body. In the end, it's a political body with a life of its own and, as such, the advice to policy makers that the IPCC releases no longer gives primacy to scientific reasoning. It actually gives primacy to political advice."[396]

Here is the crux of the matter—political interests have taken immature and incomplete science, and inflated it into a global crisis. We need to be mindful of the environment. Mankind should eliminate excessive emissions of any type. But, we do not need to follow the tainted advice of bureaucrats and politicians who have turned scientific conjecture into a crusade in favor of their own opinions and a jihad against those whose ideas they dislike.

Chapter 15 Prophets of Doom

"The whole aim of practical politics is to keep the populace alarmed—and hence clamorous to be led to safety—by menacing it with an endless series of hobgoblins, all of them imaginary."

— *H. L. Mencken*

The main problem with translating the "science speak" of the IPCC reports into everyday language is that scientists are cautious by nature and not prone to making unequivocal, blanket predictions. You will never hear "by next Tuesday the ocean will rise 10 feet, three hurricanes will strike Florida, and the temperature in New York will be 123°F." Instead you get long-term possibilities with the odds of them actually happening dependent on other events taking place, like an average temperature rise of so many degrees. The temperature rise is likewise dependent on still other events and also stated with "degrees of certainty," meaning more probabilities. It is when the bureaucrats, government officials and news media get involved that probabilities become absolute predictions of disaster.

Before proceeding, let us restate some findings of *The Resilient Earth* thus far:

- Earth is warming slowly—about 1.8°F (1°C) per century.

- The greenhouse effect is real and CO_2 is a greenhouse gas.

- CO_2 is much less potent than methane or NO_x and is far less plentiful than the primary GHG, water vapor.

- Based on empirical data, carbon dioxide is probably responsible for only a quarter of the past century's temperature increase.

- The theory behind how Earth's climate functions is incomplete and incapable of explaining all observed climate variation.

- Climate history data are inherently inexact and are constantly being revised. Error margins for historical temperature measurements are of the same magnitude as the change observed over the past century.

- The IPCC's predictions of future temperature increase are not based on empirical data but rather on the output of complex computer models that are even less complete than scientists' theoretical understanding of Earth's climate.

- New theories explaining climate change are constantly being proposed, though they have been mostly ignored by mainstream climatologists.

The IPCC is not the only scientific organization to make pronouncements regarding global climate change. In recent years, several major scientific bodies in the United States and around the world have issued similar statements. Such blanket statements tend to downplay legitimate dissenting opinions. Before examining global warming coverage in the non-scientific media, it is instructive to briefly consider the scientific literature.

In 2004, Naomi Oreskes, a historian, published a study based on article abstracts of 928 papers published in refereed scientific journals. All the papers were published between 1993 and 2003, and listed in the Institute for Scientific Information (ISI) citation database with the keywords "climate change." The papers were divided into six categories: explicit endorsement of the consensus position, evaluation of impacts, mitigation proposals, methods, paleoclimate analysis, and rejection of the consensus position.

According to the study, 75% of the papers fell into the first three categories, either explicitly or implicitly accepting the consensus view; 25% dealt with methods or paleoclimate, taking no position on current anthropogenic climate change. Authors might believe that current climate change is natural but, according to Oreskes, none of the papers argued that point. The study concluded, "this analysis shows that scientists publishing in the peer-reviewed literature agree with IPCC."[397]

It should be noted that the definition of the "consensus" position used for this study does not require agreement that man is the primary cause of warming. It also does not require belief in or support for catastrophic global warming of the type so luridly reported in the media. This weakened version of the IPCC claims allowed the report to find higher levels of agreement than actually exists for the more radical IPCC conclusions. Even so, Oreskes' work seems to strongly reinforce the claim of consensus among scientists with regard to global warming. The study has been repeatedly cited, though some of its data are nearly 15 years old and its conclusions somewhat dated.

Since the Oreskes' study, Klaus-Martin Schulte updated this research using more recent publications. Using the same ISI database and search terms as Oreskes, Schulte examined papers published from 2004 to February 2007. Of 528 total papers on climate change during this period, only 38 (7%) gave an explicit endorsement of the IPCC backed consensus position.

If one considers "implicit" endorsement—accepting the consensus position without explicitly saying so—support for the IPCC position rises to 45%. The largest category (48%) are neutral papers, refusing to either accept or reject the hypothesis, while 32 papers (6%) reject the IPCC's assertions outright. Of all papers published during this period, only a single one makes any reference to climate change leading to catastrophic results. This does not meet the definition of "consensus."

The jury is still out on human-caused global warming, at least for the IPCC's version of it. Surveying the scientific literature cannot muster a majority, let alone consensus, supporting the IPCC's position. To quote Dr. Schulte, "If unanimity existed in the peer-reviewed literature between 1993 and 2003—which I have reason to doubt—it certainly no longer exists today."[398] It also bears repeating, even if consensus existed, it would not constitute scientific proof. Non-scientists who advocate the IPCC position unfailingly fall back on the "there is consensus" argument because they do not possess the technical knowledge to support their beliefs. Keeping this in mind, we present a survey of global warming as reported by the media, politicians and other non-scientists.

The Media Reports on Global Warming

The IPCC case for global warming and the predicted effects are fairly troubling, but the claims themselves are made in cautious language, befitting a report written by scientists and edited by bureaucrats. What is more alarming is the way this matter is reported in the press.

Since science isn't usually front page stuff with newspapers and magazines, the advent of the global warming crisis has thrust a number of new voices forward and caused some of the established, mainstream reporters to bone up on their climate science. Scientists rarely make bold, solid predictions and are not practiced in the art of the sound bite. This leads to news interpreted, and made more exciting, by reporters and editors. Here are some examples:

- "The Planet NASA Needs to Explore — As momentum gathers to reinvigorate human space missions to the moon and Mars, we risk hurting ourselves, and Earth, in the long run. Our planet -- not the moon or Mars -- is under significant threat from the consequences of rapid climate change. Yet the changing NASA priorities will threaten exploration here at home." Tony Haymet, Mark Abbott and Jim Luyten, *Washington Post*, 10 May, 2007.

273

- "Global Warming Expert Fears 'Refugee Crisis' — Within two or three decades, there could be one and a half billion people without enough water, according to a new report on the impacts of global warming." Bill Blakemore, ABC News, 2 April, 2007.

- "6 Ex-Chiefs of E.P.A. Urge Action on Greenhouse Gases — The panel said that the Bush administration needs to act more aggressively to limit the emission of greenhouse gases linked to climate change." Michael Janofsky, *New York Times*, 19 January, 2006.

- "About ten percent of the Earth's surface is covered by ice, most of that in the polar regions. But if enough of that ice melts, the seas will rise dramatically and the results will be calamitous. Scientists are keeping a watchful eye on the largest concentrations of ice on the planet: Greenland and Antarctica... If this worst-case scenario should occur, in the coming centuries New York could be abandoned, its famous landmarks lost to the sea." Tom Brokaw, NBC/Discovery Channel special on Global Warming, aired 16 July, 2006.

- "Angry environmentalists are denouncing the Bush administration for censoring the scientific evidence on global warming," John Roberts, CBS News Chief White House Correspondent, 19 June 2003.

- "Fluorescent Bulbs Are Known to Zap Domestic Tranquillity — Energy-Savers a Turnoff for Wives. ...one of the dimly lighted truths of the global-warming era is that fluorescent bulbs still seem to be flunking out in most American homes." Blaine Harden, *Washington Post*, 30 April, 2007.

- "The top climate scientist at NASA says the Bush administration has tried to stop him from speaking out since he gave a lecture last month calling for prompt reductions in emissions of greenhouse gases linked to global warming." *New York Times*, 29 January, 2006.

- "WASHINGTON (AP) -- A House committee launched an inquiry Tuesday over a former museum administrator's claim that the Smithsonian Institution toned down a climate change exhibit for fear of angering Congress and the Bush administration." Brett Zongker, AP News, 22 May, 2007.

- "Leonardo DiCaprio hit back at accusations of hypocrisy as he unveiled an eco-documentary he wrote, produced and narrated at the Cannes film festival. Asked after the premiere of *The 11th Hour* whether he had taken a fuel-guzzling jet on his way to the French Riviera, the *Titanic* star spat back sarcastically 'No, I took a train across the Atlantic'." ABC News Online, 20 May, 2007.

Though there are some warnings of extreme consequences, most of the reporting seems to concentrate on the political aspects of global warming. Perhaps the media feel on firmer ground when bashing government officials, bureaucratic foot dragging, misplaced agency priorities, and high-level cover-ups. Chronicling the struggles of people trying to "live green" is also popular, particularly if there is a Hollywood celebrity involved. For a decade, the image of Leonardo DiCaprio driving his hybrid Toyota Prius has defined Hollywood environmentalism.

How accurate and unbiased is the reporting? There are some disturbing signs that getting the right story may be more important to some reporters than getting the story right. Consider this statement made by Dr. David Deming, a geologist and geophysicist at the University of Oklahoma, in testimony given before the US Senate Committee on Environment and Public Works:

"In 1995, I published a short paper in the academic journal *Science*. In that study, I reviewed how borehole temperature data recorded a warming of about one degree Celsius in North America over the last 100 to 150 years. The week the article appeared, I was contacted by a reporter for National Public Radio. He offered to interview me, but only if I would state that the warming was due to human activity. When I refused to do so, he hung up on me."[399]

This seems to indicate that the media are reporting the story they want, not the truth. Deming continued, "There is an overwhelming bias today in the media regarding the issue of global warming," adding "In the past two years, this bias has bloomed into an irrational hysteria."

Dr. Deming is not alone in charging the media with inaccurate reporting. One reader of the *Baltimore Sun* wrote in regarding an article about the IPCC conclusions saying, "I'm not a scientist, but the following words used to describe what the media reports to be scientific fact are troubling: 'could,' 'probably,' 'might,' 'maybe,' and 'possible,'"[400] Such writers are often pilloried for "denying global warming" when what they are actually complaining about is the *way* global warming is being reported. Paul More, an editor at the *Sun*, dismissed the complaints of readers saying:

"The readers who criticized this round of The Sun's reporting did so, in my view, not just because they disagreed with the conclusions of the scientific panel. They were angry because they reject the reality of global warming because they see it mostly in political and ideological terms."[401]

Not all news outlets are slanting the news to raise ratings or increase circulation. The *Wall Street Journal* printed the following on their op-ed page. It was written by MIT Professor Richard Lindzen, a member of the National Academy of Science's expert panel:

"Our primary conclusion was that despite some knowledge and agreement, the science is by no means settled. We are quite confident (1) that global mean temperature is about 0.5 degrees Celsius higher than it was a century ago; (2) that atmospheric levels of carbon dioxide have risen over the past two centuries; and (3) that carbon dioxide is a greenhouse gas whose increase is likely to warm the earth (one of many, the most important being water vapor and clouds). But—and I cannot stress this enough—we are not in a position to confidently attribute past climate change to carbon dioxide or to forecast what the climate will be in the future. That is to say, contrary to media impressions, agreement with the three basic statements tells us almost nothing relevant to policy discussions."[402]

Of course, others see matters in a different light. Al Gore, speaking about the conclusions of the IPCC, said the media "have failed to report that it is the consensus and instead have chosen ... balance as bias."[403] After citing a study that showed 53 percent of topical newspaper articles offer "false balance" on global warming, he said, "I don't think that any of the editors or reporters responsible for one of these stories saying, 'It may be real, it may not be real,' is unethical. But I think they made the wrong choice, and I think the consequences are severe."

The science is complex and hard to understand, balanced reporting is biased reporting, and nothing gets in the way of a good catastrophe story. This isn't science, it's bad journalism driven by politics.

Other Voices

Along with the "official" IPCC reports and the reporting on them in various media outlets there is a third set of players in the global warming crisis. These are special interest groups, lobbying groups, and activists of all kinds that fall into the category of Non-Governmental Organizations (NGOs). Many are local or national in scope and membership, while others are international, claiming millions of members.

Among the NGOs are Greenpeace International, the World Wildlife Fund, Environmental Defense, Center for a New American Dream, Center for International Climate and Environmental Research - Oslo (CICERO), Centre for Science and Environment (CSE), Climate Change Belgium, Columbia Earth Institute, Earth Repair Foundation, The Environmental Change and Security Project (ECSP), and Friends of the Earth International, to name just a few.

What do these various defenders of Earth say about global warming? Here are some sample statements from well-known NGOs:

- "Time is running out. We need more international action now to reduce the gas emissions responsible for global warming." Catherine Pearce, Friends of the Earth International's climate campaign coordinator, 30 September 2004.[404]

- "There are many serious and troublesome problems going on in the world which will affect all present and future generations, unless humanity and all governments implement practical solutions from now on." Earth Repair Foundation.[405]

- "Without action, climate change will cause the extinction of countless species and destroy some of the world's most precious ecosystems, putting millions of people at risk." WWF.[406]

- "This problem is unlike anything seen in the past. It affects the whole planet and threatens every person living in every country on every continent. However, we can do something about it. It's not a threat coming from outer space. It is people, us, who are causing climate change by polluting the atmosphere with too much carbon dioxide and other greenhouse gases." Greenpeace International.[407]

- "Bush and Co. have betrayed the cause of climate protection up to now. We have 13 years at most to avert the worst impacts of climate destruction, and yet there is not even a glimmer of a breakthrough on the horizon at present." Greenpeace International Climate & Energy campaigner Stephanie Tunmore.[408]

- "The most important thing you can do is to get involved in the political process and get rid of all of these rotten politicians that we have in Washington D.C. Who are nothing more than corporate toadies for companies like Exxon and Southern Company, these villainous companies that consistently put their private financial interest ahead of American interest and ahead of the interest of all of humanity. This is treason and we need to start treating them now as traitors." Robert F. Kennedy Jr., Comments at Live Earth.

- "This is a glimpse into an apocalyptic future. The earth will be transformed by human induced climate change, unless action is taken soon and fast." Stephanie Tunmore, Greenpeace International Climate & Energy Campaigner commenting on the IPCC report.[409]

- "As climate change wreaks its havoc across the globe, ecosystems could disappear altogether, or they may undergo serious and irreversible changes, such as those happening to coral reefs." WWF.[410]

- "Human activities are emitting vast amounts of 'greenhouse gases' that prevent heat from escaping from the Earth's atmosphere. Scientists report that this phenomenon will increasingly lead to catastrophic natural disasters, such as more frequent and intense droughts, floods, and hurricanes; rising sea levels; and more disease outbreaks." GoVeg.com.[411]

Unlike the restrained and measured warnings from the IPCC, NGOs are a bit more strident in their claims of impending doom. They promise an "apocalyptic future" filled with "havoc," "irreversible changes," and "catastrophic natural disasters." "Time is running out" and our leaders are not doing enough or have "betrayed the cause of climate protection" altogether. Robert Kennedy Jr., a lawyer for the Natural Resources Defense Council, goes so far as to call anyone who questions global warming a "traitor."

This type of rhetoric, replete with *ad hominem* attacks and accusations of treason, is not a fitting part of any public debate, let alone a scientific one. When Marlo Lewis, of the Competitive Enterprise Institute, published an article opposing mandatory limits on carbon-dioxide emissions, arguing that Congress should not impose caps until the technology exists to produce energy that doesn't depend on carbon dioxide, he was publicly threatened by

Michael Eckhart, the president of the American Council on Renewable Energy. In Eckhart's own words:

> "Take this warning from me, Marlo. It is my intention to destroy your career as a liar. If you produce one more editorial against climate change, I will launch a campaign against your professional integrity. I will call you a liar and charlatan to the Harvard community of which you and I are members. I will call you out as a man who has been bought by Corporate America."[412]

David Roberts, a writer for Grist magazine, goes even further. His solution for those who are members of what he terms the global warming "denial industry" is: "When we've finally gotten serious about global warming, when the impacts are really hitting us and we're in a full worldwide scramble to minimize the damage, we should have war crimes trials for these bastards – some sort of climate Nuremberg."[413]

People wonder why more scientists don't speak out against the distortions and inaccurate reporting of global warming. It is because anyone brave enough, or foolish enough, to do so risks being branded a "climate criminal" by fanatical global warming "true believers." As a number of scientists have learned, questioning the "truth" about human-caused global warming can have dire consequences. They risk their grant money, their careers[414] and even their lives.[415,416]

This all makes *Weather Channel* personality Heidi Cullen's suggestion, that television meteorologists be stripped of their American Meteorological Society certification if they question predictions of catastrophic global warming, seem like tame stuff. The global warming debate has strayed far from rationality and, as Thomas Jefferson once said, "It is as useless to argue with those who have renounced the use of reason as to administer medication to the dead."

Al Gore's Convenient Calling

Albert Gore Jr. was born into privilege, the scion of an influential Tennessee family. His father served in the United States House of Representatives before WWII and, after the war, in the US Senate for two decades, so it came as no surprise when Al Jr. also chose a career in politics. Following in his father's footsteps, he served first as a Congressman, then a US Senator, and finally Vice-President of the United States during the presidency of Bill Clinton.

Reaching for the top prize in American politics, Gore was the Democratic nominee in the 2000 presidential election. It was one of the most controversial elections in American history. After a series of voting discrepancies and court challenges, Gore lost the election to George W. Bush when the US Supreme Court ruled that vote recounting had gone on long enough. Embittered by his narrow defeat, Gore left politics for private life and more academic pursuits. Asked to speak at the 2004 Democratic National Convention, the lingering resentment over his defeat was still evident when he said in his speech, "Let's make sure not only that the Supreme Court does not pick the next president, but also that this president is not the one who picks the next Supreme Court."

Al Gore was to eventually return to the public eye with a new mission. He found a higher calling in the environmental movement, a cause he had always supported. Gore claims the roots for his passionate belief, that human civilization harms the Earth, came from Rachel Carson's book *Silent Spring*. During Gore's childhood, his mother, an educated and politically active woman who took Eleanor Roosevelt as a model, read the book to the 14 year old Albert and his older sister.

Carson's book was about DDT, a pesticide that was widely used to control malarial mosquitoes. But it also had negative effects on birds' eggs, weakening their shells and threatening a number of species with extinction. As a result of Carson's book, DDT was banned around the world. Four decades later, studies have shown that the removal of DDT caused millions of deaths in countries plagued with malaria and dengue fever.

While Gore attended Harvard, he was a student of the noted oceanographer Roger Revelle, who is mentioned in Chapter 2 of this book. Dr. Revelle, the self-proclaimed father of global warming, had a profound and lasting influence on young Albert. In Gore's own words, Revelle taught him "what was happening to the atmosphere of the entire planet, and how that enormous change was being caused by human beings." Gore attributes his lifelong dedication to environmental causes to Revelle's teachings.

Mr. Gore comes by his environmentalist leanings honestly. His reemergence from the political wilderness as an environmental prophet of doom was no sudden conversion. But Gore has reinvented himself. From failed politician to a speaker on the college lecture circuit, the new Al Gore is only concerned with saving mankind from its own excesses. He now lectures widely on the topic of global warming, which he calls "the climate crisis."[417]

His power-point show on global warming is entertaining, but many of his conclusions are not supported by the science. That did not stop his lecture hall pitch from being turned into a slick documentary—basically a video of Gore presenting his talk. This film was immediately embraced by eco-conscious Hollywood. It would have been shocking if Gore's film hadn't won the Oscar™ for best documentary in 2006. Perhaps more surprising was the announcement that Al Gore and the IPCC would share the Nobel Peace Prize for 2007.

The link between proselytizing about global warming and world peace seems tenuous at best. The *Economist,* observing that two out of the previous three peace prizes went to people and organizations who had nothing to do with peace, stated: "Evidently the committee has decided to redefine the award as the Nobel Prize for Making the World a Better Place in Some Unspecified Way."[418]

At the same time that Gore's documentary was being praised by activists, it was coming under progressively harsher criticism. A UK High Court judge, in response to a lawsuit questioning the film's suitability for showing in British classrooms, ruled that *An Inconvenient Truth,* contains significant factual errors. Justice Burton said Gore makes nine statements in the film that are not supported by current mainstream science.[419] The specific errors are:

1. Mr. Gore asserted that sea levels will rise up to 20 feet due to melting ice sheets "in the near future." Justice Burton: "This is distinctly alarmist" and will only occur "after, and over, millennia."

2. Low-lying Pacific atolls have already been evacuated. Judge: There was no evidence of any evacuation having yet happened.

3. The Gulf Stream that warms the northern Atlantic would shut down. Judge: It was "very unlikely" this would be happening in the future, though a slow down was possible.

4. Graphs of CO_2 and temperature rise over the past 650,000 years showed an "exact fit." Judge: There was a connection, but "the two graphs do not establish what Mr. Gore asserts."

5. The snows of Mt. Kilimanjaro are disappearing due to global warming. Judge: It cannot be established that the recession of snows on Mt. Kilimanjaro is mainly attributable to human-induced climate change.

6. The drying up of Lake Chad is an example of a catastrophic result of global warming. Judge: There was "insufficient evidence to show that."

7. Hurricane Katrina was caused by global warming. Judge: Insufficient evidence to establish the exact cause.

8. Polar bears were found that had drowned "swimming long distances—up to 60 miles—to find the ice." Judge: Only four polar bears have been found drowned, because of a storm not the lack of ice.

9. Coral reefs are bleaching because of global warming and other factors. Judge: Separating the impacts of stresses due to climate change from other stresses, such as over-fishing and pollution, was difficult.

The judge concluded that the film could be shown, but teachers must alert students to the errors he identified.

Interestingly, the Motion Picture Association of America rated Gore's film PG, requiring parental guidance due to "thematic elements." More disturbing is Gore's picture book for children, *an inconvenient truth: the crisis of global warming—adapted for a new generation.* Supposedly a book about environmental awareness for young children, the book is heavy on images and light on substance. For example, on page 155 looms the full page image of a nuclear fireball. In very small letters the caption reads, "Test detonation of a nuclear bomb, Nevada, 1957." The opposing page shows five German soldiers in a trench. The photo caption reads: "German soldiers in World War 1, 1914." Above the German photo, on white background and in bold, large text are the words: "New technologies for fighting, such as nuclear bombs, have dramatically changed the consequences of war."

Another image of supposed global warming disaster is of stranded fishing boats in the dried out Aral Sea, in Central Asia. This image has a link to

New technologies for fighting, such as nuclear bombs, have dramatically changed the consequences of war.

Illustration 135: Misleading pictures from Al Gore's An Inconvenient Truth.

ecological problems, but not ones caused by global warming. The Aral Sea was destroyed by Soviet mismanagement. The former USSR diverted the rivers that fed the sea to other uses, another triumph of communist central planning. What do those words and those images have to due with global warming? Nothing. But they leave a horrible verbal and visual impression on young children—and there is a disturbing pattern in this.

Al Gore is a politician and politicians deceive the public all the time. They lie to get elected, it is a way of life for them. That might be the way it works for voting adults, but not with impressionable kids. Gore wore the mantle of a politician for nearly three decades. It seems that it is a hard cloak for him to abandon, because he has brought politics into science. When politics gets involved in science, the lies, opinions, and deceit, that are a politician's everyday fare, destroys scientific discipline. Gore did this trying to indoctrinate young children, a "new generation," with his horrific pictures and misleading words.

Gore made a speech to New York University Law School on September 18, 2006. His sponsors were two NGOs, the World Resources Institute and Set America Free. A few statements from his speech are presented here to demonstrate Mr. Gore's approach to the problem of global warming.

> "My purpose is not to present a comprehensive and detailed blueprint— for that is a task for our democracy as a whole—but rather to try to shine some light on a pathway through this terra incognita that lies between where we are and where we need to go. Because, if we acknowledge candidly that what we need to do is beyond the limits of our current political capacities, that really is just another way of saying that we have to urgently expand the limits of what is politically possible."

Though Mr. Gore denies having a plan for the future he has made suggestions. In his book, *An Inconvenient Truth,* he put forth a number of things everyday people can do to save the planet. While some of the suggestions are sensible, others are laughable. These include a simplistic admonition to "use a tote bag when shopping" to the fatuous "buy things that last." Gore doesn't present a detailed plan because, despite claims to have invented the Internet, he knows little about science or technology.

Gore's film, *An Inconvenient Truth,* devoted a long segment to the imminent meltdown of the Greenland and West Antarctic ice-sheets. The film predicted a global sea level increase of 20 feet (6 m) that would flood Manhattan, Shanghai, Bangladesh, and other coastal settlements. As proof of this impending disaster, Gore quoted Oreskes' essay as proving that all credible climate scientists were in agreement regarding this supposed threat. He did not point out that Oreskes' definition of climate change "consensus" did not encompass his alarmist notions.

What is more revealing is his call to urgently expand the limits of governmental power. A lifelong member of America's political left, Mr. Gore has always embraced the expansion of government control. It comes as no surprise that he sees the solution to global warming crisis, indeed all environmental problems, as a need for expanding government regulation and oversight. He truly believes that humanity's future is in jeopardy.

> "Many Americans are now seeing a bright light shining from the far side of this no-man's land that illuminates not sacrifice and danger, but instead a vision of a bright future that is better for our country in every way ... Our children have a right to hold us to a higher standard when their future—indeed the future of all human civilization—is hanging in the balance."

Presumably, the bright light is the one Mr. Gore is shining on the path to the future. Casting aside the overblown, campaign stump speech language, Gore thinks that mankind is in peril and the future looks bleak, unless bold political leadership asserts itself. Being unable to formulate an actual plan for reducing GHG emissions, or solving the world's future energy needs, he falls back on the need for moral leadership.

> "Developing countries like China and India have gained their own understanding of how threatening the climate crisis is to them, but they will never find the political will to make the necessary changes in their growing economies unless and until the United States leads the way. Our natural role is to be the pace car in the race to stop global warming."

Only a politician could think that the world is waiting for America to "lead the way" before they "find the political will" to tackle the dual problems of rising pollution rates and falling energy supplies. China, India or any other developing country will not slow their industrial growth to please the US They cannot. Their people would rise up and topple their governments if told the future had to be put on hold.

China is building one new coal-fired power plant every week with India following close behind. China and India will stop building power plants when they think they have enough of them, or a better alternative arises. A foreshadowing of this can be seen in Westinghouse's recent deal to build four new nuclear power plants in China.

Even the greener-than-thou nations of Europe have discovered they have little appetite for such economic sacrifice. To date, six EU nations have petitioned for relief from their GHG reduction goals. The natural role for the United States and the other technologically advanced nations is not to lead the way to national impoverishment through limiting growth, self denial, and industrial suicide, but rather to develop new technology that is both environmentally sound and economically competitive.

The most interesting codicil to Mr. Gore's saga as prophet of ecological doom is that, after winning an Oscar™ and publishing a best selling picture book, Gore steadfastly refuses to meet his critics head to head. A sizable and growing list of "contrarian" scientists have offered to debate the former politician in public. This list includes Bjørn Lomborg, author of *The Skeptical Environmentalist,* Dennis Avery, coauthor of the best-selling book, *Unstoppable Global Warming: Every 1,500 Years,* and Lord Monckton of Brenchley, a former adviser to Margaret Thatcher during her years as Prime Minister of the United Kingdom. Quoting from Lord Monckton's open challenge:

"The Viscount Monckton of Brenchley presents his compliments to Vice-President Albert Gore and by these presents challenges the said former Vice-President to a head-to-head, internationally-televised debate upon the question "That our effect on climate is not dangerous," to be held in the Library of the Oxford University Museum of Natural History at a date of the Vice-President's choosing.

Forasmuch as it is His Lordship who now flings down the gauntlet to the Vice-President, it shall be the Vice-President's prerogative and right to choose his weapons by specifying the form of the Great Debate. May the Truth win! Magna est veritas, et praevalet. God Bless America! God Save the Queen!" [420]

Illustration 136: Lord Monckton throws down the gauntlet.

Lord Monckton and Mr. Gore have clashed in the media before. Monckton has stated publicly, "A careful study of the substantial corpus of peer-reviewed science reveals that Mr. Gore's film, *An Inconvenient Truth*, is a foofaraw of pseudo-science, exaggerations, and errors, now being peddled to innocent schoolchildren worldwide."

It is understandable that Gore would not wish to debate a real scientist, but Lord Monckton is a fellow politician who has also been involved in the global warming policy debate. To date, Mr. Gore has demurred.

Masters of Deception

While politicians need no underlying agenda as an excuse for distorting the global warming crisis—the desire to be in control is enough for them—what motivates others is not always so clear. The following are a few examples of global warming related claims that are not what they appear to be on the surface.

The Pacific island nation of Tuvalu stands just 13 feet about the sea level at its highest point. In 2002, the government announced they might sue the United States and Australia because they have rejected the Kyoto protocol on global warming. Enele Sopoaga, Ambassador and Permanent Representative of Tuvalu to the United Nations, has stated that his nation is still considering bringing a suit "drawing in the rights of the people to exist and the injury that climate change has caused on the existence of people of Tuvalu." He continued, "It's unfortunate because, you know, a country such as the US that has—that's contributing about 25% of the CO_2 greenhouse gas emissions into the atmosphere, not to participate in this global effort, it does create some concerns."[421]

This is simply not the truth. UN reports have clearly stated that Asia far exceeds North America in GHG emissions with India alone surpassing the US.[422] The "brown clouds" of Asia may be contributing half of the global temperature increase attributed to CO_2, yet both India and China were exempted from Kyoto reductions. Add to this the claims by climate experts, that even full compliance with Kyoto will have no appreciable affect on global warming. The question becomes why is Tuvalu thinking of suing Australia and the US for not signing the Kyoto Protocol? Why doesn't Tuvalu sue China and India?

Could it be that China and India would tell them to sod off? Or that the US and Australia are rich countries against whom Tuvalu could play David vs. Goliath in the World Court? Perhaps the true reason for Tuvalu's environmental problems are poor water management and over pumping from the island's fresh water wells. This mismanagement is causing the island to sink and promoting salt water encroachment that is ruining the natives' gardens. The residents have applied for relocation money, but the government says it lacks the cash, hence the law suit. This is simply the victim culture mentality applied on an international scale.

We have already commented on reports that domestic animals may contribute more to GHG emissions than human transportation. The United Nations published a report on livestock and the environment concluding: "The livestock sector emerges as one of the top two or three most significant contributors to the most serious environmental problems, at every scale from local to global." Others are now pursuing this story for different purposes. ABC News, picked up the story in an article titled *Meat-Eaters Aiding Global Warming? New Research Suggests What You Eat as Important as What You Drive.* In it they reported that you should become a vegetarian if you want to help lower greenhouse gas emissions. They quoted researchers who liken eating red meat to driving an SUV.[423] This position would certainly be supported by PETA, the group running the GoVeg.com web site.

In the US, there have been claims that the large number of domestic cattle, estimated by the Department of Agriculture at 100,000,000 head, are a major contributor to methane emissions. While this may be so, prior to the arrival of men with fire arms, as many as 80,000,000 bison roamed the American prairie. Since the DOA considers cattle and bison as equal in terms of emissions, there has not been an appreciable increase in such emissions in North America since prehistoric times. Furthermore, unless those cattle are eating coal or drinking oil, their emissions are as carbon neutral as biodiesel and ethanol. Vegetarians and animal rights activists are simply seizing on global warming to promote their own beliefs. As one blogger put it, "Vegetarian is the New Prius."

Illustration 137: Ad from GoVeg.com

Artistic luminaries have a long tradition of decrying man's wasteful behavior. No save-the-earth rally is complete without a cadre of rock stars and Hollywood actors. Like Al Gore, the great Japanese novelist Junichiro Tanizaki (1886-1965) was also something of an ecological prophet. Tanizaki's 1933 essay, "In Praise of Shadows," is a meditation on the aesthetics of shadows in Japanese culture. In it, he railed against electric lighting, the construction of highways, and women using makeup. In an on-line article on Tanizaki, Joshua Sowin captures the essence of the artistic class's opposition to human technological progress:

> "Tanizaki was too astute and realistic to be an optimist. And his fears were accurate—what he denounces has become progressively worse. Technology is not ushering us into some kind of techno-utopia like so many seem to believe. Our progress has caused immeasurable social and ecological destruction along with its many advantages. Yes, we live longer, richer, and with more gadgets. But if life is robbed of rewarding and meaningful work, community, stability, silence, health, and wilderness, can our progress really be considered progress?"[424]

Optimism is foolish and unrealistic? Living longer, richer, and with more gadgets is not progress? Seek out those who live shorter, poorer, and with nothing—ask them their opinion. The mindset of idle intellectuals, avaricious politicians, and society's spoiled, pampered celebrities becomes clear—our science-based, technological civilization is inherently evil. It is easy for those who have an excess of everything to criticize the masses for wanting a portion of the good life. But there are those who take these opinions to their final, inescapable conclusion.

There is an organization called VHEMT that is dedicated to the goal of voluntary human extinction. Their motto is "may we live long and die out." Describing human population growth as an "inexorable horror," Voluntary Human Extinction Movement head and founder Les Knight has said "as long as there's one breeding pair of homo sapiens on the planet, there's too great a threat to the biosphere."[425] Knight believes that man has done more harm to nature than good, and most environmentalists would agree. But VHEMT takes environmental extremism to its limits by saying mankind's primary concern should not be its own survival. Claiming to be non-violent, they suggest people simply stop having children and allow our species to die out. It would be easy to dismiss Knight and his compatriots as part of the lunatic fringe, people who simply hate mankind, but they are not alone.

Illustration 138: VHMET's logo: sign up and die out.

David Suzuki, Canadian broadcaster and population control activist, hosted American ant expert Edward O. Wilson on his 1999 television series, *From Naked Ape to Super Species*. Wilson, a fellow population control extremist, proclaimed: "if all humanity disappeared the rest of life would benefit enormously ... If the ants were all to disappear, the results would be close to catastrophic." Wilson added that if humans disappeared, "the forests would grow back, the whole Earth would green up, the ocean would teem, and so on."[426] To some environmentalists, ants are preferable to people.

So far, humanity has shown little interest in volunteering for extinction, prompting suggestions of other, less radical courses of action from ecological activists. A suggestion that surfaces frequently is for mankind to turn its back on high technology living and return to a simpler, more natural lifestyle.

The notion that life was better for human beings at some time in the past has surfaced before. This type of thinking goes back to Rousseau[†] and the myth of the noble savage. Primitivism, an 18th century response to the stress of "modern" living, idealized savage man as uncorrupted by the influences of civilization. Rousseau thought that people in a state of nature did not know good and evil, and that "as every advance made by the human species removes it still farther from its primitive state, the more discoveries we make, the more we deprive ourselves of the means of making the most important of all."[427] Natural equality disappeared "from the moment one man began to stand in need of another."

In the 1930s, George Orwell railed against "cranks" in his book, *The Road to Wigan Pier*. English advocates of the simple life in the 1930s, like Edward Carpenter and Leslie Paul, can be regarded as forerunners of modern ecological anti-capitalism. Quoting from an article in *Cabinet Magazine*, "In the spirit of *The Road to Wigan Pier*, today's anti-capitalism could be said to draw towards it with magnetic force every tree-hugger, organic fruitarian, solar-powered scooter rider, water-birth enthusiast, Tantric-sex practitioner, world-music listener, teepee-dweller, hemp-trouser wearer, and Ayurvedic massage addict."[428] Crank culture is alive and well in the 21st century.

In 2001, the International Forum on Globalization's *Teach-In on Technology and Globalization,* held at New York's Hunter College, was attended by some 1,400 activists. According to one report, "The speakers included an all-star cast of technophobes and other rebels against the future, featuring proud self-declared Luddites such as Kirkpatrick Sale, Jeremy Rifkin, Jerry Mander, Andrew Kimbrell, Paul Hawken, Pat Roy Mooney, Mae-Wan Ho, and Vandana Shiva."[429] During the conference, speakers called for civil justice, environmental justice, and green peace. They urged civil society

---

† Jean-Jacques Rousseau (1712-1778) Swiss philosopher of the Enlightenment.

groups to join together to create "an action program to control or halt progress in the development of technology."

They call themselves "progressives" but the last thing they desire is progress. Rich Hayes demanded "an immediate global ban on human reproductive cloning, an immediate global ban on manipulating genes that we pass on to our children, and accountable and effective regulation of all other human genetic technologies." Jeremy Rifkin called for "a strict global moratorium, no release of GMOs (genetically modified organisms) into the environment." Stephanie Mills advocated the "precautionary principle," the idea that before any new development in science and technology can be used, it must be shown to have no negative impact. "New technologies should be presumed guilty until proven innocent," Mills declared.

It is not just science and technology these activists loathe, dismantling the global economy is also high on their agenda. For these later-day Luddites, free trade is anathema and shopping online causes excess consumption by being "too easy." Jerry Mander, head of the IFG and organizer of the conference, recommended that countries return to the import substitution model of economic development, which was abandoned after it bankrupted most of Africa and Latin America in the 1960s and 1970s. These people find nothing good in human technological civilization and long for a simpler, pre-technological world. To quote again from the article by *Reason* magazine's Ronald Bailey:

> "Whether willfully or out of sheer ignorance, the congregants in Manhattan this past weekend dismiss any and all evidence that the human race has progressed over the past 100 years, much less the past 1,000; the longer life expectancies, higher standards of living, and cleaner environments that are everywhere becoming the rule and not the exception for the masses have seemingly made no impression (nor have the economic forces that make such things possible). The hopeful future of humanity freed from disease, disability, hunger, ignorance, poverty, and inequity depends on beating back the forces of know-nothing reaction such as those assembled at this weekend's Teach-In. The struggle for the future begins now."

Our Global Civilization

People's lives prior to the Industrial Revolution show the folly of longing for "simpler times." Medieval France, blessed with an abundance of arable countryside, had a 14th century population density of 100 people per square mile. The British Isles were less populous, having little more than 40 people

290

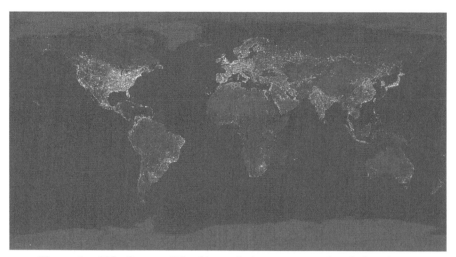

Illustration 139: Image of Earth's city lights was created with data from the Defense Meteorological Satellite Program. Source NASA.

per sq. mile.[430] Modern France has more than twice as many people and the U.K.'s population density is now 640 people per sq. mile. Japan's medieval population was around 6 million in 1280 compared with 127 million today.[431] The total world population in 1250 AD was around 400 million, today there are 6,600 million people on Earth.[432] There were not only fewer people on Earth, people's lives were much harder than most people's today. The people of Medieval times had to toil from sunrise to sunset to survive, always only a failed harvest away from starvation. How large a population Medieval technology could support is uncertain, but one conclusion is inescapable: Without modern technology, Earth could not support its current human population.

Despite the exhortations of ecological Luddites, most people enjoy the benefits of our high-tech, global civilization. While a return to simpler, non-technological times is the anti-technologists' fantasy, most people in the less developed parts of the world long for the benefits of modern civilization. Those who say, "we can keep the good things, like medical care and cellphones, and dump everything else," do not understand how interconnected people have become. The benefits of science and technology cannot be chosen or rejected on an individual basis—technological civilization is a package deal.

Even the simplest modern conveniences require the common efforts of millions of people around the world to produce. Not just those who design and assemble the immediate product, but all those who supply parts and raw materials, and all those who build the machines that manufacture the parts and mine the raw materials as well. Every human is connected to the rest of humanity by the web of industry and commerce. No nation on Earth can exist in isolation. Those that have tried, for political or other reasons, have brought nothing but misery and despair upon their people. As members of a global civilization, we all stand in need of one another.

The simplest way to alleviate human misery and suffering is to bring modern technology to underdeveloped areas. What impoverished areas need most is energy, and the most beneficial form of energy is electricity. Electrification frees people from burning biomass, wood and dung that denudes the land and poisons the air. It allows refrigerated food storage, powers pumps for clean drinking water, and lights homes and classrooms. It would be criminal, bordering on racist, for developed nations to deny the impoverished peoples of the world a future. As shown in Illustration 139, Africa is literally the dark continent, but not by choice.

The leaders of developing countries are aware of global warming, but they have more immediate concerns. Though he is "extremely concerned about the consequences, the adverse affects of climate change," Juan Rafael Elvira Quesada, Mexico's environment minister, adds "We have always to bear in mind that half our population is at the poverty line." Xie Zhenhua, vice chairman of China's national development and reform commission, states the developing world's priorities clearly: "For a developing country, the main task is to reduce poverty."[433] As long as older, dirty technologies are widely available at lower cost, developing nations will always choose the quickest path to technological prosperity. For these reasons, attempts to force developing nations to adopt greener, but more expensive technologies will fail.

Illustration 140: The Cuyahoga River burning in 1969. Source NOAA.

The developed nations have made environmental mistakes in the past—pollution and accidental extinctions among them—but learning is part of growing up. A half a century ago, the air over major cities in the industrialized nations was a visible menace to human health, much like the atmosphere in Asia has become today. Pollution was so pervasive that life in many rivers and lakes was vanishing. One US river, the Cuyahoga in Ohio, famously caught fire on several occasions between 1936 and 1969. Today, air and water quality in industrial nations are markedly improved. In the times ahead, humanity's continued success requires that our technology and attitudes toward nature continue to mature. If the developing nations of the world are to avoid repeating the worst mistakes of the developed nations, cleaner technologies must be made available and affordable. Progress will not be denied—the path to the future does not lie in the past.

This book is about science, about gaining understanding. It is about analyzing the true extent of global warming, anticipating the possible effects it might cause, and offering rational solutions to fix real, not imaginary, problems. After wading through the media morass, the next three chapters will return to the realm of science.

Chapter 16 will examine the possible effects of climate change, based on the scientific evidence presented. Chapter 17 will examine the IPCC's suggested cures, called mitigation strategies, to find out which ones are useful and which ones are not. Then Chapter 18 will present what science and engineering suggest as the best plan for the future. As we will discover, things are not nearly as bad as the activists, pundits and politicians would have us believe.

Chapter 16 The Worst That Could Happen

"For myself I am an optimist—it does not seem to be much use being anything else."

—*Winston Churchill*

Much has been written and said regarding the dire effects of global warming. Just as the IPCC temperature predictions are wildly overstated, the predictions of environmental disaster have been blown far out of proportion. Cormac McCarthy won a Pulitzer Prize for his book "The Road." We quote part of the dust jacket as follows:

> "A father and his son walk alone through burned America. Nothing moves in the ravaged landscape save the ash on the wind. It is cold enough to crack stones, and when the snow falls it is gray. The sky is dark. Their destination is the coast, although they don't know what, if anything, awaits them there. They have nothing; just a pistol to defend themselves against lawless bands that stalk the road, the clothes they are wearing, a cart of scavenged food and each other."

McCarthy paints a grim picture of man's future; dismal and dangerous. This form of post apocalypse tale has been a staple of the science fiction and horror genres for decades. But the future doesn't have to be a dystopian hell, returning to a time when life would be a "struggle of all against all," as Hobbs[†] said, "solitary, poor, nasty, brutish and short." We think that humankind has a much brighter future than that. If, that is, we are willing to learn what nature has to teach us.

What is the worst that could happen? Recall that the predicted affects of global warming are predicated on varying levels of temperature rise. Illustration 141 shows the increase in temperature predicted by the IPCC's GCM based on increasing atmospheric CO_2. The increasing width of the temperature ranges for a given CO_2 level reflects growing uncertainty in the model predictions. The temperatures reflect global mean change above preindustrial levels. The central black line was calculated using "best estimate" climate sensitivity of 3°C. The upper bound line reflects climate sensitivity of 4.5°C, and the lower bound line 2°C. The Roman numerals labeling the graph segments are the scenario numbers, as described in the IPCC Working Group III report. The shaded regions show the ranges of

---

† Thomas Hobbes (1588-1679) English philosopher, author of the famous book *Leviathan*.

predictions corresponding to the stabilization scenario categories I to VI. The data for the graph are drawn from AR4 WGI, Chapter 10.8.

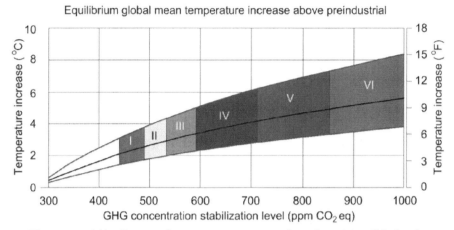

Illustration 141: Projected temperature increase based on rising CO_2 levels. Source IPCC.

The IPCC's best estimate increase for a "low scenario" is 1.8°C with a likely range of 1.1 to 2.9°C. Their best estimate for a "high scenario" is 4.0°C with a likely range of 2.4 to 6.4°C. Combining the bottom of the low scenario with the top of the high scenario gives the total range of possible increase as 1.1 to 6.4°C (2.0 to 11.5°F). Keeping this range of possible increase in mind, here is an examination of the major disasters the IPCC predicts as a result of global warming.

A Rising Tide

Sea levels could rise, they have changed significantly in the past. According to the IPCC AR4, it is estimated that sea level rise will be 7 to 15 inches (18-38 cm) in a low scenario and 10 to 23 inches (26-59 cm) in a high scenario. This is based on multiple models, which all exclude ice sheet flow due to a lack of reliable published data. They also think it likely that there will be increased occurrence of extreme high tides over the coming century. With high confidence, the report predicts that coastlines will be exposed to increasing risks such as erosion and that "Many millions more people are projected to be flooded every year due to sea-level rise by the 2080s." But sea levels are always changing and scientists have had little success predicting change in the past.

These relatively modest levels of increase stand in stark comparison to warnings issued by environmentalists, such as Al Gore. Gore has warned that "a massive destabilization may now be underway deep within the second largest accumulation of ice on the planet, enough ice to raise sea level 20 feet worldwide if it broke up and slipped into the sea."[434] While it is true that a total melting of the Greenland Ice Sheet (GIS) could theoretically cause a 20 ft (7 m) increase in world sea level, Gore's warning fails to mention how long melting the GIS would take.

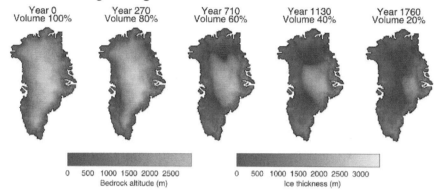

Illustration 142: Timeline for melting of the Greenland ice sheet. Source IPCC.

The role of the GIS in climate change is not clear. Researchers have to account for the loss of ice around the edges of the sheet, while also measuring the buildup of ice in Greenland's interior. The influence of short-term weather cycles adds to the complexity. It is this complexity and incomplete understanding, a reflection of the totality of climate science on a smaller scale, that causes scientists to admit that their ice sheet models do not accurately reflect reality (page 267). Based on their best guess warming rates, the IPCC has generated model predictions for ice sheet melting in Greenland (Illustration 142).

Notice that even after 270 years, 80% of the GIS remains. According to these predictions, it will take 1,000 years to melt half of Greenland's ice. Mr. Gore's warning about a 20 foot sea level rise due to the melting of the Greenland Ice Sheet is a bit premature.

As previously mentioned, tectonic activity is constantly causing local sea levels to change (page 135). This complicates collecting long-term sea level information. During the late 19th and 20th centuries, tidal gauges located along the Atlantic and Pacific coastlines indicated that sea levels suddenly began rising around 1920. The rate of increase nearly doubled, going from about 0.04 inches per year to about 0.07 inches per year. Naturally, this change has previously been attributed to climate change, but new data have changed that interpretation.

Evidence gathered by the US National Oceanic and Atmospheric Administration suggests that this change may be partly explained by the movement of large amounts of water. Atmospheric records suggest that movement of large mid-ocean water mounds, called gyres, redistributed water from the centers of the Pacific and Atlantic oceans to coastal areas. This shift explains the sudden increase in the rate at which sea levels changed as recorded by coastal instruments in the 1920s. Since tidal gauges only measure sea levels along the coasts, they could not have detected the drop in levels at the oceans' centers. Satellite measurements have only recently become sufficiently accurate to detect these types of changes. NOAA's Laury Miller states, "my guess is that it will be 20 or 30 years before we are able to identify how fast sea-level rises are truly accelerating."[435]

What about the larger concentration of ice in Antarctica? The IPCC Summary for Policy Makers reports: "Antarctic sea ice extent continues to show inter-annual variability and localized changes but no statistically significant average trends, consistent with the lack of warming reflected in atmospheric temperatures averaged across the region." It further notes that "current global model studies project that the Antarctic ice sheet will remain too cold for widespread surface melting and is expected to gain in mass due to increased snowfall."

As for the Arctic pack-ice, it is irrelevant to ocean levels. The ice at the North Pole covers water, not land. Since the ice floats, even a 100% melting of the Arctic ice will not change ocean levels, just as ice melting in a glass of water does not cause the glass to overflow.

Nils-Axel Mörner, head of paleo-geophysics at Stockholm University, has been studying the subject of sea-level rise for 35 years. He has been observing changes in ocean levels in the Maldives, one of the first places expected to disappear under the waves, and found no evidence that they are in peril. Satellite altimetry data collected over the past two decades tells the same story. His best estimate is for a rise of a couple of inches by the end of this century.[436]

Worsening Weather

Weather is caused by the exchange of heat among land, sea and air. The IPCC thinks it very likely that there will be an increased frequency of warm spells, heat waves and events of heavy rainfall due to global warming. They also call for an increase in areas affected by droughts, intensity of tropical cyclones, which include hurricanes and typhoons. These predictions have led

Illustration 143: Hurricane Gordon from orbit, source NASA.

to breathless news anchors attributing any storm activity to global warming, even though there has been no detectable trend in storm activity for over a hundred years.

Cyclonic storms, called hurricanes in the northern hemisphere, are heat engines, driven by the thermal energy released when rain condenses from water vapor. When water condenses, it undergoes what physicists call a *phase change*, a transformation from one state to another. When water evaporates, it absorbs heat energy. To change back into liquid form, the absorbed heat must be released to the surrounding environment. Hurricanes are powered by moist, warm air that is created by the tropical ocean. As the moisture-laden air rises it forms thunderstorms, which are whipped into a swirling pattern by Earth's rotation.

A tropical storm becomes a hurricane when its winds reach 74 mph (120 kph). The Atlantic hurricane season begins in June and ends in November, while the East Pacific hurricane season peaks during July through September. There are six Atlantic hurricanes during an average year. Over an average three-year period, five hurricanes strike the United States coastline, coming on shore anywhere from Texas to Maine.[437]

The claims about increasing monetary damage from hurricanes are true, there is more storm damage now than a century ago. But this is because population growth has disproportionately been in littoral (coastal) areas. Florida alone has 50 times the number of people it had in 1900, mostly living on the coasts. It is easy to see how damage can increase without an increase in storm frequency or intensity.

Bill Gray, Professor Emeritus of Atmospheric Science at Colorado State University, is often called America's most reliable hurricane forecaster and the world's most famous hurricane expert. A pioneer in the science of forecasting hurricanes, Gray's predictions have been used by insurance companies to calculate premiums since 1983. Gray, now retired, has publicly denounced attempts to link hurricane activity to global warming. According to him, "this is one of the greatest hoaxes ever perpetrated on the American people."

In a paper published after the 2006 hurricane season, Dr. Gray stated that historical records "indicate that Atlantic and global tropical cyclone activity over the last century and particularly over the last 30 years has not increased despite the global warming that has occurred over the last century and the last three decades."[438] Gray goes on to explain in detail:

"The most reliable long-period hurricane records we have are the measurements of US landfalling tropical cyclones since 1900. Although global mean ocean and Atlantic surface temperatures have increased by about 0.4°C between these two 50-year periods (1900-1949 compared with 1956-2005), the frequency of US landfall numbers actually shows a slight downward trend for the later period. If we chose to make a similar comparison between US landfall from the earlier 30-year period of 1900-1929 when global mean surface temperatures were estimated to be about 0.5°C colder than they have been the last 30 years (1976-2005), we find exactly the same US hurricane landfall numbers (54 to 54) and major hurricane landfall numbers (21 to 21)."

Historical records do not indicate a rising trend for either hurricane frequency or strength (see Illustration 144). Even so, global warming experts continue to confidently predict increasing tropical storm activity. To understand why this is not true, the factors that contribute to storm activity need to be examined.

According to NOAA, the conditions that determine active and inactive Atlantic hurricane seasons are largely controlled by recurring rainfall patterns along the equator. These patterns are linked to two dominant climate phenomena: The El Niño Southern Oscillation (El Niño and La Niña) cycle, and the tropical multi-decadal signal. The latter is a set of atmospheric and

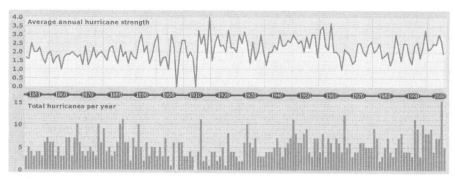

Illustration 144: Number of hurricanes and average strength for North America, 1851-2006. Source Associated Press.

oceanic conditions known to produce active hurricane eras with 25-40 year alternating periods of active/inactive hurricane seasons. It is strongly related to monsoon rainfall patterns over western Africa and the Amazon Basin, and to Atlantic Ocean temperatures.[439] NOAA's official paper on recent tropical storm activity states, "NOAA research shows that the tropical multi-decadal signal is causing the increased Atlantic hurricane activity since 1995, and is not related to greenhouse warming."

While the frequency of hurricanes has not been noticeably affected by global warming, most experts expect the strength, or intensity, of tropical storms to increase. This is because the energy that drives cyclonic storms comes from warm surface water—the higher the sea surface temperature (SST) the more intense the storm. That is the theory, but so far there is no indication that this is happening either. There has been a noticeable shift in rainfall patterns over the past century. The increase in rainfall in some areas has been credited with increased forest growth in some areas (page). A worldwide average increase is harder to detect because of the impact of the same climate cycles that affect the hurricane pattern.

The other severe weather phenomena frequently seen in North America and elsewhere, are tornadoes. In *An Inconvenient Truth,* Al Gore makes the claim that 2004 was the most active year for tornadoes ever in the United States, and that there has been a steady trend in increasing tornadoes as the globe has warmed. Again, the data presented by climate change activists are not accurate. During the time from 1950 to 2000, the technology and network for detecting tornadoes improved vastly. According to NOAA: "With increased national doppler radar coverage, increasing population, and greater attention to tornado reporting, there has been an increase in the number of tornado reports over the past several decades. This can create a misleading appearance of an increasing trend in tornado frequency."

A better understanding of the tornado activity trend in the US can be gained by looking at the most violent tornadoes in the past. These strong to violent tornadoes, rated from category F3 to F5 on the Fujita scale, are more likely to have been reported in the past, thus providing a more accurate indication of historical tornado frequency. As the bar chart in Illustration 145 indicates, there has been no discernible trend in the strongest tornadoes over the past 55 years.

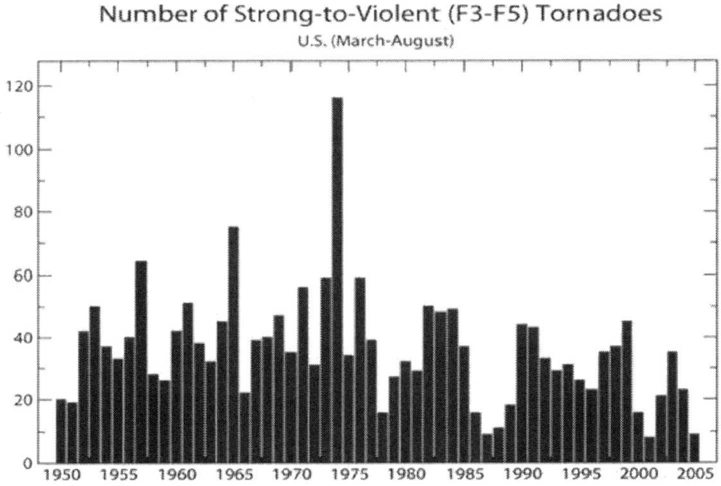

Number of Strong-to-Violent (F3-F5) Tornadoes
U.S. (March-August)

Illustration 145: Frequency of strong tornadoes since 1950.
Source NOAA.

Climatologists need to show that other, related aspects of climate such as winds and moisture levels are also changing as a result of GHG emissions. Climate scientists at Lawrence Livermore National Laboratory in California, have studied satellite measurements of atmospheric water content. They found that total water vapor over the oceans had increased since satellite records became available in 1988,[440] but whether this was due to human activity remains an open question.

Scientists at NASA's Jet Propulsion Laboratory and Goddard Space Flight Center used five years of rainfall observations collected by the Tropical Rainfall Measuring Mission satellite to identify the strongest influences on global rainfall patterns. The two major factors in global precipitation are seasonal variation and the effect of El Niño-la Niña cycles that occur over a period of several years.

As stated earlier, changes in weather and rainfall patterns are cyclic. Some areas become dryer and others become wetter. In terms of crop production, areas in the temperate latitudes have been getting wetter and having better harvests. Areas around the equator have experienced drought conditions. But overall, moderate global warming should mean an increase in global average rainfall. Given the world's ever-increasing population, this might not be a bad thing at all.

Mass Extinctions

One reason so many environment organizations have jumped on the human-caused global warming bandwagon is the reported link to species loss. The poster children for this concern are polar region animals, specifically, penguins and polar bears. There is no doubt that these two species are being affected by environmental changes. But since climate is always changing, it is hard to say whether the changes are due to global warming or other factors. Regardless of the source of change, the question becomes, are penguins and polar bears at risk?

Penguins are aquatic, flightless birds that live almost exclusively in the Southern Hemisphere. There are between 17 and 20 living species of penguin, depending on which authority is consulted. The largest living species is the Emperor penguin, *Aptenodytes forsteri,* which average 42 inches (1.1 m) in height and weigh about 75 lbs. (35 kg). Penguins are popular around the world, noted for their tuxedo-like coloration, waddling gait and curiosity about humans.

Penguins have been on Earth much longer than people. According to fossil evidence, the earliest penguins appeared between 130-65 mya, originating in the core of the ancient supercontinent Gondwanaland. At that time, Antarctica was still attached to both Australia and South America, with New Zealand close by. By 70 mya, the part of Gondwanaland where penguins originated had drifted much further south. This movement carried penguins to a cooler climate than the one in which they evolved.[441]

During the late Eocene and early Oligocene (40-30 mya), several species of enormous penguins appeared. These man-sized, prehistoric giants stood as tall as 6 feet (1.8 m). The interesting thing about these penguins is that their remains have been found as far north as Peru—they didn't need ice to survive.[442]

Over time, continued continental drift and global cooling helped transform the penguin into the cold loving birds we have come to know. In Illustration 146, the climatic transformation of Antarctica from a temperate continent to an ice covered land is depicted. Starting 40 mya Antarctica had no ice cover (a), but, as the circumpolar current formed 25 mya it became partially (b–c), and then fully covered in ice (d–f). The locations of the oldest and biggest

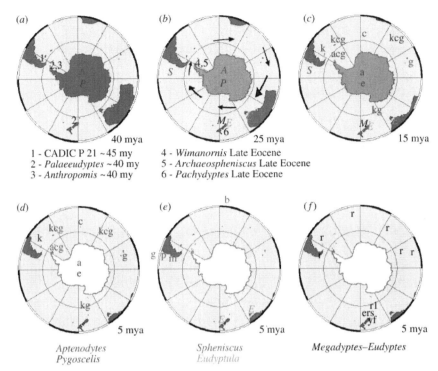

1 - CADIC P 21 ~45 my   4 - *Wimanornis* Late Eocene
2 - *Palaeeudyptes* ~40 my   5 - *Archaeospheniscus* Late Eocene
3 - *Anthropornis* ~40 my   6 - *Pachydyptes* Late Eocene

Aptenodytes   *Spheniscus*   *Megadyptes–Eudyptes*
Pygoscelis   *Eudyptula*

Illustration 146: Penguin evolution and radiation in Antarctica starting 40 mya.
Source Baker, et al, 2006.

penguin fossils are numbered 1-6. As Antarctica became ice-encrusted, modern penguins expanded to islands nearby, and later to the southern continents.

Penguins only reached the cooler tropical waters surrounding the Galapagos Islands about 4 mya. It is thought that the warm water around the equator has prevented them from spreading to the Northern Hemisphere.[443] Clearly, penguins have been able to adapt to changing environmental conditions in the past, and there is good evidence that they remain robust and flexible today.

Adélie penguins, *Pygoscelis adeliae,* are the smallest of the penguins living on the Antarctic continent. Adults average about 28 inches (70 cm) tall and weigh about 8 to 9 lbs. (4 kg) —much smaller than their larger cousins, the emperor penguins. Named after the wife of French explorer Dumont d'Urville, in the 1830s, they are the most studied of all the penguin species. There are over 2.5 million breeding pairs living in the Antarctic region. These penguins nest and breed on rocky, ice-free beaches in large colonies, numbering tens of thousands of birds. Though Tux, the mascot of the open source *Linux* computer operating system, does not represent a single species of penguin, many believe that he resembles an Adélie Penguin.

Studies of the eating habits of Adélie penguins have uncovered some surprising results. It seems that the diet of Adélie penguins in Antarctica changed significantly about 200 years ago.[444] Scientists have attributed the shift in diet to whaling and other hunting in the region during the 19th and 20th centuries. The population of krill in southern seas exploded after Antarctic fur seals and krill-eating whales were hunted nearly to extinction. Man's removal of krill's natural predators enabled the tiny crustaceans to proliferate.

Illustration 147: Tux, the Linux mascot, by Larry Ewing.

Chemical analyses of the penguins' eggshells indicate that the birds' primary diet shifted from fish to krill, previously a secondary food source. Why an abundance of krill would cause penguins to suddenly change from eating fish is a source of speculation among scientists. Charles H. Peterson, a marine ecologist at the University of North Carolina, commented: "It's a cosmic irony of foodweb ecology that a rare species is only rare because it's kept in check by predators," adding "Maybe krill was one of [the penguins'] favorite foods all along."[445] The Adélies seem to be thriving on their new diet, penguin populations have risen by 40% over the past twenty years.[446]

Krill are amazingly prolific, when measured by biomass they are the most successful animal on the planet, totaling some 725 million tons. There have been recent warnings that krill are threatened, not only by global warming but by human harvesting. Fishing fleets were projected to catch 764,000 tons in 2008.[447] This "over fishing," a catch totaling 0.1% of the world's krill, stands in comparison with the 300 million tons of krill eaten by penguins, fish, seals, and whales each year.

If penguins are the stars of the Antarctic, then polar bears are the headliners for the Arctic. Modern bears have been around for a much shorter time than penguins. Bears are large, omnivorous mammals that first evolved around 5 mya, though some of their distant relatives were in Asia during the late Eocene, 37 mya. Modern bears are a widespread, successful species, appearing in a variety of habitats throughout the Northern and parts of the

Southern Hemispheres. The white bear of the Arctic is an even newer resident of planet Earth.

Our knowledge of the development of polar bears is well-documented by fossil transitions. Scientists theorize that between 100,000 to 250,000 years ago, during the mid-Pleistocene, a number of brown bears (*Ursos arctos*) became isolated by glaciers. While many probably perished on the ice, they did not all die out. The survivors offspring underwent a rapid series of evolutionary changes in order to survive. Some think this was possible because of the small population, and extreme selection pressure. The end result was a new species of bear adapted to harsh Arctic conditions—the polar bear.

Illustration 148: A mother polar bear with cubs. Source www.firstpeople.us.

The first "polar bear," *Ursus maritimus tyrannus,* was essentially a brown bear subspecies, with brown bear dimensions and teeth. Over the next 20,000 years, body size reduced and the skull elongated. As late as 10,000 years ago, polar bears still had a high frequency of brown-bear-type molars. Only recently have they developed what biologists call polar-bear-type teeth: smaller molars and stouter canines that reflect a more carnivorous diet.[448]

There are claims that polar bear numbers have been plummeting because of melting polar sea ice and that the polar bear will soon go extinct. This claim is hard to substantiate, as *New Scientist* stated:

"There are thought to be between 20,000 and 25,000 polar bears in 19 population groups around the Arctic. While polar bear numbers are increasing in two of these populations, two others are definitely in decline. We don't really know how the rest of the populations are faring, so the truth is that no one can say for sure how overall numbers are changing."[449]

Many experts think that hunting and threats to habitat are much more important factors than global temperatures. While estimates of polar bear losses due to global warming are about 15 animals a year, hunting losses are 49 bears per year.[450] There are few accurate numbers available, and there are many reasons for polar bear populations to decline—but are they?

Mitchell Taylor, a polar bear biologist with the Canadian government, confirms what Inuit hunters have said for a long time: polar bears who live along the southeast coast of Baffin Island, in northern Quebec, and the northern coast of Labrador are healthy, and growing in numbers. "The Inuit were right. There aren't just a few more bears. There are a hell of a lot more bears," Taylor said, in an interview. Writing in the Toronto Star, in 2006, he stated: "Of the thirteen populations of polar bears in Canada, eleven are stable or are increasing in number. They are not going extinct, or even appear to be affected at present."

Global warming might actually be helping polar bears in that area. A reduction in ice cover creates better habitat for seals, which are the bears' main food, while on land, blueberries, which the bears adore, become more plentiful. Taylor says he's seen bears so full of blueberries they waddle. "Life may be good," said Taylor, "but good news about polar bear populations does not seem to be welcomed by the Center for Biological Diversity. It is just silly to predict the demise of polar bears in 25 years based on media-assisted hysteria."[451]

What can we learn from the penguin and the polar bear, species that, like Homo sapiens, are products of the ice age? For one thing, they continue to adapt to changes in their environment. Penguins opportunistically changed from fish to krill as krill became more abundant. Polar bears are now building their winter dens on land instead of on pack ice. Both polar bears and penguins instinctively know what humans native to Earth's coldest regions also understand—you have to adapt to survive.

Ecological activists are always saying we should learn from the wisdom of indigenous peoples. During an NPR story on the residents of Barrow, Alaska, the northern-most village in the US, geologist Richard Glenn was interviewed. When asked about the impact of climate change on the Eskimo's way of life Glenn, who is half Inuapiat, said "yeah it's changing, but it's our job to know these changes, and so we live with change." He added, "I think MTV has more affect on us than global climate change."[452]

Species go extinct all of the time—the ultimate fate of *all* species is extinction. Instead of agonizing over the fate of the penguin and the polar bear, we should worry about the fate of our species. If we don't learn from nature—which dictates that the future always belongs to the adaptable—humans won't outlive penguins or polar bears, let alone evolutionary superstars like the cockroach and the crocodile.

Plague, Pestilence and Famine

Climate scientists have noted that severe weather in the climate record is closely linked to calamities in the past. Disastrous floods in China, in 1331 and 1332, killed 7 million people and rank among the worst weather disasters in history. Those floods have been linked to subsequent epidemics of the bubonic plague both in China and Europe. In Europe and China, in the 14th century, the population was cut in half by plague, famine and war. Storms also increased during these intervals. Climate history is clear, it suggests that times of global cooling, not warming, are more perilous. Yet, warnings of plague, pestilence and famine caused by global warming are posted around the world.

The diseases most frequently projected to spread because of global warming are called vector-born diseases. These diseases are not transmitted to new hosts directly, instead they depend on a carrier organism called a vector. Most prominent among vector-born diseases is malaria, a sickness caused by parasites carried by mosquitoes. Each year malaria infects 500 million people, causing more than a million deaths. According to Pim Martens, writing in *American Scientist:*

> "The spread of this disease is limited by conditions that favor the disease vector (the malarial mosquito Anopheles) and the protozoan parasite (Plasmodium). The malarial mosquito is most comfortable at temperatures of approximately 20 to 30 degrees centigrade and at a relative humidity of at least 60 percent."[453]

This seems a cogent argument, linking the spread of disease to global warming. But Martens' statement isn't totally accurate. Although today, malaria is thought of as a tropical disease, historically malaria has been found most everywhere around the world. According to entomologist Paul Reiter, the first major reported outbreak of malaria occurred in Philadelphia, in the 1780s. Endemic malaria reached 68°N latitude in Europe during the 19th century, where the summer mean temperature rarely exceeded 16°C.[454] As recently as the 1880s, malaria was a serious problem throughout all of North America and was present as far north as Finland. It is only since 1950 that malaria has been eliminated from Europe and North America and from large areas in Central and South America. History proves that malaria can flourish in colder regions when the mosquitoes that carry the disease are not kept under control.

Reports that malaria is on the rise throughout Africa, South America and elsewhere have been attributed to a decrease in DDT use by those countries, not global warming.[455] Tony McMichael, Professor of Epidemiology at the London School of Hygiene and Tropical Medicine, said of the future threat from diseases such as malaria, yellow fever and dengue fever, "There may or may not have been recent influence of climate change but if there has been, we certainly can't detect it against the background noise."[456] The most disturbing thing about the vector-borne disease/global warming link is just how ill-informed the so-called experts were who first made these claims in the IPCC reports. As Reiter told the BBC, "The bibliographies of the nine lead authors of the health section show that between them they had only published six research papers on vector-borne diseases."[457]

Sidney Shindell, Professor Emeritus at the Medical College of Wisconsin, has fought disease in Western Europe, North Africa, the Middle East, the Caribbean, Central Africa and the Pacific Rim—thirty-five countries in all. After studying disease patterns around the world for over 50 years, he was asked to evaluate the potential risk of plague from global warming as part of a study by the American Council on Science and Health (ACHS). The ACSH based its work on the assumption that the IPCC's predictions are correct. Summarizing the report's conclusions, Dr. Shindell stated:

> "The global burden of disease is formidable. Well understood public health measures could significantly decrease the current incidence of premature death, but resources for applying these measures are currently inadequate. Thus, work toward increasing these resources is prudent, regardless of the prospect for climate change."[458]

Dr. Shindell concluded that "nearly all of the potential adverse health effects of the projected climate change are significant, real-life problems today that have long persisted under stable climatic conditions." In short, global warming will not appreciably change the world's health outlook and any preventative steps we might wish to take, should be taken anyway. No increase in plague is foreseen.

The threat of pestilence conjures images of swarming locusts, stripping crops bare and causing disaster on a Biblical scale. The news media breathlessly reported on the vast swarms of locusts that were eating Africa in 2004, only to have the ravenous insects inconveniently disappear, as they always do, after swarming. In 2005, the *Independent* reported, "Freak swarms of locusts devouring vineyards in and around the northern Italian province of Alessandria... threatening this year's production of a venerable wine." The locals had no doubt the locust outbreak was linked to global-warming.

While it is true that warmer temperatures and wetter weather may be beneficial to some pests, they also favor the natural predators that feed on them. If natural pests have been spreading across the world in recent years, it is more likely the result of invasive species (page 99). Around the world Formosan termites, African bees, fire ants, kudzu and zebra mussels are on the rise. This can be blamed on human beings, but not on global warming.

The IPCC admits that rising levels of CO_2 and increased precipitation can lead to increased global food production. This is projected for temperature rises of 1-3°C, which are most probable. This prediction comes with medium confidence—a 50/50 chance of being correct. William Kininmonth, speaking at a global warming debate at the University of Tasmania, summed up the situation saying, "In the context of pestilence, famine and warfare it is ludicrous to suggest that global warming might be considered the biggest threat that humankind faces in the 21st century." We agree. Even with six and a half billion people on the planet, there is more than enough food to feed us all. The harder problem is distributing that food to the hungry around the world. Famine, at least famine caused by global warming, doesn't seem to be a problem.

When examined in detail, all the predictions of disaster prove false— exaggerations seized on by activists and the media. The ocean will not rise suddenly, there will be no abnormal increase in hurricanes and tornadoes, and species will die out no faster because of global warming. The level of misery in the world due to plague, pestilence and famine will not increase. There are even indications that the warming may be beneficial.

A Sudden Shock to the System

Faced with a diminishing prospect of catastrophic temperature rise, environmental alarmists have come up with something new to terrify the public. The new terror is the threat of a sudden halt of the "ocean conveyor belt," casting the northern hemisphere into a sudden ice age. A mechanism that can cause large sudden change, triggered without notice by minute changes in temperature, is called a *tipping point.*

While a number of "tipping points" have been postulated, the one that gets the most attention is a shutdown of the ocean current that moderates temperatures in northern Europe. This theory was first proposed by Wally Broecker to explain the onset of the Younger Dryas cold snap at the beginning of the Holocene (page 139). There is good evidence for periods of rapid climate change linked to the ocean thermohaline circulation, also known as the meridional overturning current (MOC). But there are a number of reasons to think that such circulation shutdowns are not as easy to trigger as the alarmists suppose.

The proposed cause of the Dryas periods is the sudden dumping of huge quantities of fresh water into northern ocean regions. There is geological and climate proxy data that support this hypothesis. Along with Lake Agassiz, which triggered the Younger Dryas by emptying into the Arctic sea, evidence pointing to the existence of other glacial lakes has been found. Geologist J. Harlen Bretz, theorized that titanic floods, originating at a colossal lake in the area of Montana, were responsible for etching the Channeled Scablands of eastern Washington state. First proposed in 1923, Bretz's theory took decades to gain acceptance, but today, the existence of glacial Lake Missoula, and other ice-dammed bodies of freshwater at the end of the last glacial period, is widely held. Part of the supporting evidence consists of sand beds hundreds of feet thick, a thousand miles off the northern Pacific coast, carried there by the flood waters.

Ice-dammed glacial lakes releasing huge floods are now seen to be a normal part of the transition from glacial to interglacial conditions. High-resolution sonar images of the English channel have revealed a deep scar in the seafloor limestone. It is thought that a torrent of water, 400,000 years ago, sliced through a natural chalk ridge at what is now the Dover straits, forming the English Channel.[459] Prior to this event, England's southern coast was reachable from France by a broad land bridge. A similar flood widened and deepened the Channel during the Eemian interglacial, 125,000 years ago. Given the magnitude of these cataclysmic events, it is understandable that they could disrupt normal ocean circulation. And that is the point—there are no gigantic, ice-dammed glacial lakes on Earth today, waiting to trigger a sudden shift in climate.

All the solid examples of ocean current disruption are related to significant forcing events. The conditions necessary to cause such disruptions do not exist today. Recent studies have found that the thermohaline circulation is much more robust than previously thought. Evidence has been found that the MOC diminished significantly during the depths of the Pleistocene, but did not stop. The ocean conveyor belt still functioned with a third of the planet covered with ice. At the end of the previous glacial period, Earth's climate resumed its warming trend, despite several, sudden shocks to the system. According to the IPCC, no climate models predict "an abrupt reduction or shut-down."[460] Those who claim that disruption of the MOC can occur suddenly without a major trigger event, and that such disruptions cause "irreversible" climate change, are on very shaky ground.

Climate scientists fear that the Arctic has entered an "irreversible" phase of warming. They believe global warming is melting Arctic sea ice so rapidly that the region is beginning to absorb more heat from the sun, causing the ice to melt still further and so reinforcing a vicious cycle of melting and heating. The loss of highly reflective sea ice, which has helped to keep the climate stable for thousands of years, is another form of tipping point. Yet Earth has been ice-free in the past and has returned time and again to frozen ice age conditions.

Other tipping points involve thawing Siberian peat bogs, bleaching coral reefs, and a release of methane from the ocean floors. Some reports say the tipping point has been passed and others say it is ten years in the future. There is scant scientific evidence to support any of these claims. A sudden, catastrophic event is always possible—Earth's past is littered with them. Such events could happen again, mankind has no control over them. Add meteor strikes, the Sun suddenly swelling, and a Corillian death ray[†] to the list of Chicken Little scenarios. Such claims are more science fiction than science.

Global Warming Summarized

After examining the science, the IPCC's claims and predictions, and the possible impact realistic levels of global warming could have on our world, we conclude that there is no imminent threat—not to nature, people or human civilization. While it is possible that there will be continued mild warming in the future, this may not come to pass. Our climate may take a turn for the colder—no one really knows (those who claim they do are lying). Earth is still in an ice age, and will remain in an ice age until all the

---

† From the movie *Men In Black:* "There's always an Arquillian battle cruiser, or a Corillian death ray, or an intergalactic plague that is about to wipe out all life on this miserable little planet, and the only way these people can get on with their happy lives is that they Do... Not... Know about it!"

ice in Greenland and Antarctica have melted. This has happened before and will happen again, with or without the meddling of Homo sapiens. Weather patterns will continue to shift, ocean levels will change, storms will come and go, species will go extinct. In short, things will go on as they always have on planet Earth. If humanity wishes to survive, it must take a lesson from the penguin and the polar bear—adapt.

Given the probable moderate level of temperature rise between now and 2100, we can expect milder winters, extended growing seasons, more precipitation and a few inches of sea level increase. Most of these changes will be beneficial to mankind, while those that are not, like sea level increase, can be handled with modest expenditures and a bit of forethought. The Netherlands is already planning to update their ocean dikes to deal with any increase in sea level.

It is amusing how those who dislike technology and science the most also attribute technological civilization with powers far greater than it possesses. Humanity does not control Earth's climate. We do have an impact on the ecosystem, as do termites, blue-green algae and every other life-form on Earth. Pollution is an ongoing problem, and that includes the reckless consumption of fossil fuels that releases vast quantities of CO_2 into the atmosphere. But the truly global, earthshaking changes to our climate are beyond our control. With all the talk about human-caused global environmental disaster, it's worth remembering that the real global disasters—the mass extinctions and ice ages—had geological and astronomical, not environmental, causes. We can certainly be better stewards of our planet, but this irrational, anti-technology panic must cease—before we do the planet, and ourselves, serious harm.

The actions proposed by the IPCC and other international organizations would not only be ineffectual, they come with a high monetary and human cost. Bjørn Lomborg, former director of Denmark's Environmental Assessment Institute, has estimated that, if Kyoto was fully implemented, fewer than 4,000 people would be saved from premature deaths due to higher temperatures. Noting that the media rants about heat-deaths, while ignoring the larger problem of deaths caused by cold weather, Lomborg summarized the cost of Kyoto this way:

> "We cannot just change the thermostat where temperature impact is on the whole negative. We also end up changing it in all the other places—in the United States, Europe, Russia, China, and India. Here the effect of Kyoto on lives lost directly from temperature impacts would mean an increase in deaths of about 88,000 annually. Thus, to save four thousand people in the developing world, we end up

sacrificing more than $1 trillion, and eighty thousand people. Bad Deal."[461]

The human-caused global warming crisis, while it is not a crisis, should serve as a wake-up call for our civilization. Carbon dioxide emissions could become a major problem in the future if we don't take corrective measures starting today. But there are other forms of pollution that are as important to global warming and more damaging overall than CO_2.

Aerosols and other emissions are rapidly degrading the environment in large portions of the developing world. Development and slash-and-burn agriculture destroys ecosystems. Deforestation and destruction of ocean habitat diminish nature's capability to sequester CO_2 released into the atmosphere. All of these problems can be directly linked to ongoing human development, particularly in the poorer parts of the world. As the less developed nations catch up with the developed nations, the only humane way to ease the environmental impact is to develop affordable, clean energy technologies.

It is time we get serious about the true pending crisis—the worldwide energy shortages that will soon be upon us. Because that is what this crisis is really all about, energy. This will become very clear in the next two chapters, when we present the IPCC's mitigation strategies and then our own suggestions for the future.

This is a problem that cannot be solved by panicky immediate action, no matter how drastic. Only rational, long-term planning will preserve the environment and secure humanity's future. The best immediate course of action an individual can take is to tell the grasping politicians, the fruitcakes and fanatics, where to get off. In the meantime, treat the recent warming in global temperatures as it should be treated—as a blessing. In the words of author, Orson Scott Card,‡ "'global warming' is just another term for 'good weather.'"

---

‡ Orson Scott Card (1951-) American science fiction writer, author of the novels *Ender's Game* and *Speaker for the Dead.*

Chapter 17 Mitigation Strategies

"Skeptical scrutiny is the means, in both science and religion, by which deep insights can be winnowed from deep nonsense."

— *Carl Sagan*

Though the future impact of global warming will not be as devastating as the prophets of doom would have us believe, it does not mean humanity is off the hook. Our species does have an affect on climate and ecology of Earth. It is in our best interest to take care of the planet we live on. What is the best path forward? We will start by examining the IPCC's proposed actions to reduce human-caused greenhouse gases—in IPCC speak, these actions are called mitigation strategies.

The IPCC Suggestions

The suggestions made by the IPCC are contained in the Summary for Policy Makers report generated by Working Group III. According to the report, stabilization of atmospheric GHG levels is possible at a reasonable cost. The IPCC states that to achieve significant GHG reductions would require a "large shift in the pattern of investment, although the net additional investment required ranges from negligible to 5-10%." We think this is a significant understatement of the total cost, if all the proposed strategies are implemented.

The IPCC estimates that stabilizing atmospheric greenhouse gases between 445-535 ppm CO_2 equivalent would result in a reduction of average annual GDP growth rates of less than 0.12%. If stabilized at 535 to 590 ppm, would reduce average annual GDP growth rates by 0.1%, while stabilization at 590 to 710 ppm would reduce rates by 0.06%.[462]

They also concluded that it is often more cost-effective to invest in energy efficiency improvement than in increasing energy supply. Energy conservation through improved efficiency has been talked about for decades. This was suggested by President Jimmy Carter as a fix for the energy "crisis" of the late 1970s.

The specific suggestions made by the IPCC in the AR4 Summary for Policy Makers (SPM) are shown in Table 9. Listed are the key mitigation technologies and practices identified by the IPCC to help slow the advance of human-caused global warning. Note that these are the short-term, immediate strategies. The strategies suggested for the year 2030 are all based on speculative technologies, so we will ignore them for now. This table is taken from the SPM report.

315

Sector	Key mitigation technologies and practices
Energy	Improved supply and distribution efficiency; fuel switching from coal to gas; nuclear power; renewable heat and power (hydro-power, solar, wind, geothermal and bioenergy); combined heat and power; early applications of CCS (e.g. storage of removed CO_2 from natural gas)
Transport	More fuel efficient vehicles; hybrid vehicles; cleaner diesel vehicles; biofuels; modal shifts from road transport to rail and public transport systems; non-motorized transport (cycling, walking); land-use and transport planning
Buildings	Efficient lighting and daylighting; more efficient electrical appliances and heating and cooling devices; improved cook stoves, improved insulation; passive and active solar design for heating and cooling; alternative refrigeration fluids, recovery and recycle of fluorinated gases
Industry	More efficient end-use electrical equipment; heat and power recovery; material recycling and substitution; control of non-CO_2 gas emissions; and a wide array of process-specific technologies
Agriculture	Improved crop and grazing land management to increase soil carbon storage; restoration of cultivated peaty soils and degraded lands; improved rice cultivation techniques and livestock and manure management to reduce CH_4 emissions; improved nitrogen fertilizer application techniques to reduce N_2O emissions; dedicated energy crops to replace fossil fuel use; improved energy efficiency
Forestry	Afforestation; reforestation; forest management; reduced deforestation; harvested wood product management; use of forestry products for bioenergy to replace fossil fuel use
Waste	Landfill methane recovery; waste incineration with energy recovery; composting of organic waste; controlled waste water treatment; recycling and waste minimization

Table 9: The IPCC AR4 mitigation strategies.

There are a number of suggestions in the IPCC report that are eminently sensible; higher fuel economy using hybrids, more efficient airplanes, replacing incandescent lighting with fluorescent or LED based lighting, use of co-generation and more efficient building design are all reasonable things to pursue. However, most of these steps will take time as new, more efficient technology replaces older, worn out units, be they lights, cars, household appliances, or airplanes.

There are several energy sources mentioned in the report that, though they get great press coverage, won't prove very effective in curbing global warming. These sources can be broken into two broad categories; alternative fuels for transportation and renewable energy for generating electricity.

Biofuels

Foremost among the greener alternatives for transportation are biofuels, mainly ethanol and biodiesel. As popular as these "carbon neutral" fuels are with the media and corn belt politicians, they promise much more than they can deliver. In 2005, global ethanol production was 9.66 billion gallons, of which Brazil produced 45.2 percent and the United States 44.5 percent. American output of ethanol, based primarily on corn (maize), is rising by 30% a year. In 2006, the United States produced about 6 billion gallons of ethanol. World leader Brazil is pushing ahead as fast as it can with the expansion of its sugarcane crop, from which its ethanol is made. China has built the world's biggest ethanol plant, and plans more. Global production of biodiesel, made mostly from oilseeds, was almost one billion gallons in 2005. In Europe, Germany is the biggest producer of biodiesel, with output expanding 40-50% a year. Even so, there are problems with biofuels.

In 2006, corn was grown on 70 million acres in the US. If the entire corn crop was converted to ethanol, the amount produced would replace only 12% of US gasoline consumption. Unfortunately, only about 20 percent of each gallon of ethanol made from corn is "new" energy because current ethanol plants use fossil fuels to power part of the production process. As a result, the energy gained would be very small—the equivalent of less than 9 days of oil imports. Car tune-ups and proper tire air pressure would save more energy.[463] Biofuel proponents' solution to the marginal energy contribution is to put more land into production. Even if biofuel production could be expanded enough to make a significant dent in fossil fuel use, there is an environmental cost.

Since the corn crop is already used for human food and animal feed, the rapid expansion of biofuels proposed by some will mean that large tracts of land will have to be converted to agricultural use. The most fertile lands are

317

already dedicated to food production, so the acreage used to grow biofuel feedstock will be lower yielding marginal land. The land that would be used is currently natural grassland or forest, which would have to be cleared before planting, further diminishing the carbon reduction effect of biofuels. Increased runoff and use of fertilizers will add to the environmental damage caused by ramping up for biofuel production. Environmental impact aside, there is also a human cost for biofuels.

As demand for both food and energy increases, competition for arable land could raise food prices enough to threaten the poorer peoples of the world with undernourishment. In late 2006, the price of tortilla flour in Mexico, which gets 80 percent of its corn imports from the United States, doubled. Half of Mexico's 107 million people live in poverty and rely on tortillas as a main source of calories.[464] The Mexican people took to the streets in protest. Do we really want people competing with automobiles for food?

Recognizing that ethanol made from food crops will not make an appreciable dent in solving the fossil fuel problem, and that it comes at a high human price as well, advocates have turned to a more experimental source of biofuel—cellulosic ethanol. Instead of using corn or sugar cane, the cellulose locked up in plants' cell walls is extracted, broken down into its component sugars, and fermented into ethanol. Sources that might provide the raw material for cellulosic ethanol include trees, grasses, and waste organic matter.

The Departments of Energy and Agriculture estimates that the United States could produce 1.3 billion tons of plant matter that, if turned into cellulosic ethanol, could reduce the nation's petroleum needs by 30%.[465] To reach that goal will take more than merely gathering up agricultural waste. High cellulose crops like switchgrass, wheatgrass, and poplar, must be grown specifically for ethanol production. But the main problem is that production of ethanol from cellulose is still in the early experimental stage and has not been demonstrated on a commercial scale.

The Department of Energy (DOE) has granted $375 million to fund three new Bioenergy Research Centers to develop technology for cellulosic ethanol and other biofuels. The DOE also plans to spend $385 million over the next four years, working with commercial partners, to build six pilot plants to produce ethanol from cellulose. Even so, significant progress is not assured. Tim Donohue, a bacteriologist at the University of Wisconsin–Madison, put it this way: "I equate what we're doing to society saying, 'We're going to the moon,' or 'We're going to sequence the human genome.' To me, this is a critically grand scientific mission that we're just setting off on today."[466]Clearly, biofuels are not a panacea for the world's energy ills.

Hydroelectric

The second area of unrealistic expectations is renewable energy for generating electricity, mostly in the form of hydro-power, geothermal, wind and solar. The world's primary sources of energy are depicted in Illustration 149. Hydroelectric plants supply about 715,000 megawatts (MW), some 19% of total world electricity generation. Hydro is also the largest renewable source of electricity at this time, making up 64% of the total renewable non-carbon portfolio. Hydroelectric power plants convert the kinetic energy contained in falling water into electricity. The Sun evaporates water from the ocean, which condenses into water at higher elevations. The energy in flowing water is ultimately derived from the Sun and is constantly being renewed.

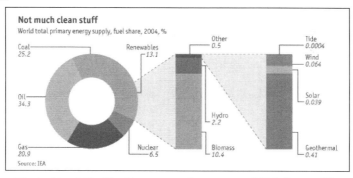

Illustration 149: World sources of energy ca. 2004, Source The Economist.

The largest hydro producing nations are China, Canada, Brazil, the United States, Russia and Norway. Currently, the largest hydroelectric power station is the Three Gorges Dam in China, with a projected maximum capacity of 22,500 MW. The largest plant in the US is Grand Coulee Dam in Washington state, with a capacity of 6,800 MW. While the generation of electric power from flowing water generates no pollution, large dams can have a significant environmental impact. Habitat is lost under the water of dam-created reservoirs, fish migration paths can be blocked, and the natural renewal cycles of rivers downstream are disrupted. Water exiting a hydroelectric turbine usually contains little suspended sediment, which can lead to scouring of riverbeds and riverbanks. Large dams can cause environmental problems both upstream and downstream.

The expansion of hydroelectric power from fresh water sources is limited by geography and the balancing of environmental and human concerns. There are those who advocate small-scale hydro-power plants that would provide electricity for a cluster of houses or a village. While this is a worthwhile solution where conditions permit, care must be taken when relying on hydro-power from small sources. This is due to the fact that smaller water sources may prove to be seasonal or highly variable over a period of years.

For example, in Peru, home to 70 percent of Earth's tropical glaciers, the glaciers in the Andes mountains have lost at least 22% of their area since 1970. According to Peru's National Resources Institute, the melting is accelerating. This threatens not only fresh water supplies but also hydroelectric plants that generate 70% of Peru's power.[467] Expanding hydro-power greatly is not possible and what is achievable in the short-term, may not be prudent in the long-term.

Geothermal

Another area of suggested expansion is electrical generation and, to a lesser degree, heating from geothermal sources. Taking advantage of the natural internal head of Earth's molten interior is a great idea, if you have access to such a heat source. It is a non-polluting, renewable and long-lasting energy source. Geothermal energy is generally harnessed in areas of volcanic activity. The Pacific Ring of Fire is a prime spot for the harnessing of geothermal activity because of significant tectonic activity (see page 134). Other areas, such as Iceland, Italy and around America's Yellowstone Park, have abundant supplies of geothermal energy.

Geothermal energy can be produced by drilling a well into the ground where thermal activity occurs. Once a well has been sunk and a well-head attached, energy can be extracted by injecting water into the ground and collecting the resulting steam and hot water, which can be used for heating or running a generator turbine.

According to the US Bureau of Land Management, the production of steam and hot water from 22 producing geothermal leases on public lands generated over 4.1 billion kilowatt-hours of electricity in 2005; enough for over 500,000 people. Royalties associated with this level of production totaled over $12 million dollars.[468] Most of this takes place in California. In the US, the potential of geothermal energy has been estimated as high as five times the current total use of electricity. Such estimations are theoretical, and current technology does not come close to allowing such levels of generation.[469]

As of 1999, 8,217 MW of electricity were being produced from 250 geothermal power plants in 22 countries around the world. One of the advantages that geothermal generation has over wind and solar power is the ability to run day and night. These plants provide reliable power for over 60 million people, mostly in developing countries. The top ten producers in 1999 are listed in Table 10.

Producing countries in 1999	Megawatts
United States	2,850
Philippines	1,848
Italy	768.5
Mexico	743
Indonesia	589.5
Japan	530
New Zealand	345
Costa Rica	120
Iceland	140
El Salvador	105

Table 10: Top 10 producers of geothermal energy in 1999. Source GEO.

Currently, this method of energy production can only be used where there is significant thermal activity close to the surface. Though geothermal reservoirs as deep as two miles (3.2 km) have been tapped by drilling wells, this still leaves geothermal as a niche player in the overall renewable energy game.

Wind Power

The current star of the renewable energy stage is wind power. During the 1970s, wind turbines produced around 200-300 kW of energy each at a cost of around $2 per kWh. Today, large turbines produce up to 2.5 MW each at a cost of 5-8 cents per kWh. In comparison, electricity generated by a coal plant costs 2-4 cents per kWh.[470] How much potential energy is there in wind? In a study published in the Journal of Geophysical Research, global wind power potential for the year 2000 was estimated to be ~72,000 GW (gigawatt). According to the study: "Even if only ~20% of this power could be captured, it could satisfy 100% of the world's energy demand for all purposes and over seven times the world's electricity needs, 1.6-1.8 TW (terawatt)."[471] But much like geothermal energy, wind power's potential cannot be realized in the real world.

There are problems with wind power that make its widespread use much less attractive. Some areas do not have sufficient wind to make installing turbines viable. Wind generators are practical where the average wind speed is 10 mph (16 km/h) or greater. Even in areas with high enough average wind speeds, fluctuations can limit generation. Too little breeze means no power, while strong winds can force the turbines to shut down to prevent damage. Most wind farms achieve, on average, only 30% of their full capacity. Worse, from a power grid management point of view, is that output can fluctuate by up to 20% of the total national wind capacity in the space of a single hour.[472] In America, there are many areas where wind generation is not cost-effective or practical, including large portions of the south (see Illustration 150).

UNITED STATES ANNUAL AVERAGE WIND POWER

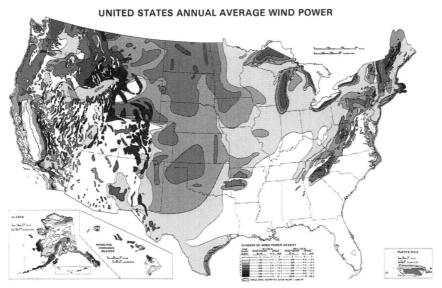

Illustration 150: U.S. wind potential—lighter areas have less potential. Source U.S. National Renewable Energy Laboratory.

Because of these limitations for ground-based wind turbines, a number of proposals have been made to elevate the problem. Among the more ambitious ideas for expanding wind power is a suggestion to loft huge flying generators, a cross between a kite and a helicopter, into the perpetual winds of the jet-stream, six miles above the ground.[473] The idea is being developed by Sky WindPower, a company based in San Diego, California. According to Ken Caldeira, a climate scientist at the Carnegie Institution, harvesting just 1% of the jet-stream's energy would produce enough power for everybody on the planet. Aiming a bit lower, a Canadian company called Magenn Power has proposed wind generators filled with helium. Spinning around its horizontal axis, like an airborne water mill, the generators would fly at altitudes up to ½ mile (1 km). Other proposals include arrays of giant kites and similar, far-fetched ideas.

Beyond the physical and engineering problems, wind farms are running into increasing opposition from the public, a form of the NIMBY (not in my back yard) phenomenon. In Britain, the *Economist* reports: "Wind power, once seen as the eco-friendly cure-all for Britain's energy problems, is attracting unprecedented criticism. The latest campaign, which unites veteran Greens and the opposition Tories, opposes a proposed installation of 27 wind turbines next to Romney Marsh in Kent, a noted bird sanctuary and beauty spot."[474] The truth of the matter is that large wind farms are ugly, despoiling the natural beauty of the landscape and endangering local bird populations. In many highly populated regions, wind farms are simply not acceptable. For this reason, many proposed wind farms are to be built in offshore waters.

Locating wind turbines at sea raises the cost of power generation by as much as 50%, but this disadvantage can be offset by more reliable wind. Offshore saltwater environments can also raise maintenance costs by corroding the towers. Repair and maintenance are usually much more difficult, and more costly, than for onshore turbines. Offshore saltwater wind turbines must be outfitted with extensive corrosion protection measures like coatings and cathodic protection, helping to raise the cost of power generation. Though building offshore doesn't compete with local people or wildlife for land use —and partially hides the unsightly windmills—offshore wind farms have run afoul of public opinion.

One area that is a prime candidate for installation of an offshore wind farm is off Cape Cod, in Massachusetts. The location is one of America's prized resort areas, including such vacation destinations as Cape Cod, Martha's Vineyard and Nantucket—all playgrounds for the wealthy and powerful. The proposed Cape Wind Energy Project, on Horseshoe Shoals in Nantucket Sound, would be America's first offshore wind farm, and would provide 74% of Cape Cod's energy needs with clean, ecologically acceptable power. Though the turbines would be hard to see except on very clear days, and even then, they'd be tiny objects on the horizon, wealthy landowners have bitterly opposed the project.

Leading the opposition to the proposed wind farm are such environmental notables as former CBS news anchor Walter Cronkite, US Senator Ted Kennedy, and his nephew Robert Kennedy, a lawyer for the Natural Resources Defense Council (NRDC). Both the NRDC and Greenpeace strongly back the project.[475] Support for environmental issues evidently evaporates when landowner's property values or the scenic views from their homes are threatened. In the face of this formidable opposition, the project seems to be going ahead, recently gaining authorization from the Massachusetts secretary of environmental affairs.[476]

The fact that wind power, one of the greenest and most advanced renewable energy technologies, can encounter such stiff resistance from people who are supposedly pro-environment, is an indication that the potential of these sources of energy may never be fully realized. This is not a reason to give up on building wind farms—just practical recognition that overly enthusiastic projections for any alternate energy technology should be viewed cautiously.

Solar Power

The last of the major renewable technologies mentioned in the IPCC reports is solar energy. In 2001, solar electricity provided less than 0.1% of the world's electricity. The potential energy present in sunlight is tremendous—after all, the source of energy for all life on Earth and the planet's climate comes from the Sun. More energy strikes the Earth as sunlight in one hour than all the energy consumed by people in a year. According to the US DOE, America could supply its entire energy needs by covering 1.6% of its land area with solar cells. For comparison, the required land area is about 10 times the rooftop area of all single-family homes and is comparable with the land area covered by the nation's federal highways.[477]

Because of day/night and time-of-day variations in insolation and cloud cover, the average electrical power produced by a solar cell over a year is about 20% of its production rating. Solar cells have a lifetime of approximately 30 years and, though they incur no fuel expenses, they do involve initial capital cost. The cost of electricity produced by solar cells is calculated by amortizing the capital cost over the lifetime of the cells. The electrical output produced diminishes over the cells lifetime, further increasing overall cost.

Unfortunately, even though there have been major advances in solar energy technology in recent decades, solar is a far less mature technology than those already mentioned. The efficiency of photovoltaic (PV) cells has increased from 6% when they were first developed to around 15%. Their cost has dropped from around $20 per watt of production capacity in the 1970s to $2.70 in 2004.[478] But that progress has not made solar cells even remotely competitive with wind or fossil fuels. Today, a 2-kilowatt capacity PV system in Tucson, Arizona, would generate about 9.4 kilowatt hours (kWh) per day, a similar size unit in New York would produce 6.2 kWh. A PV system connected to the local power grid costs about $7,000 to $10,000 per kW of capacity, before incentives.[479]

Though solar cells are the most visible means of harvesting energy from the Sun, there are other technologies available. Many residential and commercial buildings use solar energy for heating water. Other commercial installations have been built to generate electricity without using expensive PV technology. Concentrating Solar Power (CSP) uses reflectors to focus and concentrate sunlight on a specific point to boil water. The mirrors may take the form of troughs, parabolic dishes or multiple flat mirrors. The concentrated sunlight produces steam, which is used to spin turbines driving electrical generators. The most complicated part of this type of system is the mirror tracking control, which must slowly move the reflectors to keep them aimed at the Sun. The heat from excess steam can also be used to desalinate water or heat buildings. Another advantage CSP has over photovoltaic cells is that a CSP plant can continue generating power for several hours after sunset using stored thermal energy.

The largest solar power plant in the world has been operating in the Mojave desert of California since the mid-1980s. Containing 400,000 mirrors, covering a total area of 4 square miles (10.3 km^2), it is capable of generating 354 MW of electricity, enough for 900,000 homes. More plants are planned for America's desert southwest. One new CSP plant in Nevada will generate electricity for an estimated 17 cents per kWh.

The US Department of Energy (DOE) published a report in 2005 that identified 13 priority research directions with the "potential to produce revolutionary, not evolutionary, breakthroughs in materials and processes for solar energy utilization."[480] The report's overly enthusiastic tone cannot hide the fact that most of the technologies discussed are highly speculative. The report notes that progress in the proposed research could lead to: artificial "molecular machines" that turn sunlight into chemical fuel; "smart materials" based on nature's ability to transfer captured solar energy with no energy loss; self-repairing solar conversion systems; devices that absorb all the colors in the solar spectrum for energy conversion, not just a fraction; far more efficient solar cells created using nano-technology; and new materials for high-capacity, slow-release thermal storage.

The report further notes that revolutionary breakthroughs come only from basic research and that, "We must understand the fundamental principles of solar energy conversion and develop new materials that exploit them." Solar remains hundreds of times more expensive than other sources and, barring "revolutionary breakthroughs," will not be a major factor. Basing future energy, economic and environmental policy on revolutionary breakthroughs is like basing a retirement plan on winning the lottery.

The bottom line on renewable energy? We should maximize its use where it makes economic and environmental sense. But, as we saw, biofuels cannot replace fossil fuels and, when it comes to generating electricity, renewables hold little hope. The US DOE estimates that 0.007 TW[†] will be available from solar by 2020. The United Nations estimates that the remaining global, practically exploitable, hydroelectric resource is less than 0.5 TW. The cumulative energy in all the tides and ocean currents in the world amounts to less than 2 TW. The total geothermal energy at the surface of the Earth, integrated over all the land area of the continents, is 12 TW, of which only a small fraction could be practically extracted. The total amount of globally extractable wind power has been estimated by the IPCC to be 2-4 TW.

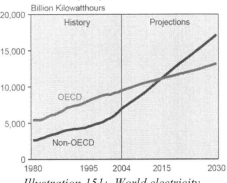

World Electric Power Generation by Region, 1980-2030
Billion Kilowatthours

Illustration 151: World electricity generation. U.S. EIA.

All these clean, renewable sources of energy would not satisfy the world's current appetite for power, roughly 13 TW of continuous energy consumption in 2005. Based on current growth (see Illustration 151), world energy consumption is projected to more than double by 2050. Further, demand will more than triple, exceeding 46 TW, by the end of the century.

Coal's False Promise

Much attention is being given to a non-renewable resource that many countries have in abundance—coal. Currently, coal is one of the major contributors to both greenhouse gas emissions and more traditional air pollution, so how can coal be the fuel of the future? This hope all hinges on an experimental and mostly non-existent technology called "carbon sequestration."

Carbon sequestration is the capture and storage of carbon dioxide and other greenhouse gases that would otherwise be emitted to the atmosphere. The greenhouse gases are to be captured at the point of emission. The captured gases can be stored in underground reservoirs, dissolved in deep oceans, converted to rock-like solid materials.

---

† TW stands for terawatt. A terawatt is 1,000,000 megawatts or 1,000,000,000,000 watts of power, the equivalent of 1.4 million horsepower.

At present, the state-of-the-art technology for existing power plants is limited to "amine absorbents." Amines are organic compounds that contain nitrogen as a key atom. Structurally, amines resemble ammonia (NH_3) with one or more of its hydrogen atoms replaced by an organic molecular group. It is from such a nitrogen-based group that amino acids, the building blocks of protein molecules, get their name. Amines are used extensively in the petroleum refining and natural gas processing industries. The process works as follows:

1. Flue gas that would normally go out the stack is bubbled through a solution of water and amines in what is called a contactor vessel.

2. The amines in the water react with the carbon dioxide in the flue gas to form an intermediate chemical called a "rich" amine. Rich amines are water soluble and stay in the water solution.

3. Some of the flue gas bubbles out of the top of the amine solution and is emitted to the air, but a portion of the carbon dioxide has reacted with the amines and remains in solution.

4. The rich amines are pumped to a vessel where they are heated to make them decompose back into regular "lean" amines and carbon dioxide gas.

5. The pure carbon dioxide gas is collected from this vessel and the regular amines are recycled to the flue gas contactor.

Of course, something must then be done with the captured CO_2. There are three basic options: Use the carbon dioxide as a value-added commodity; store the carbon dioxide in underground formations; or convert the carbon dioxide to methane, biomass, mineral carbonates, or other substances.

Currently, 98% percent of all coal-fired power plants burn their fuel in air and exhaust flue gas that contain carbon dioxide in moderate concentrations (3-12% by volume). Retrofitting carbon dioxide capture to these facilities is expensive. For pulverized coal plants, the cost of carbon dioxide capture, transport, and storage in an underground formation would add at least 70-100% to the cost of electricity.[481]

Emerging R&D technologies are attempting to lower the cost to less than a 20% increase in the cost of electricity compared to a non-capture counterpart. A new technology for coal-fired power plants, integrated gasification and combined cycle (IGCC), has much lower cost for carbon dioxide capture and storage because of inherent characteristics of the process. Equipping an IGCC plant for capture and storage would add at least 30% to the cost of electricity.

A different approach is being investigated by researchers from the Ohio Coal Research Center at the University of Ohio. They are among several groups of scientists trying to capture coal's excess CO_2 using photosynthesis. This approach feeds algae on carbon dioxide exhaust and sunlight, resulting in biomass. It is estimated that algae could recapture 75% of the CO_2 emitted by a coal-fired power plant, at the same time producing 8,000 gallons (30,000 liters) of biodiesel and enough leftover plant material to make (9,000 liters) of ethanol.[482] This totally defeats the purpose of carbon sequestration and biofuel use. If the carbon captured is eventually released into the atmosphere by burning the biodiesel and ethanol, it will have the same effect as releasing it directly. This scheme amounts to a carbon dioxide shell game, not solving the emissions problem, but obscuring it with technological slight of hand.

Other concepts for converting carbon dioxide to different chemicals, especially fuels, are in the very early stages of research but disposal of the carbon remains the main stumbling block. To quote from the US DOE report on solar energy:

> The challenge for carbon sequestration is finding secure storage for the 25 billion metric tons of CO_2 produced annually on Earth. At atmospheric pressure, this yearly global emission of CO_2 would occupy 12,500 km^3, equal to the volume of Lake Superior; it is 600 times the amount of CO_2 injected every year into oil wells to spur production, 100 times the amount of natural gas the industry draws in and out of geologic storage in the United States each year to smooth seasonal demand, and 20,000 times the amount of CO_2 stored annually in Norway's Sleipner reservoir. Beyond finding storage volume, carbon sequestration also must prevent leakage. A 1% leak rate would nullify the sequestration effort in a century, far too short a time to have lasting impact on climate change. Although many scientists are optimistic, the success of carbon sequestration on the required scale for sufficiently long times has not yet been demonstrated.[483]

In short, no one knows if carbon sequestration will even work. This has led to suggestions that CO_2 emissions be offset by planting new forests. The DOE estimates that about 220,000 acres would be required to offset emissions from an average size power plant. Given competition for land use from food production and possibly biofuels, this option seems untenable.

Meanwhile, worldwide use of coal as a primary energy source continues to grow. With oil and natural gas prices expected to continue rising, coal is an attractive fuel for nations with ample coal resources. Despite government policies aimed at reducing coal use, its share of world energy consumption is projected to increase further (see Illustration 152). China is building a new coal plant every week to ten days and will continue to do so for the foreseeable future. Germany is faced

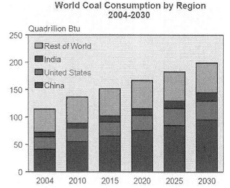

Illustration 152: World coal consumption. Source U.S. EIA.

with aging existing coal plants, green opposition to retaining nuclear power stations, and politically unreliable supplies of natural gas from Russia. Their only option is to build new coal generating plants, even though advanced sequestration technology is not yet available. The US, with a 200 to 400 year supply of indigenous coal, is also tempted by the allure of anthracite. But coal has a dark side.

In 1952, London experienced a period of thick smoggy conditions that came to be called the "Killer Fog." On December 5th, a toxic mix of dense fog and sooty black coal smoke enveloped the city, causing residents to gasp for air. Soon, the roads were littered with abandoned cars as visibility fell to less than a foot. Before long, Londoners started dying. A study in the journal *Environmental Health Perspectives*, estimated that as many as 12,000 may have been killed before the fog lifted, four days later.[484]

Since requiring non-smoking fuels be used within the city, London is no longer threatened by killer fog, but other cities around the world are not so fortunate. In developing countries such as Mexico, India and China, rising emissions from trucks and automobiles combined with soot and sulfur from burning wood, dung and coal have created conditions similar to those that afflicted London. In China, the sulfur dioxide produced by burning coal poses an immediate threat to people's health. In China alone, the infamous "brown clouds" contribute to an estimated 400,000 premature deaths a year. It also causes acid rain that poisons lakes, rivers, forests and crops across Asia and around the world.[485]

With its economy expanding at an annual rate of 10%, China already uses more coal than the United States, the European Union and Japan combined. With China rapidly adding new coal-fired plants, each with an operational

Illustration 153: Killer Fog hits London.

lifetime of 75 years, the brown clouds can only be expected to grow worse in the future. And China is not alone—India is right behind China in stepping up its construction of coal-fired power plants.

What is worse, around the world huge underground coal fires are pointlessly burning millions of tons of coal each year. Mega-fires burning in India, China and elsewhere in Asia spew out huge volumes of air pollutants, force local residents to relocate, and ruin the land above them. Coal fires burn throughout the Chinese coal mining region, an area that stretches 3000 miles (5000 km) from east to west and about 450 miles (750 km) from north to south. Chinese fires alone consume 120 million tons of coal each year, equivalent to the annual coal production of Pennsylvania, Ohio, and Illinois combined.[486] In India, coal fires also rage. Though naturally-occurring coal fires can be dated back to the Pleistocene, mining activity only aggravates the situation by making new fires more probable.[487]

Suffocating smog is not the only way that coal kills. Each year, the mining of coal takes a terrible toll among coal miners. In the US, 688 coal miners died between 1990 and 2006. In 2006, US coal mining deaths soared to a 10-year high, reversing an 80-year trend of fatality reduction.[488] In China, official government statistics estimate that 250,000 people have died in the mines since 1949. More than 2,000 perished in 2007 alone.[489] No one knows how many people coal has killed, but the truth is plain to see—coal is the most harmful and deadly energy source on the planet.

Methane Ice

There is another, new source of hydrocarbon fuel being talked about for the future—gas hydrates. Gas hydrates, also known as *clathrates*, are actually natural methane-water ices, which form under conditions of high pressure and low temperature in many areas worldwide. A hydrate is a crystalline solid consisting of gas molecules, usually methane, each surrounded by a cage of water molecules. Methane hydrate looks very much like water ice. It is stable in ocean floor sediments at water depths greater than 300 meters and, where it occurs, it is known to cement loose sediments in a surface layer up to several hundred meters thick. Since their discovery, scientists have been investigating seafloor hydrates as a future source of energy.

As with all fossil fuel resources, it is hard to estimate just how much methane is trapped in clathrates worldwide. Deposits have been identified off the coasts of every continent and several of the larger lakes in Central Asia are cold enough to permit clathrate formation. In theory, this undersea source of methane could greatly expand the world's supply of relatively clean-burning natural gas. Deposits may surpass 100 times proven natural gas reserves and twice all other sources of hydrocarbon-based fuels combined.

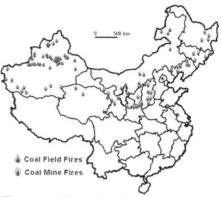

Illustration 154: Distribution of coal fires across China from satellite data.

A Japanese collaboration has drilled about 30 wells, with a time-line to start production and distribution of methane from hydrates by 2016. China is reported to have recovered the first methane-bearing samples from the South China Sea. In the US, a consortium of government agencies and petroleum companies has been drilling for clathrates in the Gulf of Mexico with some success.

It now appears that these vast stockpiles of frozen seafloor methane are more unstable than previously thought. The sudden release of methane from seafloor clathrates may have been responsible for sudden global warming events in the past. Scientists have blamed tsunamis resulting from sudden sea floor movements for several catastrophic events. Around 8,000 years ago, the Storegga submarine landslide resulted in the movement of 700 cubic miles (3000 km³) of ocean floor sediments from an area rich in gas hydrates off the western coast of Norway. This movement generated a tsunami that inundated the coasts of Scotland and Norway. Greenland ice core records show that, following the event, atmospheric methane concentrations increased by 80-100 parts per billion. This increase corresponds to methane releases from the slide debris at an estimated rate of 20-25 megatons each year for several hundred years following the slide.[490]

The release of large volumes of methane from sea floor deposits have been suggested as the cause for warming anomalies like the PETM and even the Permian-Triassic Extinction. A sudden release of large volumes of methane would have a dramatic impact on greenhouse warming, since methane is 20 times as potent a GHG as CO_2. Methane rapidly decomposes in the presence of oxygen so a methane release would show up in geological records as a sudden increase in CO_2 levels. Such a release could deplete oxygen in the ocean and lower atmospheric levels as well. This has led to speculation that a warming trend could trigger a devastating release of methane that would cause a spike in global temperatures or even worse—a worldwide extinction event.

We have to ask ourselves, is developing yet another form of hydrocarbon-based energy worthwhile? Particularly when poking and prodding underseas clathrate deposits may result in devastating tsunamis or catastrophic release of methane that could cause real global warming. We should let this sleeping dog lie.

Renewable Energy Redux

If our goal is to significantly reduce emission of carbon dioxide, use of fossil fuels must be severely curtailed. Illustration 155 provides a breakdown of US energy consumption and CO_2 emissions by fuel type and use. Petroleum is the largest source of CO_2 emissions, accounting for 43% of the total, while

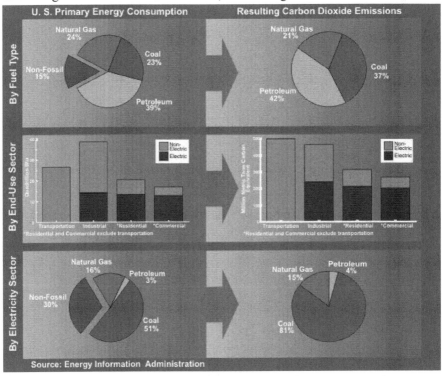

Illustration 155: US energy consumption and CO2 emissions by sector and fuel type. EIA.

coal is second with 37%. The majority of petroleum fuel use is in the transportation sector. This is of particular concern because, as we have shown, there are no suitable renewable replacements for liquid fossil fuels. Since world automobile use is predicted to continue rising, a successful energy strategy for the future must solve the problem of automotive emissions.

It is no surprise that coal is used in more stationary applications than petroleum. In the US, it accounts for 51% of electrical generation and an astounding 81% of CO_2 emissions from power generation. Emanating from fewer locations than petroleum emissions, pollution from coal is a major part of the GHG problem. Because it is inexpensive and abundant, replacing coal as a major energy source will be as difficult as rethinking the automobile. Thus, the challenge for the future is twofold—redesign the automobile and replace coal in electrical generation.

While the preceding examination of potential renewable and existing energy sources may seem discouraging, things are not as bad as they appear. Within the bounds of economic and physical possibility, renewables may be able to provide as much as 25% of the world's energy needs using current technology. What is inadvisable is forcing the adoption of renewable sources that would prove unreliable or excessively costly. This will cause consumer backlash that could damage the appropriate use of renewable energy for decades to come.

Diversifying our sources of energy is a good idea for many reasons, not just to reduce GHG emissions and other forms of pollution. Major fossil fuel deposits are located within the territories of many of the world's more unsavory regimes. Oil money often props up repressive governments and bankrolls international terrorism. Lessening the world's dependency on fossil fuels would diminished these bad actors' income stream—a beneficial side-effect of reducing pollution.

Renewable energy from hydroelectric, geothermal and wind power should be pursued. Solar also has a place but, without significant reduction in cost, will remain a promise for the future. But other steps can be taken now. The intelligent use of energy holds a potential as great as that of current renewable technology. Moreover, other sources of non-polluting energy are available. Building on the contributions of renewables and energy conservation, the next chapter will explain why the future is not as bleak as many fear—technologies already available can be expanded, and world CO_2 emissions can be greatly diminished.

Chapter 18 A Plan for the Future

"When it comes to the future, there are three kinds of people: those who let it happen, those who make it happen, and those who wonder what happened."

—John M. Richardson, Jr

As we have seen, there is little prospect of solving the problem of greenhouse gas emissions using existing renewable energy sources and technologies. The world demand for energy continues to rise and the sources that are best positioned to fill that demand are the ones that are most damaging to Earth's environment (see Illustration 156).

Notice that the largest fuel source is labeled "Liquids," meaning oil. The area with the greatest potential for reducing oil consumption is the transportation sector.

Getting from Point A to Point B

Parts of the transportation industry are already moving towards higher fuel efficiency, with commensurate reduction in GHG emissions. Airlines and airplane manufacturers have taken steps with only minimal prodding by governments. For them, it is a matter of simple economics, the cost of jet fuel is rising and accounting for an ever larger portion of airline operating budgets. For most airlines, fuel is the second largest expense category behind labor. Increase in fuel consumption over the past 15 years was 52.2% from domestic operations and 111.8% from international. When these figures are combined the net total increase is 68.7%.[491]

Even with this increased fuel consumption, aviation is the most carbon-efficient form of transportation generating only 2-3% of total current CO_2 emissions and 13% of CO_2 emissions from all transportation sources. Aircraft account for just 3% of the world's annual petroleum usage and modern aircraft are more than

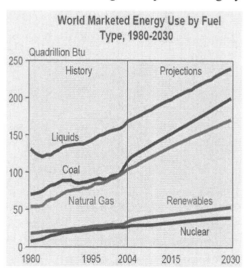

Illustration 156: Growing dependence on fossil fuels. Source U.S. EIA.

335

three times as efficient as today's average car, with fuel efficiencies of 67 passenger-miles per gallon. Next generation aircraft, like the Boeing 787 and Airbus's planned A350 XWB, will increase fuel efficiency to 78 passenger-miles to the gallon, much higher than any modern compact car.[492]

Airbus, which participates in the European Commission's Clean Sky initiative, has a program that should speed development of new technology that helps reduce air transport's impact on the environment. The seven year, €1.7 billion initiative started early in 2007 is expected to reduce CO_2 emissions by 50% through reduced fuel consumption and cut NO_x emissions by 80%. Clean Sky also addresses greener aircraft life cycles, which includes evaluating maintenance and aircraft disposal.

Fuel savings are possible even for older planes. The amount of fuel they burn can be reduced through changes to air traffic control approach patterns and delaying engine start until cleared to an active runway. Airlines operating Boeing 737s, in Europe, have been given approval to use an optimized landing approach that significantly reduces the amount of fuel used during approach operations. Changes in approach patterns reduce CO_2 and NO_x emissions by roughly 20% compared to previous arrival procedures.[493]

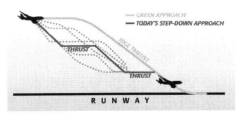

Illustration 157: CDA keeps aircraft higher longer and has them descend at near-idle power. Credit: NAVERUS and AVTECH

In the US, the use of continuous descent approaches can save airlines up to $30 million in annual fuel costs. All of these savings can be realized without increasing the air traffic controller workload.[494] This is an example of how being greener and more efficient benefits industry and the environment.

Today, there are 600 million automobiles on the road worldwide, and that number is forecast to double to 1.2 billion in the next 30 years. The environmental impact of those 600 million vehicles greatly exceeds the impact of the world's 22,000 airliners. Consequently, one of the most promising technologies for the future is that of hybrid electric power for automobiles. The first hybrids became available a decade ago and hybrid models are now offered by most major auto manufacturers. Early hybrids added electric motor-generators to existing internal combustion (IC) based drive trains. The electric motors boosted performance by allowing the use of smaller IC engines. The generator function allowed energy to be recovered when breaking or slowing down—called *regenerative braking*. These features greatly enhanced mileage in stop and go driving.

Unfortunately, none of the early or existing hybrid models allowed their battery packs to be recharged from external sources. All the energy to drive the vehicles still had to come from their IC engines. Also, some hybrids only use electric motors to assist their IC engine and could not run on electric power alone. These vehicles are called *power assist hybrids*. Honda's hybrids, including the Insight, use this type of design. However, starting with the 2006 Civic, some newer model Honda's can run on electricity alone.

Other models, such as the Toyota Prius, are *full hybrids*. This means that the car can move under electric power when it is going slowly, but when speed or acceleration increases, the gasoline motor kicks in automatically. A full hybrid, sometimes called a strong hybrid, is a vehicle that can run on just the IC engine, just the batteries, or a combination of both. A number of enterprising, and ecology-minded hobbyists have taken full hybrid models, expanded their battery capacity and added external recharges—creating *plugin hybrids*. We think that plugin hybrids are the future of the automobile.

The primary advantage of being able to plug in a hybrid is that some of the vehicle's power can come from sources other than oil. A plugin hybrid can be powered by wind, solar, hydro or nuclear power, reducing a portion of its emissions to zero. Another advantage is that, at least for full hybrids, short trips at moderate speeds can be accomplished without ever turning on the IC engine. This would enable many people to effectively drive fully electric vehicles, particularly in large cities and urban areas where air pollution is heavily concentrated. The drawback with electric cars has always been their short range due to limited battery capacity. Plugin hybrids overcome that disadvantage by also having an IC engine ready to kick in whenever needed. They can be electric for local trips and gas powered for long trips.

The Chevrolet Volt, from General Motors (GM), is a hybrid car but it is radically different than any on the road today. While current hybrids, such as the Prius, are *parallel hybrids*—meaning they have a small electric motor that moves the car when it is going slowly, but when speed or acceleration increases, a gasoline motor kicks in—the Volt is a *series hybrid*. It has a powerful 161 horsepower (16 kWh) electric motor that is the only engine that powers the car. This engine is capable of moving the car from 0 to 60 in 8 seconds, and reaching a top speed of 120 mph.

The electric engine gets its energy from a powerful high-voltage battery pack that can store enough energy to drive the car up to 40 miles under standard driving conditions. That battery pack is recharged by plugging the car into a standard home wall outlet. A full-charge cycle takes about 6 hours. Based on current prices, electricity costs should amount to a gas equivalent price of 50 cents per gallon. Studies suggest that 78% of drivers travel less than 40 miles (65 km) per day, making it possible for many people to use no fossil fuel at all. Imagine such a car charging its batteries with electricity generated by nuclear power. We would have a vehicle that leaves a zero carbon footprint.

General Motors is still in the early stages of production planning for this car and the official release date is expected to be late 2010. As of now, GM is not soliciting customers, but an activist web site, GM-Volt.com, is asking people to join a waiting list to help ensure that the Volt makes it to production. But even if this particular car never gets produced, interest in hybrids is strong and growing stronger around the world.

In Europe, at the Frankfurt Motor Show, Volvo unveiled a concept car called the ReCharge. Based on their existing C30 model, the ReCharge is a series hybrid like the Volt. Volvo claims an all-electric range of up to 65 miles. Power is stored in a lithium-polymer battery pack and the car can drive up to 30 miles on a one hour charge. There is an electric motor at each wheel providing all-wheel-drive (AWD) and enhanced traction. An electrical generating unit called an auxiliary power unit (APU) provides power when the batteries run low. The APU, a 1.6 liter gas engine driving an electric generator, is not directly connected to the wheels.

Operating costs for the ReCharge are estimated to be 80% lower than a similar gas-powered vehicle when using battery power alone. For a 90 mile (150 km) drive starting with a full charge, the car will require less than 0.75 gallons of fuel, giving the car an effective fuel economy of 124 mpg. At least on the automotive front the future looks bright, even exciting.

There are two main obstacles to full hybrid production; the high cost of batteries and battery life. The familiar 12 volt lead-acid batteries found in cars cost $40 to $50 per kWh (one gallon of gasoline is equivalent to 8.8 kWh of stored electricity). Nickel-metal-hydride batteries, used in portable electronics, cost $350 per kWh. Newer lithium-ion cells used in the same application cost around $450/kWh. For automotive applications, nickel-metal-hydride batteries cost $700/kWh. Several companies around the world are working on building less expensive lithium-based batteries that will last for 10 years and cost under $300/kWh. Prices are expected to drop dramatically when the batteries enter mass production.[495]

From a technological point of view, the move to full series hybrids means that auto manufacturers will be well-positioned to take advantage of future breakthroughs. This is because the APU can be easily replaced with any other source of electrical power. If fuel-cells become reliable and cost effective, the IC engine can be swapped for a fuel-cell-based power pack. If battery technology advances to the point where energy density and fast recharge time make gas engines obsolete, the APU can be replaced with more battery storage. From a development point of view, the change to hybrid technology allows manufacturers to refine electrical drive trains while awaiting future APU advances. Who knows, in the future we may be driving hybrid cars powered by "mister fusion" generators, like in the movie *Back To The Future II*.

Other modes of transportation need to be improved as well. Trucks are moving to so-called clean diesel, but that is only a small improvement. Both trucks and trains can benefit from hybrid technology. The main obstacle to developing hybrid trains, trucks, and buses is the significantly higher energy levels required. The amount of electrical power generated when decelerating a train or large truck can exceed current battery's recharge capacity. Some companies are developing super capacitors that can handle the high regenerative braking currents, allowing batteries to be charged at a more leisurely pace.

Another developing technology that could replace chemical batteries in hybrid trains or trucks is based on an old idea—flywheels. Not heavy wheels of metal spinning at a few thousand rpm, but high tech carbon composite flywheels, suspended in vacuum by magnetic fields whirling at 50,000 rpm. Some day, departing trains could get their initial energy for free, each time saving the equivalent of several days' worth of electricity usage by an average US household.

Engineers at the University of Texas at Austin are developing an improved flywheel that can store enough energy to accelerate a passenger train up to cruise speed. Such a locomotive flywheel could weigh 5,000 pounds and, when fully charged, its rim may move at the speed of sound.[496] The researchers' experimental flywheel is a cylinder 4.9 ft (1.5 m) in diameter and 4.2 ft (1.2 m) tall. When spinning at full speed, it is designed to store 133 kWh of energy. Flywheel systems can pack more energy than batteries of comparable weight. They can also last decades with little or no maintenance and do not degrade over time as chemical batteries can. It is estimated that a flywheel-based hybrid locomotive could attain a 15% increase in efficiency on a route such as New York to Boston. Light rail commuter trains, which stop and start frequently, could save even more.

Flywheels can provide a boost for smaller vehicles as well. A Dutch company, called Centre for Concepts in Mechatronics (CCM), has developed and tested a flywheel-powered hybrid bus. The prototype bus incorporated a small car engine to keep the flywheel spinning. Running at constant speed, the engine's fuel efficiency was maximized. The flywheel stored up to 3 kWh, running on conventional ball bearings in a vacuum. When needed, the flywheel could supply bursts of 300 kilowatts, the equivalent of about 400 horsepower. According to the developers, the bus "ran like a Porsche," and had 35% better mileage than a comparable-size conventional bus.[497]

CCM's flywheel technology is currently used in a tram from Fraunhofer Gesellshaft, and a light rail vehicle from Siemens. The company has also contracted with French engineering giant Alstom to develop a wireless tram that would recharge its flywheel at passenger stops. But flywheels are not limited to mobile applications.

One of the impediments to wider use of intermittent electrical power sources, such as wind and solar, is that power availability cannot be matched with demand. With solar power, energy for nighttime use must be stored during the day, and with wind power, if there is more energy available than is needed, excess energy is wasted without storage. Several countries have built energy storage facilities, mostly to cushion peak demand spikes. The conventional way this is done is to use excess power to pump water back upstream, into a dam's reservoir. When more power is needed later, the water flows back through the dam's hydroelectric generators. Unfortunately, this technology can only be used where geography allows.

Flywheels have no such geographic restrictions. Flywheels are one type of storage device, collectively known as power-quality units. These devices are used to dampen fluctuations in the power grid's frequency, current or voltage. With the addition of wind and solar generation, the role for such technology becomes even more important. In the US, several companies are already producing commercial flywheels for power-quality applications. Beacon Power, based in Wilmington, Massachusetts, manufactures 9-kWh units used by telecommunications companies to stabilize power in remote locations. Beacon is developing larger, 25-kWh flywheels meant to be installed in arrays of as many as 200. Such arrays could provide 20-megawatt bursts of power to stabilize a grid or store excess energy for later use.

Beyond adding new forms of green power, and storage devices to make the most of their spotty power output, national power grids will need a major overhaul. In Europe, there is already talk of linking the output of wind generators, scattered about the continent, to the huge hydroelectric reservoirs in Norway's fjords.[498] But moving huge amounts of power over long distances raises another problem, transmission loss.

Most existing long distance power lines are based on alternating current (AC). This is because AC is more efficient than the alternative, direct current (DC), over short to medium distances. On average, transmission and distribution (T&D) losses between 6% and 8% are considered normal. According to data from the Energy Information Administration, net generation in the US came to over 3.9 billion megawatt hours (MWh) in 2005 while retail power sales during that year were about 3.6 billion MWh. T&D losses amounted to 239 million MWh, or 6.1% of net generation. Using the national average retail price of electricity for 2005, T&D losses cost the US economy around $19.5 billion.

Over long distances, losses are even higher, sparking new interest in DC power transmission. Most of the transmission lines that make up the North American transmission grid are high-voltage alternating current (HVAC) lines. DC transmission offers great advantages over HVAC, as much as 25% lower line losses. DC is also capable of two to five times the capacity of an AC line at similar voltage, with improved ability to control the flow of power. Any nation or group of nations that wishes to take full advantage of renewable energy in the future needs to plan their power grid on continent-spanning scales. The power grid is not glamorous, usually only attracting attention when it fails, but politicians must start paying attention to power infrastructure, or the future will be filled with brown-outs and rolling power failures across wide regions.

Building for the Long Run

To fully replace older vehicles with new, more efficient ones will take some time. In the US, the average car has a median lifetime of 17 years and Boeing expects 21.4% of the aircraft in service by 2025 to be "holdovers" from today.[499] Other parts of the world's energy-consuming infrastructure last even longer. The average lifetime of a detached house in Japan is 40 years, in the US, it is around 100 years, and even higher in Europe.[500] New solar cells have been developed that look similar to traditional shingles, so retrofitting a house doesn't mean adding a large, ugly solar panel to the roof. Of course, existing structures can be made more efficient by adding insulation and upgrading heating and air conditioning equipment, but sizable savings are best found in new commercial buildings.

New office buildings are being constructed that use zero net energy from the commercial power grid, yet provide normal office lighting levels with

341

comfortable heating and cooling. These so-called z-squared buildings are just now starting to appear on the city skyline. One such building is the new San Jose headquarters of Integrated Design Associates (IDeAs). The IDeAs building, designed by Scott Shell of EHDD Architecture, makes innovative use of natural lighting and generates more electricity than it uses during the day from an array of solar cells on its roof.[501] It is one of a handful of buildings that meet the platinum energy rating of the US Green Building Council (USGBC).

To promote more efficient building design, the USGBC has developed the Leadership in Energy and Environmental Design (LEED) Green Building Rating System,[TM] a nationally accepted benchmark for the design, construction, and operation of high performance green buildings. LEED promotes a whole-building approach to sustainability by recognizing performance in five key areas of human and environmental health: sustainable site development, water savings, energy efficiency, materials selection, and indoor environmental quality. Depending on location, the payback period for a building with built-in solar power ranges between 20 and 35 years. While it used to cost 15% more to construct a green building, today the added cost is only 1-3%.[502] With buildings accounting for 18% of US energy consumption, we need to build more efficient commercial structures as well as more clean power plants.

Another change in building practices that can benefit the environment seems counter-intuitive at first glance—using more wood. Cutting down trees for building material seems anything but green, but it can be. When trees are turned into lumber, the carbon stored in the wood is taken out of circulation unless the wood decays or is burned. If the wood comes from a managed forest, where new trees are planted to replace the ones harvested for building materials, more and more CO_2 is removed from the atmosphere. This is a better use of forest material than cellulosic ethanol, which returns the carbon in the wood back to the atmosphere when the fuel is burned. It is also far better than using concrete. The production of cement, the primary component of concrete, accounts for 5 to 10 percent of the world's total CO_2 emissions.[503] The warm, natural look of wood is better for the environment than cold concrete and steel.

The New Nuclear Age

A major non-polluting energy source that we have not discussed is also mentioned as a mitigating technology in the IPCC report discussed in the previous chapter. Because this clean, carbon-free technology is proven, readily available, and can make the greatest immediate contribution to reducing GHG emissions, we include it here in our road map for the future.

Much to the displeasure of the neo-Luddite wing of the ecology movement, the technology that has the greatest potential to free the planet from fossil fuels—a technology recommended by the IPCC—is nuclear power.

Nuclear power has long been the favorite whipping boy of the ecology movement. An irrational fear of all things nuclear firmly took root in the US in the late 1960s—but the event that shut down the American nuclear power industry was the 1979 reactor accident at Three Mile Island.

Around 4 am, on March 28, through a series of errors—including a valve that was supposed to close but didn't, and a known leak that led operators to conclude high temperatures readings were false—water escaped from the reactor core. Without a way to remove the heat, reactor temperature began to rise. Within a matter of minutes, the operators, believing that water was still circulating through the core, concluded they had a "bottled-up system."

Unable to deal with the unfolding events, by 7 am the operators called a site emergency. By 7:30 am, amid concerns that a hydrogen bubble had formed in the reactor core, a general emergency was declared. No bubble had formed, although months later cleanup crews were astonished to see how much of the core had actually melted. But the melted nuclear core had been contained and the radiation released was minimal. The plant design and safety protocols had worked, despite numerous operator mistakes.

Before the Three Mile Island incident, the United States had 104 nuclear reactors generating electricity—the most of any country in the world. After the accident at Three Mile Island and the anti-nuclear propaganda film, *The China Syndrome,* the country turned its back on nuclear power. A new reactor hasn't been built in the US since. Public opinion forced power utilities to cancel 96 new nuclear projects. If those 96 plants have been built on schedule, along with additional ones that would surely have been ordered over the past 30 years, more than half of US electrical generating capacity would produce no greenhouse gases. Instead, the last nuclear power plant built in the US was started in 1973 and America's existing plants are starting to show their age. Jane Fonda, the star of *The China Syndrome*, has done more to cause global warming than all the people driving SUVs.

Today, fear of nuclear power lingers. In an August 15, 2007 interview on NPR's Morning Edition, David Whitford, editor-at-large at *Fortune* magazine, demonstrated the media's ingrained aversion to nuclear power when he was questioned about the future of the industry. Commenting on the 50 years nuclear power has been in use, Whitford said: "During that time, there have been two terrible accidents: Three Mile Island and Chernobyl. Chernobyl was by far the worst. You had a meltdown, an explosion and 75 people, according to the UN, died as a result of the accident." He continued grudgingly, "Three Mile Island, the best hard evidence that I found, was that no one was significantly harmed."[504]

Two "terrible" accidents, one where 75 people died and a large area was contaminated with radioactivity that persists to this day, the other where no one was significantly harmed. Chernobyl was undoubtedly a disaster, but Three Mile Island was not. At TMI all of the safety features worked, disaster was averted. Yet, in the eyes of the media, the two events are equivalent. Saying that TMI was a terrible accident is like saying that an airplane that loses an engine on takeoff, but still manages to land safely, had a terrible crash. What is the attitude towards nuclear power in other countries? Germany and the Scandinavian nations, in particular, seem to suffer from the same aversion as America. Other countries take a more practical and pragmatic approach.

Worldwide, there are 439 nuclear reactors in 31 countries supplying 15% of all electricity generated. In the 50 years of commercial nuclear power generation over 12,700 reactor years of service have been logged—in that time, the only fatal accident at a commercial plant was Chernobyl (Illustration 158). Despite being the only country to have suffered the devastating effects of nuclear weapons in wartime, Japan has embraced the peaceful use of nuclear energy. Today, nuclear power accounts for about 34% of the country's total electricity production. In April 2006, the Institute of Energy Economics of Japan forecast that, by 2030, nuclear energy's portion will increase to 41%. By that time, ten new nuclear power stations will come on-line.

In France, unlike America, nuclear energy is accepted and even popular. A country that, like Japan, has few natural energy resources, France's decision to launch a large nuclear program dates back to 1973. When events in the Middle East led to the quadrupling of oil prices by OPEC, the French faced what they refer to as the "oil shock." At that time, France generated most of its electricity by burning oil and had no choice but to go nuclear. Today, France has 56 working nuclear plants, generating 76% of its electricity.[505]

Some argue that safety issues and nuclear waste are too intractable to even consider nuclear power. *C'est de la merde,* as the French say. The

Illustration 158: Cumulative reactor hours, showing TMI and Chernobyl accidents. Source IAEA.

only catastrophic nuclear accident occurred at an incredibly primitive, poorly designed, and haphazardly run reactor in the old Soviet Union—a reactor that no modern nation would allow to be built on its soil today. In America, and the other technically advanced nations, nuclear power has proven to be among the safest energy sources available. In the US, this has been achieved with reactors designed and built in the 1960s. Compare a modern automobile with a car from the 1960s. Though the cars of the '60s seemed advanced and modern at the time, today they are amusing, if crude and smelly, antiques. Similarly, nuclear power technology has improved tremendously in the past four decades. Newer, more modern reactor designs are even safer and more efficient than their safe and reliable predecessors.

As for the fear of radiation coming from nuclear power plants, note that radiation is found everywhere in nature. Radon gas is found in rock structures and house basements from New England to Georgia. Our bodies contain radioactive isotopes of carbon, potassium and other elements— fortunately at low concentrations. The burning of coal releases radioactive materials that naturally occur in coal. People who live near coal-fired power plants are exposed to higher levels of radiation than those living near nuclear plants.

Creating a sustained nuclear fission reaction is not simple. A nuclear power plant is a large, complex, high-precision piece of engineering, carefully designed to create conditions that allow the splitting of uranium atoms. To keep such a plant operating requires the involvement of many scientists and technicians, who constantly monitor and adjust the plant mechanisms. Because creating fission is so hard, many people think that nuclear power is not natural—that nuclear fission is something unnatural that man has unwisely created. That is why it came as a great surprise to most when French physicist Francis Perrin announced that nature had preceded humans, creating the world's first nuclear reactors almost two billion years ago.[506]

In 1972, fifteen natural fission reactors were found in three different ore deposits at the Oklo mine in Gabon, West Africa. Collectively known as the Oklo Fossil Reactors, these natural atomic piles lie deep under African soil. About 1.7 billion years ago, natural conditions prompted underground nuclear reactions to take place. Scientists from around the world have studied the rock at Oklo to gain an understanding of how this could happen. Scientists believe that water filtering down through crevices in the rock played a key role. Without water, it would have been nearly impossible for natural reactors to sustain chain reactions.[507]

In 1956, while at the University of Arkansas, Dr. Paul Kuroda described the conditions under which a natural nuclear reactor could occur. When the Oklo reactors were discovered, the conditions found there were very similar to his predictions. Acting much as it does in human built reactors, water served as a moderator, slowing the neutrons emitted by radioactive uranium atoms. Without being slowed, these neutrons would collide with other atoms with so much energy, they would bounce off. Slowing the neutrons allowed them to be absorbed by the atoms they collided with. The added neutrons destabilized any uranium atoms that absorbed them, causing the atoms to split. This created two lighter atoms while releasing heat and more neutrons, which split even more atoms—a chain reaction was created.

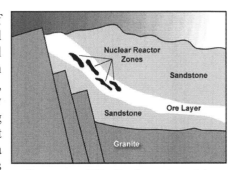

Illustration 159: Configuration of the Oklo natural nuclear reactors. Source U.S DOE.

When the heat from the reactions became too great, the moderating water turned to steam, which was incapable of slowing the neutrons. The chain reaction shut down until the reaction zone cooled and liquid water could return. Then the process would begin again. It is thought that the natural nuclear reactors operated continuously for 150 million years, releasing heat energy at an average power of 100 kilowatts.[508]

When these underground nuclear reactions ceased, nature showed that it could effectively contain the radioactive wastes they created. The radioactive remains of natural nuclear reactions that took place 1.7 billion years ago in Africa never moved far beyond their place of origin, deep underground. They remain contained in the sedimentary rocks, which kept them from being dissolved or spread by groundwater. Nature's example makes scientists confident that human built storage sites, like the one being constructed at Yucca Mountain, can contain the radioactive wastes with similar effectiveness. We have learned from nature's example, nuclear power and the waste it generates can be safe if handled properly.

Which is more manageable; 6,500,000,000 tons of CO_2, emitted by hundreds of millions of sources worldwide, or a few thousand tons of waste from a few hundred locations? There are a number of highly excitable "ecologists" who will tell you this "deadly" waste must be stored safely for 10,000 years —some even say 1,000,000 years. Trying to permanently bury nuclear waste for longer than human beings have been on Earth is utter foolishness.

Though nature has safely stored nuclear waste for more than 1.5 billion years, it has been pointed out by real world engineers that permanently interring waste material and hoping for the best is not the optimum solution. A better approach is to leave the storage facility unsealed and constantly monitor the condition of the stored material. This allows action to be taken as needed in the future. This adaptable solution also allows for the use of improved disposal technology as it becomes available. In truth, we need only safely store the nuclear waste from decommissioned plants for a few hundred years, knowing that future engineers will revisit the problem when warranted.

An even better option is to recycle spent nuclear fuel, reclaiming the highly radioactive waste products including plutonium that are produced by fission reactions. France has been doing this all along, greatly reducing the amount of waste their plants produce. A new multinational association called the Global Nuclear Energy Partnership (GNEP) is promoting an improved recycling process that does not separate plutonium from the other highly radioactive waste products. This prevents the plutonium's use in nuclear weapons while allowing the waste to be used as fuel for so called "fast" reactors. At the same time the volume and toxicity of the waste are reduced, more energy is extracted from the spent fuel. After this process, most of the remaining waste would need to be stored for only a few hundred years. William Hannum, a nuclear physicist formerly at Argon National Laboratory, has said that this process is so efficient that "for all practical purposes, the uranium would be inexhaustible."[509]

The final argument used against nuclear power is that it is too expensive and only exists because of government subsidies. Nuclear power plants do cost more than oil or gas fired plants, but once they are built, their operating costs are minimal while the cost of fossil fuels is high and bound to go higher. The Energy Information Agency (EIA) calculates that the average wholesale cost of electricity in the US is 5 cents a kWh. The Nuclear Energy Institute estimates the average generating cost for nuclear power is 1.7 cents per kWh. Why then, do people say nuclear costs too much? The main factors are delays caused by regulations and law suits.

Construction of the Shoreham reactor in New York was begun in 1973, but the plant never entered service. This was due to strong local opposition that brought about delay after delay, preventing the plant from ever producing usable power. Because of interest costs over time, the price tag for the plant rose from an initial $70 million to $6 billion when the plant was finally decommissioned in 1994.[510] In other countries, such as France, where both the people and the politicians accept the benefits of nuclear power, delays are minimal and costs much lower.

As we have seen, there is a choice to be made in energy production. Renewable sources will not meet present day needs, let alone tomorrow's energy demands. Coal and other fossil fuel use should be reduced, not increased. Solar and perhaps fusion lie in the distant future, but for now, the only proven, clean, rational choice is nuclear power. Just as cars, cellphones, and computers improve with time, so does nuclear reactor technology. Our ability to build more efficient nuclear plants, contain radioactive waste, and lower energy costs will only grow in the future, particularly if we follow the GNEP recycling scheme.

Many environmentalists remain opposed to nuclear power, even in the face of the supposedly more imminent threat of global warming. Several green luminaries such as Patrick More, a founder of Greenpeace, James Lovelock, originator of the *Gaia Hypotheses* and president of the Marine Biological Association, and Stewart Brand, creator of *The Whole Earth Catalog*, have changed their stance on nuclear power and now embrace it. Nature shows us the way—it is time for those who reject nuclear power to overcome their fears and get serious about reducing GHG emissions.

A Step Farther Out

If the desire is for truly clean power in the future, perhaps we should look to an idea from the past. Four decades ago, the idea of a solar power satellite in stationary orbit was hotly debated. The proposed stationary solar power satellite (SSPS) would collect energy from the Sun using an array of lightweight mirrors, use that energy to generate electricity, and then beam the energy to Earth as microwaves. Optimistic assessments thought the SSPS would be cost-competitive with nuclear power, but the technology of the time would have been tested to the breaking point.

More recently, NASA rechristened the idea Space Solar Power (SSP) and formed a study group to reevaluate the concept. After studying the problem from 1995-2001, John C. Mankins, who headed up the project, concluded that SSP is viable and could be operating by the 2020s. In contrast to existing renewable power sources, solar power from space would provide an uninterrupted, 24-hour supply of carbon-free power. Another estimate, from the George C. Marshall Institute, found that an SSP system could meet Europe's power needs by 2030 with a production cost of about $.05/kWh.[511]

The SPS system is composed of a space segment, the Solar Power Satellite (SPS), and a ground power receiving site. The ground station uses an antenna array to receive and convert the microwave power beam back into electricity. The receiving device is called a *rectenna,* for rectifying antenna, and would be about 6 by 9 miles (10 by 15 km) in area.

Conceptually, an SPS is very simple. It is a large satellite designed as an electric power plant orbiting in the Geostationary Earth Orbit (GEO). It consists of three main components: a solar energy collector to convert sunlight into electricity, a DC-to-microwave converter, and a large antenna array to beam the power to the ground. The solar collector can be either photovoltaic cells or a solar thermal turbine. The DC-to-microwave converter of the SPS can be either a microwave tube system and/or a semiconductor system. The third component is a sizable microwave antenna array. Because the microwave beam is so wide, the power density of the beam would only be around 0.024 W/m^2, low enough to not harm wildlife.

Such a satellite would have a large surface area, dwarfing the current International Space Station. Estimates for a gigawatt received capacity SPS using solar cells require an array area of approximately 1.25 by 3 miles. This would produce 2 GW DC power at the satellite. Optimum size of the transmitting phased array is around a mile across and the optimum microwave power of a few GW at 2.45 GHz. The DC to microwave RF conversion efficiency, including all losses (e.g. from phase shifters, power circuits, and isolators), is around 80%. The beam collection efficiency back on Earth, would be about 90%. Absorption by the atmosphere less than 2%. Such a system would deliver a bit more than 1 GW of power to the grid.[512]

Advanced solar cells convert 40% of the light energy they receive into electricity. DARPA funded research is projected to increase efficiency to 50%. Increased efficiency makes solar cells more productive, but they will remain expensive for the foreseeable future. If expensive solar cell arrays are to be used, the best place for them would be in orbit, above Earth's atmosphere. As we have seen, most of the Sun's energy striking Earth is either reflected back into space or absorbed by the atmosphere. Placing solar cells outside the atmosphere allows the reception of full strength, full spectrum solar radiation. Another advantage is that, due to the tilt of Earth's axis, a power station placed in GEO is in nearly perpetual sunshine, only occasionally passing through Earth's shadow. Unlike earthbound solar cells, which are in darkness 50% of the time, SPS cells are in the light 98% of the time.

Unfortunately, NASA disbanded the SSP research group headed by Mankins. As Martin Hoffert, Professor Emeritus of Physics at New York University, put it "we don't do energy at NASA." The US DOE is also not interested in the SSP concept because they don't do aerospace. While we are generally not in favor of large government programs, such an experimental enterprise, costing an estimated $100 billion to develop and build the initial satellite, is more than any commercial company can take on. Certainly, this would be a space program with tangible benefits here on Earth.

Plan Summary

Satisfying the world's need for clean, non-polluting energy is a gigantic problem—but not an insolvable one. The future will no doubt provide technological breakthroughs, though what form these will take, and when they will emerge, cannot be predicted. In the meantime, we need to apply the best available technology to solve the problem. Already significant advances are being made in aviation, automotive, and power storage technologies. Wind, geothermal and hydroelectric generation can help ease the demand for green energy, but will not be able to solve the problem on their own. Solar can help as well, but barring dramatic reductions in cost, it will remain a minor energy source. Even with optimistically high efficiency improvements (10-20%), the hard truth is the world must turn to nuclear power—at least as a bridging technology until better solutions become available. Here are the main points of our plan:

- Use renewable energy where economically viable. This includes hydroelectric, geothermal and wind power.

- Aggressively pursue the development of hybrid transportation technologies. This includes trains, buses, trucks and automobiles.

- Build only energy efficient new buildings and homes. Utilize both passive and active solar heating and power. Use renewable building resources such as timber from managed forests.

- Overhaul national and continental power grids. Switch to DC transmission and add off-peak storage systems to make the most productive use of variability in wind and solar power.

- Actively work on improving solar power technology, both on Earth and in space.

- Rapidly expand nuclear power capacity. Adopt safe recycling of spent nuclear fuel and advanced reactors.

These are things we can and should do today. There are also things we should not pursue. These include biofuels, "clean" coal and tapping methane ice deposits. The potential return on these technologies are poor, or the potential for environmental disaster exceeds any possible gain.

One of the actions that Kyoto and many others have promoted is creation of so called cap-and-trade carbon trading schemes. Under this type of emissions limiting system, the government sets a maximum amount of allowed pollution, the cap. Polluters are issued credits that permit them to emit a fixed amount of carbon. If a carbon credit holder reduces their emissions, they can resell or trade their excess credits to other, dirtier enterprises. This type of scheme has a spotty history for effectiveness. The US sulfur emissions program has worked fairly well, but the EU emissions trading scheme, instituted in 2005, has been a disappointment.[513]

One problem with current cap-and-trade systems is that they only targeted the electricity industry, which accounts for less than 40% of emissions. There are proposals to expand this, which have gained support from General Electric, DuPont, Alcoa, and major energy companies. When socialist leaning politicians and big business are on the same side of a proposal, the public should be highly suspicious. Turning emissions reduction into a commodity trading opportunity for speculators is not the most effective way to limit CO_2. This creates opportunities for Enron-style market manipulation and abuse.

A more equitable proposition is the carbon tax, proposed by a number of policymakers and advocates. A carbon tax simply imposes a tax for polluting based on the amount emitted, thus encouraging polluters to clean up and entrepreneurs to come up with alternatives. Estimates of how high such a tax should be, based on calculated future damage caused by global warming, range from $2 to $14 dollars per ton of CO_2.[514] For consumers, this would mean an increase of 2¢ to 14¢ per gallon of gasoline, and a maximum of 0.72¢ per kWh of electricity.

It is important that energy costs are increased to consumers in such a way that they can take individual action to avoid added expense. By increasing energy costs, the consumer is encouraged to buy more efficient household appliances, upgrade their heating and air conditioning, and buy more efficient automobiles. In America, politicians from both the right and the left have voiced support for the tax. According to former Bush advisor, N. Gregory Mankiw: "Basic economics tells us that when you tax something, you normally get less of it. So if we want to reduce global emissions of carbon, we need a global carbon tax. Q.E.D."[515]

Phasing in a carbon tax, starting at a reasonable initial rate, say $5 per ton, and then increasing the tax by $1 per ton per year, avoids a disruptive initial shock while removing uncertainty regarding the future price of energy. The thing markets hate the most is uncertainty—the carbon tax is predictable. It doesn't require the creation of a new carbon credit trading market, and it can be collected by existing state and federal agencies. It is straightforward and much harder to manipulate by special interests than the politicized process of allocating carbon credits. While the added costs under cap-and-trade go to companies, utilities and traders, the added costs under a carbon tax would go to the government. The added tax revenue could be used to promote reforestation, wetlands recovery, and flood protection, and to fund alternate energy research. But there are dangers—there have been warnings from economists that raising such a tax too high, above $30 per ton, would be wasteful and could damage the world economy.[516]

Former US president Bill Clinton has stated: "If Wal-Mart really does sell 100 million compact fluorescent bulbs and people buy them, and screw them in, and use them, it will have the CO_2 impact of taking 700,000 cars off the road."[517] Unfortunately, Wal-Mart is one of the firms that can do no right in the eyes of ecological/social activists. Mr. Clinton's own labor secretary, Robert Reich, criticized him for praising firms that find ways of making money from being green. Evidently saving Earth, while making a profit, is evil. This kind of politically correct nonsense hurts humanity and the planet.

We suggest that the best way forward is to invent new "cool green" technologies that are not only less polluting than current technology, but more cost effective as well. When the benefits of new technologies are clear, the world will readily embrace the products derived from them. This is the case with the Boeing 787 Dreamliner that is back-ordered beyond 2015. When everyone in Europe, America and Japan is zipping around in clean, efficient, high performance hybrids, consumers in developing nations will want them too. Current gas guzzlers will be as obsolete as muscle cars from the 1960s. The key to getting the developing world to shift from old, dirty technologies lies not in agreements like Kyoto, but in producing attractive new technologies that people in developing countries will want.

Enlightened self-interest is the most powerful motivational force on Earth. If for no other reason than our personal health and well-being, we should cut down on CO_2 and other pollutants, recycle our garbage, and build more nuclear power plants with a long-range goal of cleaner energy and more efficient transportation. Compared with coal and oil, nuclear power plants and hybrid electric cars are a cleaner and much healthier option. Fourth generation reactors are being developed that safely operate at temperatures of 700-900°C. These reactors can split hydrogen from water thermo-chemically, as well as generate electricity, providing hydrogen to power fuel cells in cars and homes. There is a bright future ahead, if we have the courage to make it happen. The US DOE estimates that America will need to build a new nuclear power plant every other day to satisfy our growing energy appetite. We need to get started—or we will all be left wondering what happened to our once beautiful world.

Chapter 19 The Fate of Planet Earth

"Nature never deceives us; it is always we who deceive ourselves."

— *Jean Jacques Rousseau*

Imagine an interstellar spaceship, seeking out new habitable worlds among the stars, moving silently through the void. Approaching a Sun-like star system from the emptiness of deep space, an Earth-sized planet is spotted within the habitable zone, the band of orbits where planetary temperatures are favorable to life. Though this star is about 4% dimmer than our Sun, the planet in question has livable surface temperatures. On closer examination a possible explanation for the unexpectedly warm temperatures is found—the atmospheric carbon dioxide level is 15 times that of Earth. The planet's day lasts about twenty-two and a half hours and its year 389 days. Across the surface of the planet, unfamiliar continents are scattered; one large and several smaller ones, mostly in the southern hemisphere. The land is devoid of complex life, inhabited only by patches of fungi and bacterial mats—there is no soil, only rock. But the shallow seas teem with life; marine worms,

Illustration 160: A visit to an alien planet. Original art by D. L. Hoffman.

mollusks, and arthropods. The hard-shelled arthropods appear in a bewildering number of forms; some graze the bottom ooze, feeding on fungi and bacteria, while others swim free, predators sporting shells with spikes and horns. A habitable, world but not a welcoming one for our kind—the ship moves on.

Another world is found circling a similar star. Again, the continents are unfamiliar, with the major land mass straddling the south polar region. Temperatures are warm, averaging 14°F (8°C) higher than modern-day Earth. The mix of atmospheric gases is different as well, with 10 times as much CO_2 as back home. On land, alien plants reach toward the yellow sunlight. Forests of 30 ft trees cover the continents, but these are like no trees we would recognize. The tall stalks are ribbed like giant pieces of celery and at the top, in the place of branches with leaves or needles, are bifurcating limbs that look like bottle brushes. The forest floor is littered with limbs that have been shed as the plants grow higher. Among the debris, scuttle wingless hexapods, partly aquatic scorpions, and spider-like arachnids. The seas are filled with large armored fish and other lobe finned creatures that use stubby legs to venture onto the land. A strange, alien world—again the ship moves on.

The next world brings the usual unrecognizable continents and something new, there is ice at the poles of this planet. In fact, a third of the plant's landmass is buried under thick glacial ice and the average global temperature is several degrees lower than our home world. Even so, the equatorial zone contains a number of warm shallow seas and expansive tropical swampland. The atmosphere contains carbon dioxide at levels comparable to modern-day Earth but the oxygen level is 35%, nearly twice what we are used to breathing. Throughout the peat swamps insects rule. Flying insects with unfolding wings flit about, resembling dragonflies with three foot wingspans. On the ground, spider-like creatures 30 inches across and 6 foot long millipedes hunt for prey. We move on, looking for a planet not in the grip of an ice age and without a bug problem.

Another world circling a slightly brighter star is found. This world also has most of its dry land concentrated in a single, gigantic continent. But this continent sprawls from the planet's south pole most of the way to the north polar region. There is no visible ice at either pole and the average global temperature is about 9°F (5°C) warmer than Earth. As usual, the higher temperature comes with higher CO_2 levels, but oxygen levels are close to what we consider normal. The land is warm and arid with signs of extensive volcanism. Reptilian life-forms dominate both the oceans and the land. In the air, pterosaurs glide on leathery wings. This is a dinosaur planet. Not wishing to contest ownership of the planet with the "terrible lizards," we look further ahead.

The last planet we visit looks vaguely familiar, with scattered continents whose shapes are almost, but not quite recognizable. The temperatures are warm, 5-7°F (3-4°C) warmer than home—it is a temperate world with no ice caps. There are forests covering the continent at the south pole and surrounding the shores of the northern sea. Elsewhere, there are broad savannas and grasslands. Life on land is dominated by animals that are recognizably mammalian, but unlike the animals we are used to. Giant ground sloths, the size of modern day elephants, and other animals that have no modern-day equivalent roam the plains. Predators stalk their prey with saber-like upper canines, large front limbs and strong clawed feet. Gigantic herbivores with body weights of up to five tons, six-horned rhinoceros and creatures that could be considered giant horned bunnies are but a few of the strange beasts that inhabit this menagerie. Humans could live comfortably on this world, but it is not our own.

Finishing its voyage, our ship returns to present day Earth, our familiar and comfortable home. Any of the strange and wondrous worlds we visited would make a good setting for a science fiction novel or a *Star Wars* movie. But perhaps you have guessed the truth: Our starship was voyaging in time, not space. Each of these alien worlds—the world of the arthropods, the world of bottle-brush tree forests, the worlds ruled by insects, dinosaurs and giant mammals—is Earth. These strange planets correspond to Earth 500, 400, 300, 200, and 50 million years in the past.

We embarked on this imaginary voyage to illustrate a point—that there is no single, right climate or ecology for our planet. If you go back to Earth's youngest days, over 4.5 billion years ago, and skip forward through time in 100 million year jumps, each jump would reveal a different, alien world. For the first two billion years, only worlds with toxic atmospheres would be found. For the next two billion years, though oxygen would be present, the air would still be unbreathable. Only since the advent of complex life, around 545 million years ago, would the world be livable. Even then, as we have just described, Earth's past was startlingly different from today's familiar world.

Science has revealed an ancient and ever-changing Earth, a world shaped and molded by the life it nurtured. We believe we know the ultimate fate of our planet and perhaps our species, Homo sapiens sapiens. But the more fascinating thought is what happens between now and a billion years in the future. How long will man—an infant species in the sweep of geologic time —continue to walk this planet?

We know that nature is a tireless sculptor, and that Earth is its unfinished masterpiece. As long as life exists on Earth, nature will continue to change our planet, shaping it to better fit life's needs. But what nature builds, nature can destroy. Continents move, ecosystems disappear, and species go extinct. Humans, like all life, are but transitory occupants of this blue planet. If we stay on this planet, nature will decide the fate of our species. It could be extinction by fire or ice, or a plague or something still unimagined.

The single cell microscopic organisms, that were the most complicated life on Earth for a billion years, ultimately led to a species that included Archimedes, Copernicus, Newton, Einstein, Bohr and other great scientists. There was a time, not so long ago, when people lived constantly on the edge of impending disaster. Plague, pestilence and famine lurked around every corner. Technology has freed most of humanity from those fears and can do so for the rest of us if given a chance. Science has taught us not to fear the unknown, not to live in dread of the future. We like to think that nature has more secrets to reveal.

Almost a half century ago, we built a machine and visited our nearest neighbor, the Moon, a quarter million miles away. We discovered our Moon is a lifeless, barren rock covered in craters and dust. We made a few trips there and then lost interest. How could that be, our nearest neighbor is lifeless, while our planet teams with life? Can life really be so rare, so fragile, human ignorance can destroy it?

Hurricanes, volcanoes, earthquakes and tsunamis provide frequent evidence of nature's might. Other forms of destruction have not been observed within human memory and remain more hypothetical. The threat of a massive release of methane from seabed deposits has never been experienced by people, but something similar, though on a smaller scale, has taken place.

In the African nation of Cameroon, there are two lakes with a deadly secret. Lake Monoun and Lake Nyos both harbor large volumes of CO_2 in their deepest layers. On occasion, disturbances of unknown origin have upset these lakes, triggering a subsequent release of carbon dioxide. One outburst from Lake Nyos claimed victims as far away as 17 miles. Two million people live along the shores of Lake Kivu, in Rwanda, which contains similar gas deposits—the potential for a larger disaster is quite real. Given these examples of sudden gas release, it is possible to imagine the devastation that a deep-sea eruption could cause. Clearly, more research is needed to assess the stability of oceanic clathrate deposits before we disturb them.

The last sizable meteor impact on Earth caused the 1908 Tunguska event in the Russian wilderness. An air-burst explosion of a large meteoroid or comet fragment, at an altitude of 3 to 6 miles above Earth's surface, leveled trees over an 800 sq. mile area. The energy of the blast was estimated to be between 10 and 20 megatons of TNT—1,000 times more powerful than the bomb dropped on Hiroshima. This powerful blast was produced by an object 200 feet in diameter.

In 2029, the asteroid Apophis is scheduled to pass within 30,000 miles of Earth—close enough to be seen with the naked eye. Estimated to be 800 feet in diameter, Apophis is much larger than the Tunguska object. Just a slight perturbation of Apophis' orbit could cause it to collide with our planet, unleashing an explosive force of 850 megatons. Apophis is expected to miss Earth, but astronomers will tell you it is not a matter of if Earth will be hit again, it's a matter of when. The only way to save our species—and every other existing species on Earth—from such a collision is to develop the technology to travel far out into the solar system to deflect marauding space debris.

There are those who say we are destroying nature and ourselves with technology, laying waste to the planet that gave birth to our species. Global warming is just the latest scare tactic of those who fear the future, who distrust the science and technology that has built our modern world. They claim to be saving Earth, saving nature, but in truth, they only want to hide in the past—taking the rest of us with them whether we wish to or not. They think that mindless, uncaring nature is good and people are evil. We don't think that is true.

Life has one defining characteristic—it spreads to new habitats, filling every niche and cranny. Humans are the end product of nature and our technology is therefore also a product of nature. Technology offers the possibility that our kind will venture to the other planets of the solar system and, eventually, the stars. If we do, we will spread life to other far-off lands, fulfilling nature's design and our destiny.

Some say that reaching for the stars is wrong, that we should instead look inward. We should only use "appropriate technology" and return to a simpler, more "natural" way of life. If we do, then we know how life will ultimately end—eventually our nurturing Sun will grow old and die. But not before swelling into a red giant, incinerating our blue-green oasis in space and whatever life remains on it. Billions of years into the future, if our species is lucky enough to survive that long, the last of our kind will look upon the red, swelling visage of the Sun and see life's final extinction. Even the resilient Earth and tenacious life cannot escape the death throws of our star. After all, nature is what it is and it is only we who deceive ourselves.

But there is another path. That path consists of embracing science and technology, of not being afraid of where they might take us. We should look upon the fossil fuels found in the earth much as the package of nutrients in a plant seed—there to give new life a jump-start. But the time has come to move on, to give up the use of fossil fuels, and embrace new energy sources that may eventually take us to the stars. To get there, our species must grow up and cast off the last vestiges of superstitious nature worship. In 100 million years, if a starship approaches Earth from the depths of space, what will it find? A world radically different from our world, that much is certain. But will the children of men still walk the planet? Will it be our progeny that pilot the ship on a visit to humanity's ancient home? Or will we have turned inward, awaiting the fate that nature decrees? Nature is what it is, but we have a choice.

Afterward

Since we finished writing The Resilient Earth in early December, 2007, a number of events have occurred here on planet Earth. First, a number of reports have come in from around the globe proclaiming the winter of 2007-2008 to be one of the coldest on record. While one cold year does not imply a new ice age any more than a really hot year proves global warming, consider the following headlines:

- **South America Has Coldest Winter in a 90 Years--**Residents in Argentina and Brazil are wondering if this winter will ever end. Buenos Aires recorded this Thursday (November 15th) the lowest November temperature in 90 years. Temperature in the Downtown weather station reached 2.5C. Since records began more than a century ago, only two days had colder lows in November. (The Telegraph, 17 November, 2007)

- **Record snowfalls mean big meltdown**—Locations such as Madison, Wis., and Concord, N.H., endured their snowiest winter since records began, and parts of the western USA also saw a much snowier-than-average winter, according to NOAA's National Climatic Data Center. (USA Today, 9 March, 2008)

- **China Suffers Coldest Winter in 100 Years**—Millions remained stranded in China on Monday ahead of the biggest holiday of the year as parts of the country suffered their coldest winter in a century.(Reuters, 5 February, 2008)

- **Brrr... Tajikistan Crisis!! Coldest Winter in 25 Years!** Record cold has forced 90% of the factories to close down in Tajikistan. (Reuters, 10 February, 2008)

As a consequence, despite hysterical warnings about the melting ice caps and endangered polar bears, it looks like the ice at both poles is doing just fine. Maybe too fine for those same polar bears that are supposedly endangered by melting polar ice:

- **Antarctica Ice Cap Growth Reaches Record High Levels**— The Southern Hemisphere sea ice area narrowly surpassed the previous historic maximum of 16.03 million sq. km to 16.17 million sq. km. (The Cryosphere Today, 11 October, 2007)

- **Ice between Canada and Greenland reaches highest level in 15 years** (Greenland's Sermitsiak News – 12 February, 2008)

359

- **Too Much Ice = Polar Bears Starving?** (Global Warming Politics – 15 February, 2008)

Second, it seems the link between a quiescent Sun and colder climate is being proven out in front of our eyes:

- **Report:** Sun's 'disturbingly quiet' cycle prompts fear of global COOLING (February 8, 2008 - Investor's Business Daily)

- **Solar data suggest our concerns should be about global cooling** – (By Geologist David Archibald of Summa Development Limited in Australia – March 2008 Scientific Paper)

- **Researchers Predict Another Ice Age**— Sunspots have all but vanished in recent years. (Daily Tech – 9 February, 2008)

Perhaps the climate experts who wish to blame all climate change on CO_2 need to review the works of Marsh, Svensmark and Shaviv regarding the effects of cosmic rays on low level cloud cover.

Third, some scientists are beginning to question the reliability of computer models in predicting future climate change.

- **Climate models overheat Antarctica, new study finds**-- Computer analyses of global climate have consistently overstated warming in Antarctica, concludes new research by scientists at the National Center for Atmospheric Research (NCAR) and Ohio State University...the authors caution that model projections of future Antarctic climate may be unreliable.(National Center for Atmospheric Research, 8 May, 2008)

- **In a New Climate Model, Short-Term Cooling in a Warmer World**—One of the first attempts to look ahead a decade, using computer simulations and measurements of ocean temperatures, predicts a slight cooling of Europe and North America, probably related to shifting currents and patterns in the oceans. The model is a rough replica of conditions, the scientists said. While it reliably reproduced climate patterns in Europe and North America, the model could not replicate patterns over central Africa, for example.(The New Your Times, 1 May, 2008)

- **Climate prediction: No model for success**—preliminary research at the Leibniz Institute of Marine Sciences in Kiel, Germany, suggested that natural variations in sea temperatures will cancel out the decade's 0.3C global average rise predicted by the IPCC, before emissions start to warm the Earth again after 2015. IPCC authors said this was not incompatible with their models; but the German research provoked some skeptics to ask whether models could be believed at all. (BBC News, 6 May, 2008)

This model induced confusion has scientists publicly hedging their bets on the bolder predictions of the IPCC and other eco-alarmists.

Finally, world food prices have risen sharply with the blame being placed squarely on efforts to increased production of biofuels. Oil prices have climbed to $138 per barrel yielding $4 a gallon gas in the US. Domestic American auto manufacturers are closing plants that build SUVs and pickup trucks while hurrying to bring new plugin hybrids to market. Airlines are taking old, less efficient planes out of service in order to save on fuel costs. Even so, the airline industry is hemorrhaging red ink, with debt increasing past $7 billion.

Around the world, orders for nuclear power plants are swelling. Countries that had been planning to close nuclear plants are reversing their decisions or ordering new replacements. There are even signs that a few politicians are beginning to realize that the world is facing an energy crisis that is going to totally eclipse the global warming scare of the past decade. What does this all mean?

Consider again the example of the Alvarez theory that an impact by an astronomical object caused the extinction of the dinosaurs. At first this idea was viewed as radical and a contradiction of the consensus view of slow geologic and evolutionary change. Over time, the scientific community came to accept the reality of the Chicxulub impact, displacing the older dogma. Because it was an exciting story many TV shows and magazine articles describing the event were produced. But the scientific story didn't stop there.

As it turns out, other evidence shows that the impact was not the only cause of the KT extinction. It was a factor, but not the only one. Similarly, the theory that anthropogenic CO_2 emissions are the only driver of global warming—the simplified view presented in most every news story or special program on global warming—is proving to be incorrect. Human emissions are a factor, but not the only factor—perhaps not even the major factor in climate change. The news media, capricious at best, is showing signs of moving on to the next crisis while scientists continue to do what they always do, argue among themselves.

Perhaps sanity will win out after all. The Global Warming Crisis will fade as have all the other eco-fads, replaced by the next world threatening crisis to catch the eye of the news media and pandering politicians. Calmer heads will prevail, allowing scientists and engineers to find workable, realistic solutions to the dual problems of pollution and energy production.

Alphabetical Index

[1] *Violent Past: Young sun withstood a supernova blast*, R. Cowen, Science News, May 26, 2007, vol. 171 p 323.

[2] *Climate Change 2007: The Physical Science Basis, Summary for Policymakers*, Working Group I to the Fourth Assessment Report of the Intergovernmental Panel on Climate Change, Paris, February 2007.

[3] *Astronomy 161: An Introduction to Solar System Astronomy,Lecture 36: Worlds in Comparison:The Terrestrial Planets*, Richard Pogge, 2006: http://www.astronomy.ohio-state.edu/~pogge/Ast161/Unit6/compare.html

[4] *New Thinking on the Death of Sun-Like Stars*, Charles Q. Choi, April 2007: http://www.space.com/scienceastronomy/070409_dust_clouds.html

[5] *Climate Control*, William B. Gail, IEEE Spectrum, Vol 44, no. 5 (NA), May 2007, pp 20-25.

[6] *Laurie and Sheryl Go to School,* Sheryl Crow, The Huffington Post, April 19, 2007: http://www.huffingtonpost.com/sheryl-crow/laurie-and-sheryl-go-to-s_b_46320.html

[7] *Fight Global Warming by Going Vegetarian*, GoVeg.com web site: http://www.goveg.com/environment-globalwarming.asp

[8] *Climate Change 2007: The Physical Science Basis, Summary for Policymakers*, Working Group I to the Fourth Assessment Report of the Intergovernmental Panel on Climate Change, Paris, February 2007.

[9] *The Kyoto Protocol to the United Nations Framework Convention on Climate Change*, Wikipedia, http://en.wikipedia.org/wiki/Kyoto_Protocol.

[10] *China passes U.S. as top CO_2 polluter,* Associated Press, Jun 21, 2007: http://www.tdn.com/articles/2007/06/21/biz/news03.txt

[11] *The Asian Brown Cloud: Climate and Other Environmental Impacts,* UNEP, Nairobi, 2002, p. 3.

[12] *What Goes Up*, S. Perkins, *Science News,* 8 Sept. 2007, vol 172, pp 152-154.

[13] Kyoto and beyond: Kyoto Protocol FAQs, CBC News, Feb. 14, 2007: http://www.cbc.ca/news/background/kyoto/

[14] *Problems with the Protocol,* Harvard Magazine, November-December 2002: http://www.harvardmagazine.com/on-line/1102199.html

[15] *G8 greenhouse gas emissions rise; U.S. not worst.* Alister Doyle, Environment Correspondent, Reuters, 31 May 2007.

[16] *Zero emissions needed to avert 'dangerous' warming,* Catherine Brahic, *New Scientist* news service, 11 October 2007.

[17] *Rethinking the Kyoto Emissions Targets,* M. Babiker and R. Eckaus, MIT Joint Program on the Science and Policy of Global Change, Report 65, August 2000.

[18] Hermann Grunder, Director, Argonne National Laboratory, May 14, 2004: http://www.anl.gov/Media_Center/News/2004/news040514.htm

[19] *Human Ancestors Hall: Homo sapiens*, Smithsonian Institution, 2006: http://www.mnh.si.edu/anthro/humanorigins/ha/sap.htm

[20] *On the Shoulders of Giants*, David Herring, Earth Observatory, NASA website: http://earthobservatory.nasa.gov/Library/Giants/Arrhenius/

[21] Ibid.

[22] *Carbon Dioxide and the Climate*, G. N. Plass, 1956, American Scientist 44, p. 302-16.

[23] *Effect of Carbon Dioxide Variations on Climate*, G. N. Plass, 1956, American J. Physics 24, p. 376-87.

[24] *Restoring the Quality of our Environment: Report of the Environmental Pollution Panel, President's Science Advisory Committee. R. Revelle, W. Broecker, H. Craig, C.D. Keeling, J. Smagorinsky*, Washington, DC: The White House; 1965. pp. 111–133.

[25] *History of Global Warming, the political climate*, PBS NOW: http://www.pbs.org/now/science/climatechange.html

[26] *Climate Change 2007: The Physical Science Basis, Summary for Policymakers*, Working Group I to the Fourth Assessment Report of the Intergovernmental Panel on Climate Change, Paris, February 2007.

[27]Address by Dr R K Pachauri, Chairman of the Intergovernmental Panel on Climate Change (IPCC) at the High Level Segment, Montreal, Canada, 7 December 2005. http://www.ipcc.ch/press/sp-07122005.htm

[28] *Uncertainties, in Guidance Papers on the Cross Cutting Issues of the Third Assessment Report of the IPCC*, Moss, R., and S. Schneider, 2000, R. Pachauri ed., Intergovernmental Panel on Climate Change (IPCC), Geneva.

[29] *IPCC Workshop on Describing Scientific Uncertainties in Climate Change to Support Analysis of Risk and of Options: Workshop report.* M. R. Manning, et al (Eds), 2004, Intergovernmental Panel on Climate Change, Geneva.

[30] *Guidance Notes for Lead Authors of the IPCC Fourth Assessment Report on Addressing Uncertainties,* IPCC, July 2005.

[31] *The Treatment of Uncertainties in the Fourth IPCC Assessment Report*, Martin R. Manning, IPCC Working Group I Technical Support Unit, the National Oceanic and Atmospheric Administration, 2007.

[32] Ibid, p 21.

[33] *Climate Change 2007: Impacts, Adaptation and Vulnerability, Summary for Policymakers,* formally approved at the 8th Session of Working Group II of the IPCC, Brussels, April 2007.

[34] *Climate Change 2007: The Physical Science Basis, Summary for Policymakers*, AR4 WG I, Paris, February 2007.

[35] *The Skeptical Environmentalist*, Bjørn Lomborg, 2001, p266, Cambridge University Press, ISBN 0 521 80447 7.

[36] *Useless Arithmetic: why environmental scientists can't predict the future*, Orrin H. Pilkey and Linda Pilkey-Jarvis, 2006, Columbia University Press, ISBN 978-0-231-13212-1, pXIII.

[37] Ibid.

[38] *Climate myths: We can't trust computer models*, NewScientist.com news service, 16 May 2007. http://environment.newscientist.com/channel/earth/climate-change/dn11649

[39] Address by Dr R K Pachauri, Chairman of the Intergovernmental Panel on Climate Change (IPCC) at the High Level Segment, Montreal, Canada, 7 December 2005. http://www.ipcc.ch/press/sp-07122005.htm

[40] *Climate Change 2007: Mitigation of Climate Change, Summary for Policymakers.* Working Group III contribution to the Intergovernmental Panel on Climate Change Fourth Assessment Report, 4 May 2007.

[41] *Frozen Earth, the once and future story of ice ages*, Doug Macdougall, 2004, the University of California Press, pp 240.

[42] *Scientists Map 'New Frontier' Deep Within Ocean*, Marine Technolgy Reporter, 10/02/2006.

[43] *Regional signatures of changing landscape and climate of northern central Siberia in the Holocene*, V.L. Koshkarova and A.D. Koshkarov, 2004, Russian Geology and Geophysics 45 (6): 672-685.

[44] *A highly unstable Holocene climate in the subpolar North Atlantic: evidence from diatoms*, C. Andersen, N. Koç, M. Moros, 2004, Quaternary Science Reviews 23 (2004) 2155–2166.

[45] *Holocene climate in the Atlantic sector of the Southern Ocean: Controlled by insolation or oceanic circulation?* S.H.H. Nielsen, N. Koç, and Xavier Crosta, Geology, April 2004.

[46] *Holocene Climate Variability in Antarctica Based on 11 Ice-Core Isotopic Records*, Valérie Masson, et al, *Quaternary Research* 54, 348–358 (2000).

[47] *Antarctic climate cooling and terrestrial ecosystem response*, P. T. Doran, et al, *Nature*, 2002 Jan 31;415(6871):517-20.

[48] *Trends in the length of the southern Ocean sea-ice season. 1979-99*. C.L. Parkinson, 2002, Annals of Glaciology '34: 435-40

[49] *Snowfall-Driven Growth in East Antarctic Ice Sheet Mitigates Recent Sea-Level Rise*, Curt H. Davis, Yonghong Li, et al, Science Express, 2005.

[50] *$\delta^{18}O$, Sr/Ca and Mg/Ca records of Porites lutea corals from Leizhou Peninsula, northern South China Sea, and their applicability as paleoclimatic indicators*, K. F. Yu, et al., 2005, Palaeoecology 218: 57-73.

[51] *Climate and the Collapse of Maya Civilization*, Gerald H. Haug, et al, *Science* 14 March 2003: Vol. 299. no. 5613, pp. 1731 – 1735.

[52] *Link Between Solar Cycle And Climate Is Blowin' In The Wind*, Science Daily, April 12, 1999.

[53] *Persistent solar influence on North Atlantic climate during the Holocene*. G. Bond, et al, 2001, Science 294: 2130-2136.

[54] *Solar Cycles, Not CO₂, Determine Climate*, Z. Jaworowsky, 21st Century Science & Technology, winter 2003-2004.

[55] *Possible solar origin of the 1,470-year glacial climate cycle demonstrated in a coupled model*, Holger Braun, et al, Nature, Vol 438,10 November 2005.

[56] *Climate and the collapse of Maya civilization*, L.C. Peterson and G.H. Haug, 2005. *American Scientist* 93, p. 322-329.

[57] T*he Little Ice Age: How Climate Made History, 1300-1850*. Brian M. Fagan, 2001, Basic Books.

[58] *Ice age to warming - and back?* Peter N. Spotts, Christian Science Monitor, March 18, 2004. http://www.csmonitor.com/2004/0318/p13s01-sten.html

[59] *Viking Greenland.* K.J. Krogh, Nat. Museum of Denmark, 1967.

[60] The history of the Chamonix Glaciers, Chamonix dot Net: http://www.chamonix.net/english/mountaineering/histofchamglaciers.htm

[61] *The Little Ice Age, Brian M.* Fagan, 2001, Basic Books, pp 120-125.

[62] *Did "Little Ice Age" Create Stradivarius Violins' Famous Tone?* John Pickrell, National Geographic News January 7, 2004. http://news.nationalgeographic.com/news/2004/01/0107_040107_violin.html

[63] *Climate in Art*, Hans Neuberger, Weather, vol 25, 1970, pp 46-56.

[64] *Climate Change: The IPCC Scientific Assessment*, J.T. Houghton, (Ed.). 1990, Cambridge University Press, Cambridge, UK.

[65] *Global-scale temperature patterns and climate forcing over the past six centuries* ,Mann, M.E., Bradley, R.S. and Hughes, M.K. 1998, Nature 392: 779-787.

[66] *Northern Hemisphere temperatures during the past millennium: Inferences, uncertainties, and limitations*, Mann, M.E., Bradley, R.S. and Hughes, M.K. 1999, Geophysical Research Letters 26: 759-762.

[67] *Counterpoint: Climate Change Responses*, Bob Carter, Monday 11 April 2005, Radio National, Presented by Michael Duffy, http://www.abc.net.au/rn/talks/counterpoint/stories/s1339366.htm

[68] *Mercury in a Spanish peat bog: Archive of climate change and atmospheric metal deposition,* A. Martinez-Cortizas, et al., 1999, *Science* **284**: 939-942.

[69] *Greenhouse warming or Little Ice Age demise: a critical problem for climatology*, S. B. Idso, 1988, Theo. and Applied Climatology 39: 54-56.

[70] *Climate change prediction,* W.S. Broecker, 1999, Science 283: 179.

[71] *Glaciers that speak in tongues and other tales of global warming*, W.S. Broecker, 2001, Natural History 110 (8): 60-69.

[72] *Centennial-scale Holocene climate variability revealed by a high-resolution speleotherm $\delta^{18}O$ record from SW Ireland*, F. McDermott, D.P. Mattey and C. Hawkesworth, 2001, *Science* 294: 1328-1331.

[73] *Highly variable Northern Hemisphere temperatures reconstructed from low- and high-resolution proxy data,* A. Moberg, D.M. Sonechkin, K. Holmgren, N.M. Datsenko and W. Karlén, 2005, *Nature* **443**: 613-617.

[74] *Reconstructing Past Climate from Noisy Data,* Hans von Storch, et al, *Science* 22 October 2004: Vol. 306. no. 5696, pp. 679 – 682.

[75] *Behind the Hockey Stick,* David Appell, Scientific American, March 2005.

[76] *Blowing hot and cold,* K.R. Briffa and T.J. Osborn, 2002, Science 295: 2227-2228.

[77] *The Mann Et Al. Northern Hemisphere "Hockey Stick" Climate Index: A Tail of Due Diligence,* Ross McKitrick, in *Shattered Consensus,* Patrick J. Michaels ed., Rownan & Littlefield, 2005, pp. 20-49.

[78] *Hockey sticks, principal components, and spurious significance,* S. McIntyre and R. McKitrick, Geophysical Research Letters, vol 32, 2005.

[79] *Another Ice Age?* Time Magazine, Monday, Nov. 13 1972.

[80] Ibid.

[81] Ibid.

[82] *The Cooling World,* Peter Gwynne, Newsweek, April 28, 1975.

[83] *Science Panel Calls Global Warming 'Unequivocal',* E. Rosenthal and A. C. Revkin, New York Times, February 3, 2007.

[84] *Tropical storms stepping up with climate change,* NewScientist.com news service, 30 July 2007: http://environment.newscientist.com/article.ns?id=dn12377

[85] *Extreme weather brings flood chaos round the world,* NewScientist.com news service, 30 July 2007: http://environment.newscientist.com/article/dn12385-extreme-weather-brings-flood-chaos-round-the-world.html

[86] *Birth of the Earth,* Andrew Alden, About Geology web site, http://geology.about.com/od/nutshells/a/aa_earthbirth.htm.

[87] *Satellite-sized planetesimals and lunar origin,* W. K. Hartmann and D. R. Davis, Icarus 24, 504-515, 1975.

[88] The origin of the Moon, A. G. W. Cameron and W. R. Ward, 1976, Proc. Lunar Planet. Sci. Conference 7th, 120-122.

[89] *The Age of the Earth (new edition, revised),* A. Holmes, 1937, Nelson:London, p.1-263.

[90] *A geologic time scale, 1989 edition,* Harland, W.B.; Armstrong, R.L.; Cox, A.V.; Craig, L.E.; Smith, A.G.; and Smith, D.G., Cambridge University Press: Cambridge, p.1-263. ISBN 0-521-38765-5

[91] *A Geologic Time Scale 2004,* Gradstein, F.M., Ogg, J.G., and Smith, A.G., International Commission on Stratigraphy (ICS), Cambridge University Press, March 2005. Web site: http://www.stratigraphy.org/.

[92] *Archean stromatolites in Michipicoten Group siderite ore at Wawa, Ontario,* H. J. Hofmann, R. P. Sage, and E. N. Berdusco, *Economic Geology;* August 1991; v. 86; no. 5; p. 1023-1030.

[93] *Written declaration on greenhouse gas emissions by the livestock sector,* Romana Jordan Cizelj and Jan Christian Ehler, 28 March, 2007, DC\659409EN.doc.

[94] *Prokaryotes, Eukaryotes, & Viruses Tutorial*, The Biology Project, 1997, Department of Biochemistry and Molecular Biophysics, University of Arizona. http://www.biology.arizona.edu/CELL_BIO/tutorials/pev/page3.html

[95] *A Geologic Time Scale 2004*, Gradstein, F.M., Ogg, J.G., and Smith, A.G., International Commission on Stratigraphy (ICS), Cambridge University Press, March 2005. Web site: http://www.stratigraphy.org/.

[96] *Treatise on Invertebrate Paleontology, Part O, Volume 1, revised, Trilobita*, R.L. Kaesler, ed. 1997, Geological Society of America and University of Kansas Press, Lawrence, Kansas.

[97] *Trilobite systematics: The last 75 years*, R. A. Fortey, 2001, Journal of Paleontology 75:1141–1151.

[98] *Forest Primeval: The oldest know trees finally gain a crown*, S. Perkins, Science News, April 21, 2007, Vol 171, p243.

[99] *Earth System History*, Steven M. Stanley, W.H. Freeman and Company, New York, 1999.

[100] *Impact of a Permo-Carboniferous high O2 event on the terrestrial carbon cycle*, Beerling, D. J. and Berner, R. A., PNAS, November 7, 2000,vol. 97, no. 23, 12428-12432.

[101] *When Giants Had Wings and 6 Legs*, Henry Fountain, New York Times, Science Section, February 3, 2004.

[102] *The environment of vertebrate life in the late Paleozoic in North America: a paleogeographic study*, Ermine C. Case, 1919. The Carnegie Institution of Washington: Washington D.C.

[103] *Atmospheric oxygen over Phanerozoic time*, Robert A. Berner, PNAS 1999;96; 10955-10957.

[104] *Plant fossils in geological investigation - the Palaeozoic*, C.J. Cleal, (1991): - 233 pp.; Ellis Horwood, New York.

[105] *The great Paleozoic crisis - life and death in the Permian*, D.H. Erwin, (1993): - 327 pp.; Columbia Univ. Press, New York.

[106] *Extinction: How Life on Earth Nearly Ended 250 Million Years Ago*, Douglas H. Erwin, 2006, Princeton University Press.

[107] *Extinction: How Life on Earth Nearly Ended 250 Million Years Ago*, Douglas H. Erwin, 2006, Princeton University Press, pp 7.

[108] *The mid-Cretaceous earth: Palaeogeography, ocean circulation and temperature, atmospheric circulation*, C.R. Lloyd, 1982. *J. Geol.*, 90, pp. 393-413.

[109] *A warm, equable Cretaceous: The nature of the problem*, E.J. Barron, 1983. Earth Sci. Rev., 19, pp. 305-338.

[110] *A multiple proxy and model study of Cretaceous upper ocean temperatures and atmospheric CO$_2$ concentrations*, K. L. Bice, et al., 2006, Paleoceanography, 21, PA2002, doi:10.1029/2005PA001203.

[111] *Extraterrestrial cause for the Cretaceous/Tertiary extinction*, L. W. Alvarez, W. Alvarez, F. Asaro, and H. V. Michel, 1980, *Science*, v. 208, pp. 1095-1108.

[112] *T. Rex and the Crater of Doom*, W. Alvarez. 1997, Princeton University Press, Princeton, 185 p.

[113] *Chicxulub Impact Crater Provides Clues to Earth's History*, V. L. Sharpton, 1995, in *Earth in Space*, American Geophysical Union, v. 8, pp. 7.

[114] *A Geologic Time Scale 2004*, Gradstein, F.M., Ogg, J.G., and Smith, A.G., International Commission on Stratigraphy (ICS), Cambridge University Press, March 2005. Web site: http://www.stratigraphy.org/.

[115] *Subtropical Arctic Ocean temperatures during the Paleocene/Eocene thermal maximum*, A. Sluijs, et al., 2006, *Nature*, 441(7093): 610-613.

[116] *Abrupt deep-sea warming, palaeoceanographic changes and benthic extinctions at the end of the Palaeocene*, Kennett, J.P. & Stott, L.D. 1991.. *Nature*, 353: 225-229

[117] *Extreme warming of mid-latitude coastal ocean during the Paleocene-Eocene Thermal Maximum: Inferences from TEX 86 and isotope data*, J.C. Zachos, et al, 2006, *Geology*, v.34, p.737–740.

[118] Ibid.

[119] *Catastrophe and Opportunity in an Ancient Hot-House Climate*, Cindy Shellito, *Geotimes*, October, 2006.

[120] *Global Fever*, Elisabeth Nadin, Science Notes 2003, University of California, Santa Cruz. http://scicom.ucsc.edu/SciNotes/0301/index.html

[121] *Wyoming's Garden of Eden*, Kenneth D. Rose, April, 2001, Natural History.

[122] *Did mammals spread out from an Asian Eden?* Becky Ham, 2007, AAAS, on MSNBC web site, http://www.msnbc.msn.com/id/3077464/

[123] *Horns, Tusks and Flippers: The Evolution of Hoofed Mammals*, Donald R. Prothero & Robert M Schoch, 2002, The John Hopkins University Press, Baltimore MD.

[124] *A finding of Oligocene primates on the European continent*, Meike Köhler and Salvador Moyà-Solà, Proc Natl Acad Sci U S A. 1999 December 7; 96(25): 14664–14667.

[125] *Anatomy of the Popigai impact crater, Russia*, Masaitis, V. L., Naumov, M. V. & Mashchak, M. S., 1999: 1-17 *in* Dressler, B. O. & Sharpton, V. L., (eds.) 1999: Large meteorite impacts and planetary evolution II. The Geological Society of America, Boulder Special Paper 339, 1999, viii-464.

[126] *Seismic expressions of the Chesapeake Bay impact crater: Structural and morphological refinements based on seismic data*, C. W. Pong, et al. (eds.) 1999: Large meteorite impacts and planetary evolution, II, The Geological Society of America Special Paper 339, Boulder, 1999, viii-464.

[127] *Large temperature drop across the Eocene–Oligocene transition in central North America*, Alessandro Zanazzi, Matthew J. Kohn, Bruce J. MacFadden and Dennis O. Terry, *Nature* 445, 639-642, 8 February 2007.

[128] *Oligocene climate dynamics*, Bridget S. Wade & Heiko Pälike, Paleoceanography, VOL. 19, PA4019, doi:10.1029/2004PA001042, 2004

[129] *The miocene desiccation of the Mediterranean and its climatical and zoogeographical implications*, Kenneth J. Hsü, 1973, Naturwissenschaften, Volume 61, Number 4,

[130] *Chronology, causes and progression of the Messinian salinity crisis*, W. Krijgsman, F. J. Hilgen, I. Raffi, F. J. Sierro and D. S. Wilson, *Nature* **400**, 652-655,12 August 1999.

[131] *How did the Messinian Salinity Crisis end?* Nicolas Loget, Jean Van Den Driessche and Philippe Davy, Terra Nova, Volume 17 Issue 5 Page 414 - October 2005.

[132] *The Pliocene Paradox (Mechanisms for a Permanent El Niño)*, A. V. Fedorov, et al., *Science* 9 June 2006: Vol. 312. no. 5779, pp. 1485 - 1489

[133] *The Pliocene Paradox*, A. Fedorov1, A. C. Ravelo, et al.; http://www.aos.princeton.edu/WWWPUBLIC/gphlder/pliopar.pdf

[134] *Frozen Earth, the once and future story of ice ages,* Doug Macdougall, 2004, the University of California Press, pp 242.

[135] *Human Ancestors hall: Homo sapiens,* the Smithsonian Institution: http://www.mnh.si.edu/anthro/humanorigins/ha/sap.htm .

[136] *PCR-Based Identification of Hyperthermophilic Archaea of the Family Thermococcaceae,* Galina B. Slobodkina, et al., Applied and Environmental Microbiology, Sept. 2004, p. 5701–5703.

[137] *Dark Power: Pigment seems to put radiation to good use*, Davide Castelvecchi, Science News, May 26, 2007; Vol. 171, No. 21 , p. 325.

[138] *Frozen Earth, the once and future story of ice ages*, Doug Macdougall, 2004, the University of California Press, pp 15.

[139] *Louis Agassiz*, University of California Museum of Paleontology web site, http://www.ucmp.berkeley.edu/history/agassiz.html.

[140] *Paleoproterozoic snowball Earth: Extreme climatic and geochemical global change and its biological consequences*, Joseph L. Kirschvink et. al., PNAS, Vol. 97, Issue 4, 1400-1405, February 15, 2000.

[141] *Frozen Earth: the once and future story of ice ages*, Doug Macdougall, 2004, the University of California Press, pp 143.

[142] *A Neoproterozoic snowball Earth,* Hoffman, P.F., A.J. Kaufman, *et al.* 1998.. *Science* 281(Aug. 28):1342.

[143] *Snowball Earth*, P. f. Hoffman and D. P. Schrag, *Scientific America,* January 2000.

[144] *Pre-Mesozoic ice ages: their bearing on understanding the climate system*, Crowell, J.C., Geological Society of America Memoir192, 1999.

[145] *Clastic wedges and patterned ground in the Late Ordovician—Early Silurian tillites of South Africa*, B. Daily and M. R. Cooper, April 1976, *Sedimentology* 23 (2), 271–283.

[146] *Late Ordovician glaciation in southern Turkey*, O. Monod, H. Kozlu, J.-F. Ghienne, W. T. Dean, Y. Günay, A. Le Hérissé, F. Paris, M. Robardet (2003) *Terra Nova* 15 (4), 249–257.

[147] *The Late Ordovician Mass Extinction*, Peter M. Sheehan, 2001, *Annual Review of Earth and Planetary Sciences,* Vol. 29: 331-364.

[148] *Deglaciation sequences in the Permo-Carboniferous Karoo and Kalahari basins of southern Africa: a tool in the analysis of cyclic glaciomarine basin fills.* Johan Visser, 1997, *Sedimentology,* 44 (3), 507–521.

[149] *Grooves and striations on the Stanthorpe Adamellite: evidence for a possible late Middle - Late Triassic age glaciation*, A. P. Spenceley, 2001, *Australian Journal of Earth Sciences,* 48 (6), 777–784.

[150] *The Greenland-Norwegian Seaway: A key area for understanding Late Jurassic to Early Cretaceous paleoenvironments*, Jörg Mutterlose et al. *Paleoceanography,* Vol. 18, No. 1, 1010, doi:10.1029/2001PA000625, 2003

[151] *Isotopic evidence for temperature variation during the early Cretaceous (late Ryazaian-mid-Hauterivian)*, Gregory D. Price, et al., *Journal of the Geological Society,* London, Vol 157, 2000, pp 335-343.

[152] *The Great Ice Age*, Louis L. Ray, 1992, United States Geological Survey, ISBN 0-16-036025-0.

[153] *Biogeography of mammals in Southeast Asia: estimates of rates of colonization, extinction, and speciation.* L. R. Heaney, 1986. Biological *Journal of the Linnean Society,* 28:127-165.

[154] *A 17,000-year glacio-eustatic sea level record: influence of glacial melting on the Younger Dryas event and deep-sea circulation.* Fairbanks, R. G. 1989, *Nature* 342: 637-642.

[155] *Frozen Earth: the once and future story of ice ages*, Doug Macdougall, 2004, the University of California Press, pp 232.

[156] *Earth's Albedo in Decline,* Earth Observatory News, 2005, NASA.

[157] *Frozen Earth: the once and future story of ice ages*, Doug Macdougall, 2004, the University of California Press, pp 236.

[158] *The Precambrian Earth: Tempos and Events*, P.G. Eriksson, et al. (eds), Developments in Precambrian Geology, vol 12, 2004, ISBN: 978-0-444-51506-3.

[159] *Volcanic eruptions caused Permian period extinctions*, *New Scientist*, 29 January 2005, page 16

[160] *Dates from the West Siberian Basin*, Marc Belchow et al, *Science*, 7 June 2002.

[161] *Global Crisis: The Fungi Stand Alone*, R. Monastersky, Science News, March 16, 1996, Vol. 149 No. 11 p. 164.

[162] *Large Igneous Provinces and Mass Extinctions*, P.B. Wignall, 2001, Earth-Science Reviews, v. 53, p. 1-33.

[163] *Ancient Volcanoes Were A Wipe Out*, Nicola Jones, Terra Daily, July 4, 2001. http://www.spacedaily.com/news/early-earth-01f.html

[164] *Century-Scale Shifts in Early Holocene Atmospheric CO_2 Concentration,* Friederike Wagner, et al., *Science,* 18 June 1999, 284:5422, pp. 1971-1973.

[165] *The Long Summer: How climate changed civilization*, Brian Fagan, 2004, Basic Books, p 19.

[166] Ibid, pp 1-10.

[167] *Useless Arithmetic: why environmental scientists can't predict the future*, Orrin H. Pilkey and Linda Pilkey-Jarvis, 2006, Columbia University Press, ISBN 978-0-231-13212-1, p 68.

[168] *Fundamental Concepts: Extinction,* Smithsonian Institution web site: http://paleobiology.si.edu/geotime/main/foundation_life4.html

[169] *Thomas Jefferson as Scientist*, Austin H. Clark, Journal of the Washington Academy of Sciences, vol 33, no. 7, 1943.

[170] *Jefferson and Monticello*, Jack McLaughlin, 1988, Henry Holt And Company, New York.

[171] *Notes on the State of Virginia: On Big Bone Lick and the Mammoth,* Thomas Jefferson, 1781, edited by Avi Hathor. Available online: http://www.geocities.com/bigbonehistory/jefferson-notes.html?200726

[172] Ibid.

[173] Ibid.

[174] *Big Bone Lick Timeline*, compiled by Don Clare, 2005, Friends of Big Bone. http://www.friendsofbigbone.org/Contents/timeline.htm

[175] *Georges Cuvier, Fossil Bones, and Geological Catastrophes: New Translations and Interpretations of the Primary Texts*, Martin J. S. Rudwick, University Of Chicago Press, 1998.

[176] *The Last Human: A Guide to Twenty-Two Species of Extinct Humans*, G.J. Sawyer, V. Deak, et al., Yale, 2007.

[177] *Discovery Suggests Humans Are a Bit Neanderthal*, John Noble Wilford, *New York Times,* April 25, 1999

[178] Neanderthal DNA Shows No Interbreeding With Humans, Alan Mozes, November 15, 2006, HealthDay News: http://www.medicinenet.com/script/main/art.asp?articlekey=77566

[179] *Ordovician odyssey: Short papers for the Seventh International Symposium on the Ordovician System.* J.J. Sepkoski, 1995, Pacific Section Society for Sedimentary Geology, 393-396.

[180] *Effects of the Middle to Late Devonian spread of vascular land plants on weathering regimes, marine biota, and global climate.* T. J. Algeo, et al. (eds.). 2001 *Plants Invade the Land: Evolutionary and Environmental Approaches.pp. 213-236.* Columbia Univ. Press: New York.

[181] *Western Australia: history of discovery, Late Devonian age, and geophysical and morphometric evidence for a 120 km-diameter impact structure.* A. Y. Glikson, et al., 2005: Woodleigh, southern Carnarvon Basin, Australian Journal of Earth Sciences: Vol. 52, pp. 545-553.

[182] *Extinction: how life on earth nearly ended 250 million years ago.* Douglas H. Erwin, 2006, Princeton University Press, ISBN-13: 978-0-691-00524-9.

[183] *The End Triassic Mass Extinction.* Encyclopedia of Life Sciences, 2005, John Wiley & Sons.

[184] *Did volcanism of the Central Atlantic Magmatic Province trigger the end-Triassic mass extinction?* Pálfy, J., Research Group for Paleontology, Hungarian Academy of Sciences–Hungarian Natural History Museum.

[185] After the Dinosaurs, Donald R. Prothero, Indiana University Press, 2006, pp 30-39.

[186] *Supernova Explosion May Have Caused Mammoth Extinction,* Dan Krotz, Berkeley Lab Research news, September 23, 2005.

[187] *Northern Exposure: The inhospitable side of the galaxy?* D. Castelvecchi, Science News, April 21, 2007, Vol. 171.

[188] *Paleoindian large mammal hunters on the plains of North America,* George C. Frison, PNAS, November 24, 1998; 95(24): 14576–14583.

[189] Ibid.

[190] *The Archaeological Record of Human Impacts on Animal Populations,* Donald K. Grayson, Journal of World Prehistory, Volume 15, Number 1, March, 2001.

[191] *Supernova Explosion May Have Caused Mammoth Extinction,* Lawrence Berkeley National Laboratory press release, September 26, 2005. http://www.spaceref.com/news/viewpr.html?pid=17893

[192] *Mammoth Extinction Caused by Trees, Study Suggests,* Anne Minard, National Geographic News, May 10, 2006. http://news.nationalgeographic.com/news/2006/05/ice-age.html

[193] *Cunning Contraceptives,* Robin Taylor, Ecos 95. April-June 1998, p28.

[194] *The disappearance of Guam's wildlife.* Rodda, G. H., T. H. Fritts, and D. Chiszar. 1997. *BioScience* 47: 565-574.

[195] *What Are the Impacts of Introduced Species?* Charlene D'Avanzo and Susan Musante, Teaching Issues and Experiments in Ecology - Volume 1, January 2004.

[196] *Dodo and solitaires, myths and reality.* France Staub, Proceedings of the Royal Society of Arts & Sciences of Mauritius 6: 89-122, 1996.

[197] *Earth faces sixth mass extinction,* Anil Ananthaswamy, 18 March 2004, NewScientist.com news service: http://www.newscientist.com/article.ns?id=dn4797

[198] *The Sixth Extinction,* Virginia Morell, 2003, NationalGeographic.com: http://www.nationalgeographic.com/ngm/9902/fngm/index.html

[199] *Herschel Discovers Infrared Light,* the Cool Cosmos, Caltech: http://coolcosmos.ipac.caltech.edu//cosmic_classroom/classroom_activities/herschel_bio.html

[200] The Story of the Herschels: A Family of Astronomers, Anonymous, 1886: http://www.gutenberg.org/files/12340/12340-h/12340-h.htm

[201] John Tyndall, John van Wyhe, The Victorian Web, 2002: http://www.victorianweb.org/science/tyndall.htm

[202] John Tyndall's Research on Trace Gases and Climate, http://www.tyndall.ac.uk/general/history/JTyndall_biog_doc.pdf

[203] Ibid.

[204] *Emissions tied to global warming are on the rise*, Science News, May 19, 2007, Vol. 177, p318.

[205] *Problems with the Protocol*, Harvard Magazine, November-December 2002: http://www.harvardmagazine.com/on-line/1102199.html

[206] *Weather and Climate: Death Valley National Park*, rev 7, U.S. Park Service, 2006.

[207] *Carbon Dioxide (CO₂) Properties, Uses, Applications: CO₂ Gas and Liquid Carbon Dioxide*, Universal Industrial Gases, Inc., web site, 2003/2007, http://www.uigi.com/carbondioxide.html

[208] *Fundamentals of Physical Geography*, Michael Pidwirny, University of British Columbia Okanagan, 1999-2006. http://www.physicalgeography.net/fundamentals/9r.html

[209] *Unstable Silicon Analogues of Olefins and Ketones*, L E Gusel'nikov, N S Nametkin, V M Vdovin, *RUSS CHEM REV*, 1974, 43 (7), 620-629.

[210] *Address to the British Association for the Advancement of Science*, Reynolds, J. E. *Nature*, 48, 477 (1893).

[211] *Could Life be based on Silicon rather than Carbon?* Featured question, NASA Astrobiology Institute web site, last updated May 10, 2007 : http://nai.nasa.gov/astrobio/feat_questions/silicon_life.cfm

[212] *Volcanic carbon dioxide emission rates: White Island, New Zealand and Mt Erebus, Antarctica*, Wardell, L., and Kyle, P., 1998, EOS Transactions/Supplement, vol 79, no 10, p F926.

[213] *Gas Hazard from Natural CO₂ Emissions in Central and Southern Italy*, C. Cardellini, et al., American Geophysical Union, Fall Meeting 2006.

[214] *Fundamentals of Physical Geography*, Michael Pidwirny, U of British Columbia Okanagan, 1999-2006. http://www.physicalgeography.net/fundamentals/9r.html

[215] *Major Trends in the Mineralogy of Carbonate Skeletons Reflect Oscillations in Mid-Ocean Ridge Spreading Rates and Seawater Chemistry*, S.M. Stanley and L.A. Hardie, Ninth Annual V. M. Goldschmidt Conference, August 22-27, 1999, Cambridge, Massachusetts, abstract no. 7086

[216] *Sediments, Diagenesis, and Sedimentary Rocks*, Fred T. Mackenzie editor, Treatise on Geochemistry, Volume 7, ISBN-13: 978-0-08-043751-4.

[217] *Emission rates of CO₂ from plume measurements*, D. M. Harris, et al.., 1981, in The 1980 eruptions of Mount St. Helens, Washington, U.S. Geological Survey Professional Paper 1250, p. 3-15.

[218] *Understanding the Global Carbon Cycle*, Richard Houghton, 2007, Woods Hole Research Center: http://www.whrc.org/carbon/index.htm

[219] *The Missing Carbon Sink*, Richard Houghton, 2007, Woods Hole Research Center: http://www.whrc.org/carbon/missingc.htm

[220] *Restoring the Forests*, David G. Victor and Jesse H. Ausubel, Foreign Affairs, November/December 2000, Vol 79, Number 6, pp. 127-144.

[221] *Forest response to elevated CO_2 is conserved across a broad range of productivity.* R. J. Norby, et al., 2005. *PNAS,* **102**: 10.1073/PNAS 0509478102.

[222] *Food, Climate and Carbon Dioxide*, Sylvan H. Wittwer, CRC Press, Boca Raton, Fla., 1995.

[223] *Rain Helps Carbon Sink*, Krishna Ramanujan, NASA Earth Observatory, on-line: http://earthobservatory.nasa.gov/Study/CarbonHydrology/.

[224] *The Skeptical Environmentalist*, Bjørn Lomborg, 2001, p265, Cambridge University Press, ISBN 0 521 80447 7.

[225] *Feedbacks and the coevolution of plants and atmospheric CO_2*, David J. Beerling and Robert A. Berner, PNAS, February 1, 2005, vol. 102,no. 5, 1302-1305.

[226] *Climate swings have brought great CO_2 pulses up from the deep sea*, May 10, 2007, The Earth Institute at Columbia University. http://www.eurekalert.org/pub_releases/2007-05/teia-csh051107.php

[227] *Written declaration on greenhouse gas emissions by the livestock sector*, Romana Jordan Cizelj and Jan Christian Ehler, 28 March, 2007, DC\659409EN.doc,

[228] *Asian Forecast: Hazy, Warmer*, S. Perkins, Science News, August 4, 2007, vol. 172, p 68.

[229] *Kilimanjaro ice core records: Evidence of Holocene climate change in tropical Africa.* L. G. Thompson, et al., 2002. *Science*, 298, 589-593.

[230] *This Dynamic Earth*, J.M. Watson, U.S. Geological Survey web site: http://pubs.usgs.gov/gip/dynamic/developing.html

[231] *History of ocean basins*, H.H. Hess, in Petrologic Studies, Geological Society of America, 1962, pp 599-620.

[232] *Earth: Inside And Out*, Edmond A. Mathez ed., American Museum of Natural History, New Press, 2000.

[233] Plate Tectonics: The Book, http://www.platetectonics.com/book/index.asp

[234] *Continental thermal isostasy: 2. Application to North America*, Derrick Hasterok and David S. Chapman, Journal of Geophysical Research, VOL. 112, 2007.

[235] *Of Marine Terraces and Sand Dunes:The Landscape of San Clemente Island*, Andrew Yatsko, Pacific Coast Archaeological Society Quarterly, Vol 36, no. 1, Fall 2000.

[236] *The South Sandwich Islands: A Tasty Bit Of Info On A Sandwich Served Up By Mother Earth*, Jason Miller. Geology 1020, Nipissing University: http://www.freewebs.com/geomatter/thesandwichmeat.htm

[237] *Global Climate Modeling: The Role of the Oceanic Meridional Overturning Circulation*, LANL, 2000: http://www.ees8.lanl.gov/gcm.html

[238] *Slowdown of the meridional overturning circulation in the upper Pacific Ocean*, M. J. McPhaden and D. Zhang, Nature, 415(7), 603-608 (2002).

[239] *Frozen Earth: the once and future story of ice ages*, Doug Macdougall, 2004, the University of California Press, pp 108-109.

[240] *The Two-Mile Time Machine: Ice Cores, Abrupt Climate Change, and our Future*, Richard B. Alley, Princeton U. Press, Princeton, 2002.

[241] *The routing of meltwater from the Laurentide icesheet during the Younger Dryas cold episode*, W. S. Broecker, et al., 1989. *Nature*, 341, pp 318–321.

[242] *Does the ocean-atmosphere system have more than one stable mode of operation?* Broecker, W. S., Peteet, D. M. & Rind, D. 1986. *Nature*, 315, pp 21–26.

[243] *What If the Conveyor Were to Shut Down? Reflections on a Possible Outcome of the Great Global Experiment*, W. S. Broecker, GSA Today, January 1999, 9(1):1-7

[244] *Thousands of rubber ducks to land on British shores after 15 year journey*, Ben Clerkin, Daily News, June 27, 2007.

[245] *Atlantic Meridional Overturning Circulation During the Last Glacial Maximum*, Jean Lynch-Stieglitz, et al., *Science*, 6 April 2007, Vol. 316. no. 5821, pp. 66 – 69.

[246] *RAPID MOC: Monitoring the Atlantic Meridional Overturning Circulation at 26.5° North*, Natural Environment Research Council, 2007: http://www.bodc.ac.uk/rapidmoc/

[247] *Changing Orbit Explains Ice Ages, Scientist Says*, Reuters, 22 July 1999: http://www.space.com/scienceastronomy/planetearth/iceage_orbit_wg.html

[248] *On the secular change of the orbit of a satellite revolving about a tidally distorted planet*, G. H. Darwin, Philosophical Transactions of the Royal Society of London, vol 171, pp 713-891, 1880.

[249] *Lunar laser Ranging: A Continuing Legacy of the Apollo Program*, J. O. Dickey, et al., Science 265: 482-490, July 22, 1994.

[250] *Geological constraints on the Precambrian history of Earth's rotation and the Moon's orbit*, George E. Williams, Reviews of Geophysics, vol 38, pp 37-60, 2000.

[251] *Late-Pleistocene climates and deep-sea sediments*, Erickson, D.B., W.S. Brocker, J.L. Kulp, and G. Wollin. 1956. Science 124:385-389.

[252] *Long-term variations of daily insolation and quaternary climatic changes*, Berger, A.L. 1978. Journal of the Atmospheric Sciences 35:2362-2367.

[253] *Orbital, precessional, and insolation quantities for the Earth for —20 and +10 Myr.* Laskar, J., Joutel, F., and Robutel, P. 1993. Astronomy Astrophysics v. 270, p. 522-533.

[254] *The forms of water in clouds and rivers ice and glaciers*, J. Tyndall, 1872. International Scientific Series. The Werner Company, Akron, OH, 196 pp.

[255] *The impact of precession changes on the Arctic climate during the last interglacial-glacial transition*, Myriam Khodria, et al., *Earth and Planetary Science Letters*, Vol 236, Issues 1-2, 30 July 2005, pp 285-304.

[256] *Milankovitch climate reinforcements*, George Kukla and Joyce Gavin, *Global and Planetary Change*, Vol 40, Issues 1-2, January 2004, pp 27-48.

[257] Sunspots, The Exploratorium web site, 1998: http://www.exploratorium.edu/sunspots/history.html

[258] *Galileo's Scientific Discoveries, Cosmological Confrontations, and the Aftermath,* Stephen Mason, History of Science, vol 40, pp 377-406, 2002.

[259] *The Galileo Legend,* Thomas Lessl, New Oxford Review, June 2000, pp 27-33.

[260] *On the Shoulders of Giants: The Great Works of Physics and Astronomy,* Stephen Hawking, 2002, Running Press, p. 397.

[261] *Galileo, Astrology and the Scientific Revolution: Another Look,* H. Darrel Rutkin, HPST Colloquia presentation, November 4, 2004, Stanford University: http://www.stanford.edu/dept/HPST/colloquia0405.html

[262] *Stellar Structure and Evolution,* Peter Bodenheimer, Encyclopedia of Astronomy and Astrophysics, Robert A. Myers ed., Academic Press, 1989, pp 690-691.

[263] *Astrophysics and Stellar Astronomy,* T. L. Swihart, John Wiley & Sons, 1968, p162.

[264] *Stellar Structure and Evolution,* Peter Bodenheimer, Encyclopedia of Astronomy and Astrophysics, Robert A. Myers ed., Academic Press, 1989, pp 707-709.

[265] *The Potential of White Dwarf Cosmochronology,* G. Fontaine, P. Brassard, and P. Bergeron, Publications of the Astronomical Society of the Pacific, vol 113, no. 782, April 2001, p 410.

[266] *Stellar Structure and Evolution,* Peter Bodenheimer, Encyclopedia of Astronomy and Astrophysics, Robert A. Myers ed., Academic Press, 1989, p 715.

[267] *Age of Universe Revised, Again,* Robert R. Britt, Space.com, January 2003: http://www.space.com/scienceastronomy/age_universe_030103.html

[268] *Constant as the Sun? A Look at Solar Variability,* Peter Fox, NCAR High Altitude Observatory, May 17, 2000: http://www.ucar.edu/communications/factsheets/sun/index.html

[269] *The Little Ice Age, Brian M.* Fagan, 2001, Basic Books, p 121.

[270] *The Maunder Minimum and the Variable Sun-Earth Connection,* Steven H. Yaskell, Science, 2003, Page 134.

[271] *Estimating the sun's radiative output during the Maunder Minimum,* J. Lean, W. Livingston, A. Skumanich, and O. White, Geophys. Res. Lett. 19, 1591-1594, 1992.

[272] *Evidence on the climate impact of solar variations,* S. Baliunas and R. Jastrow, Energy, 18, 1285-1295, 1993.

[273] *Solar total irradiance variations and the global sea surface temperature record,* G.C. Reid, Journal of Geophysical Research, vol 96, pp. 2835-2844, 1991.

[274] *Dependence of global temperatures on atmospheric CO_2 and solar irradiance,* David J. Thomson, Proc. Natl. Acad. Sci. USA Vol. 94, pp. 8370-8377, August 1997.

[275] *On the Law of Distribution of Energy in the Normal Spectrum,* Max Plank, Annalen der Physik 4: p. 553, 1901.

[276] *Physics of Climate,* J. P. Peixoto and A.H. Oort, Springer, 1992, p. 118.

[277] *The dynamic greenhouse: Feedback processes that may influence future concentrations of atmospheric trace gases and climatic change,* Daniel A. Lashof, 1989, Climatic Change, vol 14 (3): 213-242.

[278] *Earth's Annual Global Mean Energy Budget,* J. T. Kiehl and Kevin E. Trenberth, Bulletin of the American Meteorology Society, Vol. 78, No. 2, February 1997, pp 197-208.

[279] *How Hot Can Venus Get?* Mark A. Bullock and David H. Grinspoon, American Astronomical Society, Proceedings of DPS 2001, November 2001.

[280] *Water Absorption Spectrum,* Martin Chaplin, July 18, 2007: http://www.lsbu.ac.uk/water/vibrat.html

[281] *A global merged land air and sea surface temperature reconstruction based on historical observations (1880-1997),* T.M. Smith and R.W. Reynolds, 2005, *J. Climate,* 18, 2021-2036.

[282] *Can increasing carbon dioxide cause climate change?* R.S. Lindzen, 1997, Proceedings of the National Academy of Sciences, v. 94, pp. 8335–8342

[283] *Comparison of Global Climate Change Simulations for $2 \times CO_2$-Induced Warming,* Kavita Kacholia and Ruth A. Reck, Climatic Change, January 1997, 53-69.

[284] *The Nile River,* Marie Parsons, Tour Egypt Magazine on line: http://www.touregypt.net/magazine/mag05012001/magf4a.htm

[285] *The Nile,* H. E. Hurst, 1952, 326 pp., Constable, London.

[286] Toussoun, J. D., 1925, Mem. Inst. Egipte 18, 366.

[287] *Does the Nile reflect solar variability?* Alexander Ruzmaikin, Joan Feynman and Yuk Yung, Proceedings Solar Activity and its Magnetic Origin, V. Bothmer, ed., Proceedings IAU Symposium No. 233, 2006.

[288] *Can solar variability influence climate?* Australian Antarctic Division web site: http://www.aad.gov.au/default.asp?casid=3563

[289] *Can Solar Activities Influence Extreme Weather over Continental Portugal? Stochastic Contrasts of Temperature and Precipitation with Sunspots and Cosmic Ray Intensity,* P. S. Lucio, Geophysical Research Abstracts, vol 7, 2005.

[290] *On The Relation Between Thunderstorm Activity And Solar Variability,* V. A. Mulloyarov, et al., Geophysical Research Abstracts, Vol. 5, 2003.

[291] Victor F. Hess, The Nobel Prize in Physics 1936, Biography: http://nobelprize.org/nobel_prizes/physics/laureates/1936/hess-bio.html

[292] *Unsolved Problems in Physics: Tasks for the Immediate Future in Cosmic Ray Studies,* Victor F. Hess, Nobel Lecture, December 12, 1936: http://nobelprize.org/nobel_prizes/physics/laureates/1936/hess-lecture.html

[293] *A selected history of expectation bias in physics,* Monwhea Jeng, American Journal of Physics, July 2006, Volume 74, Issue 7, pp. 578-583.

[294] *Heliospheric Physics and Cosmic Rays,* lecture notes by Kalevi Mursula and Ilya Usoskin, Lectured in 2003 by K. Mursula, University of Oulu.

[295] *Cosmic Rays as Electrical Particles,* Arthur H. Compton, Physical Review, Rev. 50, pp. 1119 - 1130, 1936.

[296] *Cosmic Radiation,* Peter Meyer, in *Encyclopedia of Astronomy and Astrophysics,* Academic Press, 1989, pp. 159-160.

[297] *Biographical information about VICTOR HESS* in: Current Biography Yearbook 1963, ed. Ch. Moritz, The H.W. Wilson Company, New York, 1963, p. 180-182.

[298] *Cosmic Rays,* R. A. Mewaldt, *Encyclopedia of Physics,* Macmillan, 1996.

[299] *Cosmic Radiation,* Peter Meyer, *Encyclopedia of Astronomy and Astrophysics,* Academic Press, 1989, pp. 165.

[300] *Cosmic Rays,* R. A. Mewaldt, *Encyclopedia of Physics,* Macmillan, 1996.

[301] *Supernovae,* David Branch, *Encyclopedia of Astronomy and Astrophysics,* Academic Press, 1989, pp. 733-734.

[302] *Supernova 1987A in the Large Magellanic Cloud,* W. Kunkel and B. Madore, Las Campanas Observatory, IAU Circular 4316: 1987A; N Cen 1986.

[303] *Supernova 1987A,* David Branch, in *Encyclopedia of Astronomy and Astrophysics,* Academic Press, 1989, pp. 726-727.

[304] *Cosmic Radiation,* Peter Meyer, in *Encyclopedia of Astronomy and Astrophysics,* Academic Press, 1989, pp. 163-164.

[305] *Variation of Cosmic Ray Flux and Global Cloud Coverage - a Missing Link in Solar-Climate Relationships,* Henrik Svensmark and Eigil Friis-Christensen, Journal of Atmospheric and Solar-Terrestrial Physics, 59 (11) (1997) 1225-1232

[306] *Cosmic radiation and the Weather,* E. R. Ney, 1959, *Nature,* vol 183, pp 451-453.

[307] *A laboratory study of the scavenging of sub-micron aerosols by charged raindrops,* A.K. Barlow and J. Latham, 1983, Quarterly Journal of the Royal Meteorological Society, vol 109, pp 763-770.

[308] *Length of the Solar Cycle: An Indicator of Solar Activity Closely Associated with Climate,* E. Friis-Christensen and K. Lassen, *Science,* 254, 698 (1991).

[309] *Low Cloud Properties influenced by Cosmic Rays,* Nigel Marsh and Henrik Svensmark, *Physical Review Letter,* December 4, 2000 - Volume 85, Issue 23, pp. 5004-5007.

[310] *Galactic cosmic ray and El Niño southern oscillation trends in international satellite cloud climatology project d2 low-cloud properties*, Marsh, N., H. Svensmark, J.of Geo. Research-Atmospheres, 108(D6), 2003.

[311] *Cosmic rays blamed for global warming*, Richard Gray, Sunday Telegraph, 11/02/2007.

[312] *'No Sun link' to climate change,* Richard Black, BBC News, 10 July 2007: http://news.bbc.co.uk/2/hi/science/nature/6290228.stm

[313] *New Research Adds Twist to Global Warming Debate*, Steven Milloy, Fox News, 12 Oct., 2006: http://www.foxnews.com/story/0,2933,220341,00.html

[314] *No Link Between Cosmic Rays and Global Warming,* Fraser Cain, Wired, July 03, 2007: http://blog.wired.com/wiredscience/2007/07/no-link-between.html

[315] *Cosmic ray link to global warming boosted*, Jenny Hogan, New Scientist, August 2004: http://www.newscientist.com/article.ns?id=dn6270&lpos=home1

[316] *An experiment that hints we are wrong on climate change,* The Sunday Times, February 11, 2007: http://www.timesonline.co.uk/tol/news/uk/article1363818.ece

[317] *The Chilling Stars: A new theory of climate change,* Henrik Svensmark and Nigel Calder, Icon Books, 2007, p 61.

[318] N. Shaviv, blog entry on Monday, 2006-10-23 03:10: http://www.sciencebits.com/SkyResults

[319] *A Celestial driver of Phanerozoic Climate?* N. Shaviv & J. Veizer, GSA Today 13, No. 7, 4, 2003.

[320] *Period of the Sun's Orbit around the Galaxy*, Stacy Leong, The Physics Factbook, 2002.

[321] *Encyclopedia of Science Heavens 2*. Robin Kerrod, MacMillian, 1997: 35.

[322] *Galactic Habitable Zones*, Leslie Mullen, *Astrobiology,* May 18, 2001.

[323] *The spiral structure of the Milky Way, cosmic rays, and ice age epochs on Earth,* Nir J. Shaviv, New Astronomy 8, 2003, pp 39–77.

[324] *Star Formation & Interstellar Chemistry,* Jeff Mangum, National Radio Astronomy Observatory, 2002: http://www.cv.nrao.edu/~jmangum/sfchem.shtml

[325] *The Milky Way Galaxy's Spiral Arms and Ice-Age Epochs and the Cosmic Ray Connection,* Nir Shaviv, ScienceBits, 2006: http://www.sciencebits.com/ice-ages

[326] *Empirical evidence for a nonlinear effect of galactic cosmic rays on clouds.* Harrison, R.G. and Stephenson, D.B. 2006. Proceedings of the Royal Society A: 10.1098/rspa.2005.1628.

[327] *On climate response to changes in the cosmic ray flux and radiative budget,* N. J. Shaviv, (2005), *J. Geophys. Res.*, 110, A08105, doi:10.1029/2004JA010866.

[328] *Dependence of global temperatures on atmospheric CO_2 and solar irradiance,* David J. Thomson, Proc. Natl. Acad. Sci. USA Vol. 94, pp. 8370-8377, August 1997.

[329] *Of Cosmic Rays and Dangerous Days,* Phil Berardelli, ScienceNOW Daily News, 1 August 2007: http://sciencenow.sciencemag.org/cgi/content/full/2007/801/1

[330] *Do extragalactic cosmic rays induce cycles in fossil diversity?* Mikhail V. Medvedev, Adrian L. Melott, Astro-Ph, 2007: http://arxiv.org/abs/astro-ph/0602092

[331] Blog postings by D. L. Hoffman and N. Shaviv, 6 August 2007: http://www.sciencebits.com/ice-ages

[332] *Sentient Cutlery,* J. Randi, SWIFT, 20 October, 2006.

[333] *Roger Bacon,* Theophilus Witzel, The Catholic Encyclopedia, Vol XIII, 1912, Robert Appleton Co.: http://www.newadvent.org/cathen/13111b.htm

[334] Ibid.

[335] Isaac Newton, in a letter to Robert Hooke, dated 15 February 1676.

[336] *Science Fair Central: Soup to Nuts Handbook,* Discovery School web site: http://school.discovery.com/sciencefaircentral/scifairstudio/handbook/scientificmetho d.html

[337] *Horace Walpole's Correspondence,* Lewis, S.L., Ed.. 31 Vols., Yale University Press, New Haven, 1937.

[338] *Serendipity and the three princes.* Remer, T.G. From the Peregrinaggio of 1557. University of Oklahoma Press, Norman, 1965.

[339] *"7. Simplicity",* Popper, Karl, 1992, The Logic of Scientific Discovery, 2nd edition, London: Routledge, 121-132.

[340] *The Fixation of Belief,* Charles S. Peirce, Popular Science Monthly, vol 12, November 1877, 1-15.

[341] Hermann Grunder, Director, Argonne National Laboratory, May 14, 2004: http://www.anl.gov/Media_Center/News/2004/news040514.htm

[342] *Particle Formation by Ion Nucleation in the Upper Troposphere and Lower Stratosphere,* S.-H. Lee, et al., Science, Vol. 301, 26 September 2003.

[343] *Can ozone depletion and global warming interact to produce rapid climate change?* Dennis L. Hartmann, et al., *PNAS* 2000;97;1412-1417.

[344] *Error estimates of Version 5.0 of MSU/AMSU bulk atmospheric temperatures,* J. R. Christy, et al., *J. Atmos Oceanic Tech,* vol 20, pp. 613-629.

[345] *Temperature Changes in the Bulk Atmosphere,* John Christy, in *Shattered Consensus,* P. Michaels ed., Rowman & Littelfield, 2005, p 88.

[346] *An improved in situ and satellite SST analysis for climate,* R.W. Reynolds, et al., 2002, *J. Climate,* 15, 1609-1625.

[347] *Climates & Weather Explained,* Edward Linacre and Bart Geerts, Routledge, 1997, p. 17.

[348] *Temperature Changes in the Bulk Atmosphere,* John Christy, in *Shattered Consensus,* P. Michaels ed., Rowman & Littelfield, 2005, p 77.

[349] *GISS Surface Temperature Analysis: Analysis Graphs and Plots,* Makiko Sato, NASA GISS website: http://data.giss.nasa.gov/gistemp/graphs/

[350] *Impact of urbanization and land use changes on climate,* E. Kalnay and M. Cai, 2003, *Nature,* no. 423, pp. 528-531.

[351] *Tracking Environmental Change Using Lake Sediments - Volume 3: Terrestrial, Algal, and Siliceous Indicators,* in Developments in Paleoenvironmental Research, J.P. Smol, H.J. Birks, and W.M. Last editors, Springer, 2002.

[352] *The Teeth may hold clue to vanished Viking settlements,* Sally Pobojewski, The University Record, University of Michigan, October 24, 1994. http://www.umich.edu/~urecord/9495/Oct24_94/3.htm

[353] *Cosmic background reduction in the radiocarbon measurement by scintillation spectrometry at the underground laboratory of Gran Sasso,* W. Plastino, et al., 2001, *Radiocarbon,* vol 43, pp. 157-161.

[354] *Extinction: how life on earth nearly ended 250 million years ago.* Douglas H. Erwin, 2006, Princeton University Press, pp 79-81.

[355] *Benthic Foraminifers,* U.S. Department of the Interior, U.S. Geological Survey, 1999: http://geology.er.usgs.gov/paleo/forams_b.shtml

[356] *Global Warming,* Spencer R. Weart, Harvard Univ. Press, 2003, pp 47-50.

[357] *Precise measurement of isotope ratios with a single collector mass spectrometer,* J. Schutten1, et al., *Applied Scientific Research,* Vol. 6, No. 1, January, 1957, pp 388-392.

[358] *Simulation of $\delta^{18}O$ in precipitation by the regional circulation model REMO$_{ISO}$,* Kristof Sturm, et al., *Hydrol. Process.* 19, 3425-3444 (2005).

[359] *Circulation Variability Reflected in Ice Core and Lake Records of the Southern Tropical Andes,* S. Hastenrath, D. Polzin, B. Francou, *Climatic Change,* Volume 64, Number 3, June 2004, pp. 361-375(15).

[360] *Ice core evidence of rapid air temperature increases since 1960 in alpine areas of the Wind River Range, Wyoming, United States,* David L. Naftz, et al., *J. Geophysical Research,* Vol. 107, No. D13, 4171, 2002.

[361] *Stable Isotope-Based Paleoaltimetry,* David B. Rowley and Carmala N. Garzione, *Annual Review of Earth and Planetary Sciences,* Vol. 35: 463-508, 2007.

[362] *The phase relations among atmospheric CO content, temperature and global ice volume over the past 420 ka,* Manfred Mudelsee, *Quaternary Science Reviews,* vol 20, 2001, pp 583-589.

[363] *Oldest ever ice core promises climate revelations,* New Scientist, 8 September 2003: http://environment.newscientist.com/channel/earth/climate-change/dn4121

[364] *Ice age insight,* Jay Chapman, *Geotimes* Web Extra, September 16, 2004: http://www.geotimes.org/sept04/WebExtra091604.html

[365] *Orbital and Millennial Antarctic Climate Variability over the Past 800,000 Years,* J. Jouzel et al, Science, 10 August 2007, Vol. 317. no. 5839, pp. 793 - 796.

[366] *Synchroneity of Tropical and High-Latitude Atlantic Temperatures over the Last Glacial Termination,* David W. Lea, et al., Science 5 September 2003, Vol. 301. no. 5638, pp. 1361 – 1364.

[367] *A revised +10±4ĴC magnitude of the abrupt change in Greenland temperature at the Younger Dryas termination using published GISP2 gas isotope data and air thermal diffusion constants,* Alexi M. Grachev and Jeffrey P. Severinghaus, Quaternary Science Reviews, vol 24, 2005, 513-519.

[368] *The Mann Et Al. Northern Hemisphere "Hockey Stick" Climate Index: A Tail of Due Diligence,* Ross McKitrick, in *Shattered Consensus,* P. Michaels ed., Rowman & Littelfield, 2005, pp 20-49.

[369] *Hockey Sticks, Principal Components and Spurious Significance,* S. McIntyre, R McKitrick, *Geo. Research Letters,* Vol 32(3), Feb 12 2005.

[370] *About the NAS,* The National Academy of Sciences, 2007: http://www.nasonline.org/site/PageServer?pagename=ABOUT_main_page

[371] *Surface Temperature Reconstructions for the Last 2,000 Years, Report in Brief,* June 2006, the National Academy of Sciences.

[372] *Ad Hoc Committee Report on the 'Hockey Stick' Global Climate Reconstruction,* Edward J. Wegman, et al., 2006.

[373] *Leading climatologists tell Geotimes what key earth science data they need most,* Lisa M. Pinsker, Geotimes, December 2001. http://www.geotimes.org/dec01/Feature_sidebar.htm

[374] Testimony of Michael Oppenheimer before Subcommittee on Energy and Air Quality Committee on Energy and Commerce U.S. House of Representatives, 7 March 2007.

[375] *Results of Large-Scale Weather Forecast Accuracy Study of Major Internet Weather Forecast Providers,* Eric Floehr, 2003, http://www.customweather.com/accuracy/2003study.html

[376] Ibid.

[377] *Warming Trends, Global Warming,* Weather 2000 FAQ, http://www.weather2000.com/faq/faq_warming.html

[378] *Useless Arithmetic: why environmental scientists can't predict the future,* Orrin H. Pilkey and Linda Pilkey-Jarvis, 2006, Columbia University Press, ISBN 978-0-231-13212-1, p184.

[379] *Does God Play Dice? The Mathematics of Chaos,* Ian Stewart, 2004, Blackwell, p. 141.

[380] *Limitations on Predictive Modeling in Geomorphology,* P.K. Haff, in *The Scientific Nature of Geomorphology,* C. E. Thorn and B. Rhoads, eds., John Wiley, 1996.

[381] *Limitations on Predictive Modeling in Geomorphology,* P.K. Haff, in *The Scientific Nature of Geomorphology,* C. E. Thorn and B. Rhoads, eds., John Wiley, 1996.

[382] *Error analysis of system mathematical functions,* Gaston H. Gonnet, ETH, Informatik, 2002, retrieved online 2007: http://www.inf.ethz.ch/personal/gonnet/FPAccuracy/Analysis.html

[383] *The Treatment of Uncertainties in the Fourth IPCC Assessment Report,* Martin R. Manning, IPCC Working Group I Technical Support Unit, the National Oceanic and Atmospheric Administration, 2007.

[384] *Climate Simulations for 1880-2003 with GISS Model E,* J. Hansen et al, 2007, in press.

[385] *Comparison of Global Climate Change Simulations for 2 × CO_2-Induced Warming,* Kavita Kacholia & Ruth A. Reck,Climatic Change, Jan.1997, 53-69.

[386] *Ground-water models cannot be validated,* by Leonard F. Konikow & John D. Bredehoeft, Advances in Water Resources, vol. 15, no. 1, 1992, p. 75-83.

[387] *The Hydrologist,* GSA Hydrology Division, June 1998, No. 48.

[388] *Unlicensed Engineers, Part 1,* Hendrick Tennekes, February 28, 2007, from *Climate Science: Roger Pielke Sr. Research Group Weblog:* http://climatesci.colorado.edu/2007/02/28/unlicensed-engineers-part-1-by-hendrik-tennekes/

[389] *Some fresh air in the climate debate*, Hendrick Tennekes, Amsterdam De Volkskrant, March 28, 2007.

[390] *Defining Dangerous Anthropogenic Interference: The Role of Science, the Limits of Science*, Michael Oppenheimer, Risk Analysis, Vol. 25, No. 6, 2005.

[391] *Many Colored Glass: Reflections on the Place of Life in the Universe*, Freeman Dyson, University of Virgina Press, 2007:
http://www.edge.org/documents/archive/edge219.html#dysonf

[392] *Gambling on tomorrow*, The Economist, August 28, 2007, p 69.

[393] *Useless Arithmetic: why environmental scientists can't predict the future*, Orrin H. Pilkey and Linda Pilkey-Jarvis, 2006, Columbia University Press, ISBN 978-0-231-13212-1, p84.

[394] *NASA Chief Questions Urgency of Global Warming*, Morning Edition, May 31, 2007 : http://www.npr.org/templates/story/story.php?storyId=10571499

[395] *Head of NASA Undecided on Need to Tackle Global Warming*, Steven Edwards, Wired Science, May 31, 2007, 11:32:53 AM:
http://blog.wired.com/wiredscience/2007/05/head_of_nasa_un.html

[396] *Counterpoint: Climate Change Responses*, Bob Carter, Monday 11 April 2005, Radio National, Presented by Michael Duffy,
http://www.abc.net.au/rn/talks/counterpoint/stories/s1339366.htm

[397] *Beyond The Ivory Tower: The Scientific Consensus on Climate Change*, Naomi Oreskes, *Science*, 3 December 2004, Vol. 306. no. 5702, p. 1686

[398] *Open Letter in Response to Naomi Oreskes' Criticisms*, Klaus-Martin Schulte, September 3, 2007.

[399] Statement of Dr. David Deming to the U.S. Senate Committee on Environment & Public Works. Online: http://epw.senate.gov/hearing_statements.cfm?id=266543

[400] *Sun's coverage of global warming was the right thing*, Paul Moore, Baltimore Sun, 11 February, 2007.

[401] Ibid.

[402] *The Press Gets It Wrong; Our report doesn't support the Kyoto treaty*, Richard Lindzen, The Wall Street Journal, Monday, June 11, 2001.
http://www.opinionjournal.com/editorial/feature.html?id=95000606

[403] *Gore says media miss climate message*, The Tennessean, 28 Feb., 2007.

[404] *Climate change treaty to come into force*, Friends of the Earth International website, http://www.foei.org/en/media/archive/2004/0930.html

[405] Earth Repair Foundation website: http://www.earthrepair.net

[406] Climate Change, World Wildlife Fund web site:
http://www.worldwildlife.org/climate/

[407] *How to save the climate*, Greenpeace International, Amsterdam.
http://www.greenpeace.org/raw/content/international/press/reports/how-to-save-the-climate-pers.pdf

[408] *Greenpeace: The world's climate being betrayed by G8 states*, Greenpeace press release, 16 March 2007.

[409] *New IPCC report shocking: Running out of time for action*, Greenpeace press release, 6 April 2007.

[410] *Nature at risk - the impacts of Global Warming*, WWF web site: http://www.panda.org/about_wwf/what_we_do/climate_change/problems/impacts/inde x.cfm

[411] *Fight Global Warming by Going Vegetarian*, GoVeg.com web site: http://www.goveg.com/environment-globalwarming.asp

[412] *Hot tempers on global warming*, Jeff Jacoby, *Boston Globe*, Aug 15, 2007.

[413] *Are 'climate criminals' committing 'terracide'?* Walter E. Williams, *World Net Daily*, August 8, 2007

[414] *Discover Dialogue: Meteorologist William Gray*, September 9, 2005.

[415]Scientists threatened for 'climate denial', Tom Harper, Sunday Telegraph, November 3, 2007, http://www.telegraph.co.uk/news/main.jhtml? xml=/news/2007/03/11/ngreen211.xml.

[416] Death Threats for man-made-global-warming-doesn't-exist scientist, Judi McLeod, Canada Free Press, March 12, 2007, http://www.canadafreepress.com/2007/cover031207.htm.

[417] *The Resurrection of Al Gore*, Karen Breslau, *Wired Magazine*, May, 2006.

[418] *Peace Man*, Economist.com, 12 October, 2007.

[419] Judge attacks nine errors in Al Gore's 'alarmist' climate change film, The Daily Mail, 11th October 2007: http://www.dailymail.co.uk/pages/live/articles/news/news.html? in_article_id=486969&in_page_id=1770&in_a_source

[420] *Al Gore Challenged to International TV Debate on Global Warming*, E-Wire, March 19, 2007: http://www.ewire.com/display.cfm/Wire_ID/3765

[421] *Island Nation of Tuvalu Threatened By Rising Waters Caused By Global Warming*, Amy Goodman, Democracy Now website, Tuesday, December 28th, 2004. http://www.democracynow.org/article.pl?sid=04/12/28/1511226

[422] *The Asian Brown Cloud: Climate and Other Environmental Impacts*, UNEP, Nairobi, 2002, p. 3.

[423] *Meat-Eaters Aiding Global Warming?* ABC News, April 19, 2006

[424] *In Praise of Shadows: A Meditation*, Joshua Sowin, September 23rd, 2006: http://www.fireandknowledge.org/archives/2006/09/23/in-praise-of-shadows-a-meditation/

[425] *Anti-People Group Pushes for Man's Extinction*, Michael Y. Park, Fox News, July 29, 2001: http://www.foxnews.com/story/0,2933,30834,00.html

[426] *Population Control Fanatic Advocates "Voluntary Human Extinction" as Means to "Green" the Planet*, Terry Vanderheyden, Lifesite, Nov. 16, 2005.

[427] *Discourse on the Origin and Basis of Inequality Among Men,* Jean-Jacques Rousseau, Geneva, 1755, p 43.

[428] *A Brief History of Cranks,* Paul Laity, *Cabinet Magazine,* Issue 20, Winter 2005/06 : http://www.cabinetmagazine.org/issues/20/laity.php

[429] *Rebels Against the Future: Witnessing the birth of the global anti-technology movement,* Ronald Bailey, *Reason* Online, 28 February, 2001: http://www.reason.com/news/show/34773.html

[430] *Medieval Demographics Made Easy,* S. John Ross, 1993: http://www.io.com/~sjohn/demog.htm

[431] *Japan's Medieval Population: Famine, Fertility, And Warfare in a Transformative Age,* William Wayne Farris, University of Hawaii Press, 2006.

[432] *Historical Estimates of World Population,* U.S. Census Bureau, 2007: http://www.census.gov/ipc/www/worldhis.html

[433] *Nations skeptical of U.S. climate talks,* John Heilprin, Associated Press, 26 Sept. 2007.

[434] *Al Gore Gives Policy Address at NYU on Solving the Climate Crisis,* NYU website, Sept. 18, 2006: http://www.nyu.edu/community/gore.html

[435] *Flatter oceans may have caused 1920s sea rise,* Catherine Brahic, NewScientist.com news service, August 24, 2007.

[436] *Afraid of global warming? Chill out,* Neil Collins, the London *Telegraph,* 20 September, 2004.

[437] *NOAA Hurricane Portal,* the National Oceanic & Atmospheric Administration (NOAA), U.S. Department of Commerce, 2007: http://hurricanes.noaa.gov/

[438] *Hurricanes And Climate Change: Assessing the Linkages Following the 2006 Season,* William M. Gray, Washington Round-table On Science And Public Policy.

[439] *Behind The Scenes: NOAA's North Atlantic Hurricane Seasonal Outlook,* NOAA Magazine, May 22, 2007.

[440] *Increase In Atmospheric Moisture Tied To Human Activities,* Science Daily, September 19, 2007.

[441] *Multiple gene evidence for expansion of extant penguins out of Antarctica due to global cooling,* Allan J. Baker, et al, *Proc. R. Soc. B,* vol 273, 2006.

[442] *Giant Penguins Once Roamed Peru Desert, Fossils Show,* Anne Minard, *National Geographic News,* June 25, 2007.

[443] *Multiple gene evidence for expansion of extant penguins out of Antarctica due to global cooling,* Allan J. Baker, et al, *Proc. R. Soc. B,* vol 273, 2006.

[444] *Sea change: people have affected what penguins eat,* Sid Perkins, *Science News,* July 14, 2007.

[445] Ibid.

[446] *Sea ice affects the population dynamics of Adélie penguins in Terre Adélie,* S. Jenouvrier, C. Barbraud and H. Weimerskirch, *Polar Biology,* 2005.

[447] Farmed salmon threat to Antarctic krill, New Scientist news online, 27 October, 2007.

[448] *The Evolution of the Polar Bear, Ursus maritimus Phipps.* Bjorn Kurten, Societas Pro Fauna Et Flora Fennica, 1964.

[449] *Climate myths: Polar bear numbers are increasing,* Phil McKenna, *New Scientist,* 16 May 2007.

[450] *Cool It,* Bjørn Lomborg, 2007, Knopf, p. 6.

[451] *Those Bad News Bears, Investor's Business Daily,* 28 December, 2006.

[452] *Natural Resources Key to Alaska Town's Future,* Melissa Block. *All Things Considered,* September 11, 2007.

[453] *Public Health: Expected Consequences Of Global Warming,* P. Martens, *American Scientist,* 1999 87:534.

[454] *Endemic malaria: an 'indoor' disease in northern Europe,* Lena Huldén, Larry Huldén, and Kari Heliövaara, *Malaria Journal,* 2005; 4: 19.

[455] *Malaria rising as DDT use falls, scientist says,* CNN.com, 22 Nov. 2000.

[456] *Warming "not spreading malaria,"* The BBC, September 21, 2000.

[457] Ibid.

[458] *A Briefing On Global Warming, The Other Side Of The Story,* the National Center for Policy Analysis and Competitive Enterprise Institute, 29 Sept. 1997.

[459] *Catastrophic Flooding and the Origin of the English Channel Valley System,* S. Gupta, et al, American Geophysical Union, Fall Meeting 2004, abstract #OS23C-1336.

[460] *Climate Change 2007: WGI: The Physical Science Basis,* IPCC, Cambridge University Press, 2007.

[461] *Cool It,* Bjørn Lomborg, 2007, Knopf, p. 40.

[462] *Summary for Policy Makers,* IPCC AR4, WG III, 2007.

[463] *Ethanol Hype: Corn Can't Solve Our Problem,* D. Tilman, J. Hill, The Washington Post, Sunday, March 25, 2007.

[464] *How Biofuels Could Starve the Poor,* C. Ford Runge and Benjamin Senauer, *Foreign Affairs,* May/June 2007.

[465] *Biomass as Feedstock for a Bioenergy and Bioproducts Industry: The Technical Feasibility of a Billion-Ton Annual Supply,* Perlack, R.D., et al. 2005. DOE Report #: DOE/GO-102005-2135.

[466] *Cellulose Dreams: The search for new means and materials for making ethanol,* Corinna Wu, *Science News,* Aug. 25, 2007; Vol. 172, No. 8 , p. 120.

[467] *Peru's retreating glaciers stir water-supply worries,* Leslie Josephs, The Associated Press, February 12, 2007.

[468] Geothermal, U.S. Dept. of Interior Bureau of Land Management, website: http://www.blm.gov/ca/st/en/prog/energy/geothermal.html

[469] *All About Geothermal Energy – Potential Use,* Geothermal Energy Assoc, website, 2007: http://www.geo-energy.org/aboutGE/potentialUse.asp

[470] *Sunlit uplands: Wind and solar power are flourishing, thanks to subsidies,* the *Economist,* May 31, 2007.

[471] *Evaluation of global wind power,* C. Archer and M. Jacobson, *J. of Geo Research Atmos.,* vol. 110, 2005.

[472] *Ill winds,* the *Economist,* July 29, 2004.

[473] *Getting wind farms off the ground,* the *Economist,* June 7, 2007.

[474] *Ill winds,* the *Economist,* July 29, 2004.

[475] Greenpeace Supports Cape Wind, America's First Offshore Wind Farm, Greenpeace website, 2007.

[476] *Nation's first offshore wind farm clears important hurdle,* Walter Brooks, Cape Cod Today, March 30, 2007.

[477] *Basic Research Needs For Solar Energy Utilization,* U.S. DOE, 2005, p. 10: http://www.sc.doe.gov/bes/reports/files/SEU_rpt.pdf

[478] *Sunlit uplands: Wind and solar power are flourishing, thanks to subsidies,* The *Economist,* May 31, 2007.

[479] *Solar Energy for Everyone,* Scott Shephard, 2006, Professional Builder website: http://www.housingzone.com/probuilder/article/CA6340267.html

[480] *Basic Research Needs For Solar Energy Utilization,* U.S. DOE, 2005: http://www.sc.doe.gov/bes/reports/files/SEU_rpt.pdf

[481] *Carbon Sequestration Frequently Asked Questions,National Energy Technology Laboratory,* U.S. DOE website,, 2007: http://www.netl.doe.gov/technologies/carbon_seq/faqs.html

[482] *Old Clean Coal,* The Economist *Technology Quarterly,* Sept., 2007, p. 6.

[483] *Basic Research Needs For Solar Energy Utilization,* U.S. DOE, 2005, p. 9.

[484] *The Killer Fog of '52,* John Nielsen, NPR, All Things Considered, December 11, 2002.

[485] *Pollution From Chinese Coal Casts a Global Shadow,* K. Bradsher, D. Barboza, The *New York Times,* June 11, 2006.

[486] *Underground coal fires called a "catastrophe,"* Michael Woods, *Pittsburgh Post-Gazette,* February 15, 2003.

[487] *Coal fires - A natural or man made hazard?* Anupma Prakash, 2007: http://www.gi.alaska.edu/~prakash/coalfires/coalfires.html

[488] *Coal mine deaths spike upward,* Thomas Frank, *USA Today,* January 1, 2007.

[489] *China's Coal Mines: Bottoming out,* The *Economist,* August 23, 2007.

[490] *Methane gas release from the Storegga submarine landslide linked to early Holocene climate change: a speculative hypothesis,* James E. Beget, *The Holocene,* Vol. 17, No. 3,pp 291-295, 2007.

[491] *For Aviation's Greenhouse-Gas Emissions, It's Technology Versus Growth,* David Bond, *Aviation Week & Space Technology,* August 19, 2007.

[492] *Thinking Green In Our Blue Skies,* Brian Dubie, *Aviation Week,* February 22, 2007.

[493] *GE Aviation Enables New, More Efficient Aircraft Approach Patterns,* Brad Kenney, *Industry Week,* July 5, 2007.

[494] *Benefit assessment of using continuous descent approaches at Atlanta,* I. Wilson and F. Hafner, Digital Avionics Systems Conference, 2005. DASC 2005. The 24th Volume 1, Issue , 30 Oct.-3 Nov. 2005.

[495] *Lithium Batteries Take to the Road,* John Voelcker, IEE *Spectrum,* Sept. 2007, pp 27-31.

[496] *Spinning into Control: High-tech reincarnations of an ancient way of storing energy,* Davide Castelvecchi, Science News, May 19, 2007, Vol. 171, No. 20 , p. 312.

[497] *Spinning into Control: High-tech reincarnations of an ancient way of storing energy,* Davide Castelvecchi, Science News, May 19, 2007, Vol. 171, No. 20 , p. 312.

[498] *Where the wind blows,* The Economist, July 26, 2007.

[499] *Green is for Go,* David Bond, *AW&ST,* August 2-/27, 2007, p 53.

[500] *Lifetime and Life Cycle Cost Estimation of Japanese Detached House,* Yukio Komatsu, 2nd Japan-Scandinavia Seminar on Building Technology, Masala, Finland, 25-26. May, 2000.

[501] *The Zero-Zero Hero,* Tekla S. Perry, IEEE *Spectrum,* Sept. 2007.

[502] *Green as Houses,* The Economist, 15 Sept, 2007, pp. 40-42.

[503] *Nanoengineered Concrete Could Cut Carbon Dioxide Emissions,* Science Daily, February 2, 2007.

[504] *Energy Firms Plan New Nuclear Power Plants,* NPR Morning Edition, August 15, 2007.

[505] *Why the French Like Nuclear Energy,* Jon Palfreman, PBS Frontline, 1995: http://www.pbs.org/wgbh/pages/frontline/shows/reaction/readings/french.html

[506] *A Natural Fission Reactor,* G. A. Cowan, *Sci. American,* 235:36, 1976.

[507] *The Fossil Nuclear Reactors of Oklo,* Gabon, John Smellie, *Radwaste Magazine,* Nat. Analogs, March 1995:21.

[508] *Record of Cycling of the Natural Nuclear Reactor in the Oklo/Okelobondo Area in Gabon,* A. P. Meshik, et al., *Physical Review Letters,* vol 93, No. 18. 2004.

[509] Nuclear Dawn, *The Economist Technology Quarterly,* September, 2007.

[510] Atomic Renaissance, *The Economist,* September 8, 2007, p. 71.

[511] *Space Solar Power,* Frank Morring, *AW&ST,* August 20/27, 2007, p76.

[512] *Satellite Power System: Concept Development and Evaluation Program,* DOE and NASA Reference System Report, Jan. 1979.

[513] *Lightly carbonated,* the *Economist,* 2 August, 2007

[514] *Cool It,* Bjørn Lomborg, 2007, Knopf, p. 31.

[515] *One Answer to Global Warming: A New Tax,* N. G. Mankiw, *New York Times,* September 16, 2007.

[516] *Carbon Abatement: Lessons from Second-Best Economics,* Ian W. H. Parry, Resources for the Future, 2000.

[517] *The Brand of Clinton,* The *Economist,* 22 September, 2007, p 84.